MAKING THE ENFIELD PATTERN 1853 RIFLE-MUSKET

MAKING THE ENFIELD PATTERN 1853 RIFLE-MUSKET

The Evolution of Gun Making, 1820–1860

PETER G. SMITHURST

Pen & Sword
MILITARY

AN IMPRINT OF PEN & SWORD BOOKS LTD.
YORKSHIRE - PHILADELPHIA

First published in Great Britain in 2025 by
PEN AND SWORD MILITARY
An imprint of
Pen & Sword Books Limited
Yorkshire – Philadelphia

Copyright © Peter G. Smithurst, 2025

ISBN 978 1 03610 685 0

The right of Peter G. Smithurst to be identified as Author of this work has been asserted by him in accordance with the Copyright, Designs and Patents Act 1988.

A CIP catalogue record for this book is available from the British Library.

All rights reserved. No part of this book may be reproduced, transmitted, downloaded, decompiled or reverse engineered in any form or by any means, electronic or mechanical including photocopying, recording or by any information storage and retrieval system, without permission from the Publisher in writing. No part of this book may be used or reproduced in any manner for the purpose of training artificial intelligence technologies or systems.

Typeset in Times New Roman 10/12 by
SJmagic DESIGN SERVICES, India.
Printed and bound in the UK by CPI Group (UK) Ltd.

The Publisher's authorised representative in the EU for product safety is Authorised Rep Compliance Ltd., Ground Floor, 71 Lower Baggot Street, Dublin D02 P593, Ireland.
www.arccompliance.com

For a complete list of Pen & Sword titles please contact
PEN & SWORD BOOKS LIMITED
George House, Units 12 & 13, Beevor Street, Off Pontefract Road, Barnsley, South Yorkshire, S71 1HN, England
E-mail: enquiries@pen-and-sword.co.uk
Website: www.pen-and-sword.co.uk

or
PEN AND SWORD BOOKS
1950 Lawrence Rd, Havertown, PA 19083, USA
E-mail: uspen-and-sword@casematepublishers.com
Website: www.penandswordbooks.com

Contents

Preface x
Acknowledgements xii
Chapter 1 Introduction 1
Chapter 2 Basic Technology of the Enfield Rifle 8
          2.1 Introduction 8
          2.2 Fundamentals of the percussion system of ignition 8
          2.3 The lock 9
          2.4 Operation of the lock – 'cocking' and 'firing' the gun 11
          2.5 The barrel 12
          2.6 Rifling 13
          2.7 The conundrum of a muzzle-loading rifle 14
          2.8 Solutions to the problem of muzzle-loading rifles 15
          2.9 Minié's solution 18
          2.10 Minié's system adopted and adapted 19
          2.11 The stock 20
Chapter 3 Origins and Procurement 21
          3.1 Origin of the Pattern 1853 rifle 21
          3.2 Procurement by contract 23
          3.3 Discontent amongst the contractors 25
          3.4 Discontent of the government 26
          3.5 Select Committee on Small Arms 29
          3.6 Pursuit of mechanisation 31
                 The lock 31
                 The stock 32
                 The barrel 32

	3.7	Committee on Machinery 33
	3.8	Equipping the Enfield factory 33
	3.9	Specifications of the Pattern 1853 rifle 36
	3.10	'Viewing' and inspection procedures – gauges 39
	3.11	Operations at Enfield 42
	3.12	Conclusion 45
Chapter 4		Manufacture of the Lock 47
	4.1	Overview 47
	4.2	Forging the Lockplate 49
	4.3	Machining the Lockplate 51
		Machining the outer face 51
		Machining the inner face 52
		Drilling the lockplate holes 54
		Machining the lockplate edges 58
		Gauging the lockplate 65
	4.4	The Hammer 67
		Gauging the hammer 73
	4.5	The Tumbler 74
		Gauging the tumbler 78
	4.6	The sear 79
		Gauging the sear 81
	4.7	The Bridle 82
		Gauging the bridle 84
	4.8	The Swivel 86
		Gauging the swivel 87
	4.9	The Mainspring 88
		Gauging the mainspring 89
	4.10	Sear spring and its gauging 90
	4.11	Screws and their gauging 92
	4.12	Gauging the assembled lock 94
	4.13	Some metrology of Enfield Pattern 1853 Lockplates 95
		Tower 1855 Lockplates 96
		Belgian contract 1856 Lockplates 97

CONTENTS

		Damage revealed in the metrology process 97
		Questions surrounding inspection of contract locks. 97
		Further metrology and interchangeability 99
	4.14	Conclusion 101
Chapter 5		Manufacture of the Stock 103
	5.1	Introduction 103
	5.2	The advent of mechanisation – Blanchard's developments in the United States 103
	5.3	Equipping the Enfield factory 108
	5.4	The stock 109
		Slabbing 112
		Centering 114
		Rough-turning the fore end 115
		Rough-turning the butt 117
		Spotting 119
		Bedding the barrel 121
		Hand-finishing the recess for the breech tang 125
		Sawing to length at butt and muzzle and bedding the butt plate 125
		Planing the stock 127
		Bedding the butt plate tang, three holes for the screws are bored and the two large holes tapped 129
		Bedding the lock 134
		Bedding the lockplate 138
		Bedding the screw heads 139
		Drilling the hole for the sear tang 141
		Recessing for the mainspring 142
		Bedding the trigger guard, drilling screw holes, bedding trigger plate and the stop for the ramrod 145
		Stock cut for the bands and the nose cap let-on. 148
		Stock turned between the bands 149
		Butt end and fore-end of stock between lock and bottom band finish-turned in copying lathe 151
		Stock grooved for ramrod 154

		Recessed for the ramrod spring and transverse hole for fixing pin bored 155
		Hole for ramrod bored 156
		Holes for lockplate fixing screws, for breech tang screw, for nosecap screw and pin hole for tenon of trigger guard are drilled 157
		Gauging 159
	5.5	Conclusion 160
Chapter 6		Manufacture of the Barrel Part 1 – creating the tube 161
	6.1	Introduction 161
	6.2	Specifications for the Enfield barrel 161
	6.3	Manufacture of the barrel tube 162
		Creating the 'mould' 162
		Creating the seamed tube 164
		Welding the seam 165
		Tapering the tube 167
		Straightening 170
		'Lumping' 171
	6.4	Conclusion 172
Chapter 7		Manufacture of the Barrel Part 2 – finishing the barrel 173
	7.1	Introduction 173
	7.2	Equipping the Enfield Factory 173
	7.3	Operations 175
		First rough-boring 175
		Second rough-boring 178
		Barrel cut to length 178
		Rough turned outside 179
		Grinding 181
		Fine boring 182
		Breeching 184
		Machining the breech pin 187
		Threading the breech pin 189
		Machining and threading the breech counter-bore 189
		Gauging the 'breeching' 194

CONTENTS

	The front sight and muzzle	194
	Machining the bolster (cone or nipple seat)	198
	Percussioning	200
	Polishing	201
	Rifling	202
	Rear Sight	208
	'Viewing' and gauging	208
	Internal Gauging - the bore and rifling	213
	Proving	216
	Browning	218
	7.4 Conclusion	219
Chapter 8	Conclusion	220
Appendix 1	The Percussion Cap	225
Appendix 2	Joseph Smith	227
Appendix 3	James Burton	249
Appendix 4	R.S.A.F. Enfield Orders To Greenwood & Batley, Leeds	256
Appendix 5	Enfield Expenditure Memorandum 1855	259
Appendix 6	Official Returns	262
Appendix 7	Enfield Pattern 1853 Lockplate Hole Metrology Data	265
Appendix 8	Messrs. Fox, Henderson and Co.'s Contract for rifles	268
Bibliography		273
References		279

Preface

In book 1 the origin of the concept of interchangeability was outlined along with the gradual evolution of gunmaking processes in pursuit of that objective. Over the decades from 1960 onwards there have been countless books published on the history of firearms but very few concern themselves with their production, if at all. From the British viewpoint, two books can be mentioned: *British Military Firearms 1650 – 1850*[1] and *The British Soldier's Firearm, 1850 – 1864*.[2] These were, and still are, first-class ground-breaking studies, being the first of a very few serious studies of the development and evolution of British military firearms up to the mid-19th century, but both only touch on manufacture since it was peripheral to their primary purpose and neither study that topic in depth. Roads' book, unlike his thesis,[3] the main foundation for the book, does illustrate some gauges and tooling but since he does not explain their specific functions they appear more as an afterthought. A broader perspective of English gunmaking practices can be found in *The Birmingham gun trade*[4] which gives an excellent overview but by its very nature is restricted in the detail it can provide.

In America, too, there are no comprehensive contemporary accounts of military musket or rifle manufacture prior to 1878 and only a few that touch upon individual weapons, components, or technologies.

Numerous studies concerning the early development of the interchangeable manufacture of firearms are American and written from the point of view of the economic, social, political, and administrative imperatives surrounding it. Such studies might add to a wider understanding of the subject by stressing various factors which were important in driving the quest for interchangeability in America, but they offer little or nothing of that fundamental element – the technology itself. This is also true of one otherwise excellent Ph.D. thesis, *The development of the Royal Small Arms Factory (Enfield Lock) and its influence upon mass production technology and product design c. 1820 – c. 1880*.[5] They may tell us how the 'system' operated and what the technology achieved but they do not tell us how it was achieved. They may mention techniques and machines but do not explain their operations.

It is also a period in which recent studies often blend various concepts such as 'interchangeability', 'mechanisation' and 'mass production' under a single heading – 'the American system of manufactures'. The term 'mass production' first appears to have been used in 1918 in an article by W. Freeland, *Mass production in the Winchester shops*.[6, 7] Even the label 'American System of Manufactures' is British, borrowed from the Report of the Committee on the Machinery of the United States in 1854, and has given rise to extensive misconceptions. It is noted in the introduction by the American editor, Ed Battison, a recognised technology historian, to the translated version of Gamel's *Description of the Tula Weapon factory*:

> *Here, in the United States, popular and hurried historians have built an elaborate story, much of it mythical, around a few names while ignoring the work of others.*[8]

To this might be added an observation by Somerset Maugham:

> *The faculty for myth is innate in the human nature. It seizes with avidity upon any incidents, surprising or mysterious, in the career of those who have at all distinguished themselves from their fellows and invents a legend to which it then attaches a fanatical belief.*[9]

Some of this myth is present in what is otherwise an excellent overview of the early development of the 'American System' published in the American Machinist.[10] The published papers of a symposium held in 1981, *Yankee enterprise; the rise of the American system of manufactures*[11] perpetuate the idea that the 'American system' was an American initiative. The same system was at the root of Bramah's and Maudslay's set of machinery for the manufacture of Bramah's patent lock. Alfred Hobbs, the first person to 'pick' Bramah's 'unpickable' lock in 1851, in the section of his book dealing with lock manufacture[12] makes no mention of these machines. In *English and American tool builders*[13] which has a bust of Henry Maudslay as the frontispiece and is generally assumed to be an in-depth study of machine tool development, the author also fails to acknowledge the Bramah lock making machinery. If further evidence were needed, the very fact that Brunel's and Maudslay's machines, put into operation in 1805,[14] were able to produce between 130,000 and 140,000 blocks in a year[15] is surely sufficient evidence of a manufacturing 'system' based on 'machinery' to dismiss such a claim that the American system was actually American. Considering the work in 18th century France and early 19th century Russia, whether viewed from the perspective of 'organisation' of processes, 'mechanisation' or 'interchangeability', it had precedents elsewhere.

Whilst important improvements had certainly been made in America in the execution and achievement of interchangeable manufacturing, they were improvements, adaptations, and extensions of technologies borrowed, perhaps in some cases unwittingly, or in a process that anthropologists might refer to as "parallel evolution", from a variety of developments with widely dispersed dates and geographies. Collectively these technologies were to have profound effects on many aspects of manufacturing but as will be shown, their first impact in Britain was to revolutionise gun manufacture.

Finally, I do not claim this to be 'the last word' on this topic. With any historical study it is never possible to be sure that all the facts have been gathered and as will be seen, a few questions remain unanswered.

Acknowledgements

This study would not have been possible without the help and support of numerous individuals and various institutions to whom and to which I offer my unbounded thanks.

This, book 2, is almost entirely based on work carried out within the School of Computing and Engineering of the University of Huddersfield through a fee-waiver PhD studentship and I offer my deep gratitude to the University for its support of this study and to my main supervisor, Dr. Paul Bills, and to Dr. Paul Wilcock for their support and guidance throughout.

None of that would have been possible without the assistance and cooperation of three other institutions:

Royal Armouries – being appointed first as a curator and then as Curator Emeritus after retirement, gave me unrestricted access to the world-class collections. For that privilege my thanks go to Graeme Rimer, Academic Director, during my time as a senior curator of firearms and then instigating my Emeritus status, and to Laura Bell, Director of Collections, for my continuing role as Curator Emeritus.

My thanks are also due to Jonathan Ferguson, Keeper of Firearms and Artillery; Lisa Traynor, Curator of Firearms, Stuart Ivinson, Librarian and Philip Abbott, Archivist, for their unstinting help and patience in providing 'remote assistance' on my behalf. It is worth noting that the unique collection of manufacturing gauges for the Enfield Pattern 1853 rifle have not been previously studied or published and special thanks are due to Royal Armouries for free use of images of items in their collections.

To the registrar and to Nadine Lees, Digital Media and Rights Officer at Birmingham Museum and Art Gallery, for access to the unique collections of specimen barrels and stocks representing the stages in their manufacture and for allowing me free use of many photographs of these items. These specimens have not been previously studied or published and were invaluable resources for this study.

To Andrew Appleby for a copy of the writ served on Richard Brown Roden by Thomas Wilson referred to on pages 122, 224 and in the Conclusion containing evidence from Colonel Dixon regarding conversion of Enfield rifles into Snider rifles.

Similarly, I extend thanks to the Directors of West Yorkshire Archives for the loan for photography by Royal Armouries of numerous Greenwood & Batley

ACKNOWLEDGEMENTS

drawings of various machines used and which also have not been previously studied or published.

Around 2000 a small programme of metrology was carried out through the kindness and cooperation of David Eaton, Director, and Malcolm Jackson, Metrologist, of the School of Engineering at Sheffield Hallam University, which enabled new insights to be added to this study.

I must once again thank Robert (Bob) Gordon who had similar interests and, when Professor of Geophysics and Engineering at Yale University, set me off on this path many years ago.

Last but by no means least I must thank the publishers for taking a bold leap in the dark into uncharted territories – little, if anything, has been published on the history and evolution of firearms manufacturing technologies for many decades. To treat any arms and armour simply as finished products and to explore their place in history is important but this is only the end of their stories which begin with their creation.

Chapter 1

Introduction

In the 1850s, the nature of the firearm issued to British troops, and its design, procurement, and manufacture underwent revolutionary changes. Although the Pattern 1853 Rifle began as a hybrid, it was to become a thoroughbred and shortly after its creation, the *Select Committee on Small Arms* was established in 1854 to examine its manufacture and procurement.

The *Minutes of Evidence* of this committee[1] contain testimonies from numerous persons within the gun trades and eminent engineers. There are conflicting testimonies; those who claimed that mechanisation and interchangeability was feasible and those who claimed the opposite; those who claimed that machinery was already in use and those who dismissed that idea; those who felt that a government factory was a good idea and those who were against it. Their final report, submitted on 12th May 1854, reflects these disparate views, especially in regard to the expansion of the factory at Enfield to take over the manufacture of government arms from the private contract procurement system. It was both woolly and indecisive.

However, its outcome was possibly pre-empted, for in that same year the *Committee on the Machinery of the United States*, comprising three officers, Lt. Col. Burn, Royal Artillery, Lt. Warlow, Royal Artillery and John Anderson, Ordnance Inspector of Machinery, was sent to America by the Select Committee to examine the machinery used in gun manufacture and with authority to purchase appropriate machines for use at Enfield. Their report was published in 1855 and contains details of the various contracts entered into, and while it forms the basis of a later and otherwise excellent and valuable study,[2] it only lists machines and does not touch upon their purpose or functioning.

The introduction of the Pattern 1851 and almost immediately afterwards, the Pattern 1853 rifles, coupled with the debates surrounding their manufacture and procurement, sparked a small flurry of 'popular' accounts. One of the earliest was an article in the *Illustrated London News* on the manufacture of rifle barrels[3] which illustrates, amongst other things, the grinding of barrels. A second, on the manufacture of the Enfield Pattern 1853 Rifle, appeared in 1855[4] and has some details of the barrel rolling and rifling processes but makes the mistake of titling the article the 'Minié rifle' but from its date it is clearly the rifle of 1853 which is being referred to. There was a tendency in some quarters to continue referring to the 1853 rifle as the 'Minié' which, however confusing, is not entirely incorrect since it also employed Minié's principle.

A further article, written at the time mechanisation had been introduced at Enfield, describes somewhat cursorily the manufacture of the rifle and, in this author's view, is accompanied by crude and meaningless illustrations.[5]

In his book, *The Rifle Musket*[6] the author pays attention to the barrel and its manufacture at the expense of other vital components. However, it is noted for the first time that unfortunately Jervis' illustration of the barrel welding and taper-rolling mill contains a serious error and could not possibly have produced a tapered barrel. This same error is reproduced in another publication a few years later, *The Book of Field Sports*.[7]

The most expansive articles are a series which appeared in *The Engineer*[8] and another series in the *Mechanic's Magazine*[9] dealing with the Enfield Pattern 1853 rifle specifically. Each covers, in varying detail, the manufacture of the complete rifle. The series in *The Engineer* is extensively illustrated with line drawings of a number of machines, but with preference for those used in stock manufacture. However, a feature not commented on previously is that these articles exhibit an inexplicable lack of attention to detail, possibly arising from a lack of familiarity with firearms, in that illustrations of the rifle and its components, central to its theme, are mirror images!

The *Mechanic's Magazine* articles lack illustrations which is an unfortunate oversight in dealing with a subject which would have been unfamiliar to many outside of, and no doubt to many acquainted with, the 'gun trade'.

A later article written by an engineer for engineers gives detail on some of the stock-making machinery and how various machines functioned, accompanied by excellent line drawings, but in restricting its content to 'lock bedding' and 'finish-turning the fore end', offers important, but limited, assistance.[10]

As Cooper notes in her study of Thomas Blanchard and his invention of gun-stocking machines, *Blanchard had to build skill into his machines*.[11] This is equally true of all the other skilled operations in gunmaking shown in this attenuated list:[12]

Barrel Welders
Barrel Borers
Barrel Grinders
Barrel Filers and Breechers
Barrel Rib Makers
Barrel Breech Forgers and Stampers
Barrel Smoothers – prepare the barrel for browning
Barrel Browners
Barrel Riflers
Barrel Machiners – prepare the front sight and the lump end of the barrel for the nipple
Barrel Jiggers – lump filers and break-off fitters – prepare the breech end of the barrel
Percussioners – finish the nipple seat, fit the nipple and adjust the hammer to the nipple
Stock makers
Maker-off – file the stocks to give them their proper finish, glass paper and oil them
Stockers – let in the barrel and lock and roughly shape the stock
Furniture Forgers and Casters

INTRODUCTION

- Furniture Filers
- Rod Forgers
- Rod Grinders and Polishers
- Rod Finishers
- Band Forgers and Stampers
- Band Machiners
- Band Filers
- Band Pin Makers
- Sight Stampers
- Sight Machiners
- Sight Jointers
- Sight Filers
- Sighters and Sight Adjusters
- Oddwork Makers
- Engravers (lock etc.)
- Polishers – lock and furniture
- Screwers – let in the furniture and remaining pins and screws
- Strippers – prepare the gun for rifling and proof
- Finishers – distribute the parts to the browner, polisher, maker-off, barrel smoother and when they are returned put the guns together and adjust the parts
- Lock Forgers
- Lock Machiners
- Lock Filers
- Lock Freers – adjust the working parts of the lock
- Trigger Boxers

To replace all of these by machines was a formidable task by any standards.

The only way of appreciating that cumulative achievement is through an understanding of the gradual evolution of the technology from traditional gun manufacturing methods and that is covered in book 1.

Even starting with the first recorded attempts at interchangeable firearms manufacture in France in the early 18th century, by the time of the interchangeable manufacture of the Enfield Pattern 1853 rifle it had a history extending over almost 150 years and here the later stages in that evolutionary process will be covered.

The early years of the nineteenth century were a period in which British gunmakers made significant contributions to mechanising some of the processes used in the manufacture of the barrel. Numerous experiments were made at this time to introduce mechanisation into the process of making the basic tube but few, if any, are documented in the various publications which existed, until the 1850's when interest in the new British rifles was aroused. The major innovations were Osborn's rolling mill of 1813[13] for tapering gun barrels and his later use of rolls for welding the tube,[14] coupled with Heywood's patent of 1814[15] for a rolling mill which converted a flat plate into a seamed tube ready for welding. These became the basis of military barrel making technology in Britain until late in the 19th century but were not adopted in America until 1860 when they were installed at the Springfield Armory.[16]

A major achievement in interchangeability was made in America in this period. John Hall invented a breech-loading rifle in 1811[17] which was produced at the government's Harper's Ferry Armory. Hall had proposed as early as 1816 making his rifles interchangeable and it is stated[18] that Hall employed a variety of machines *designed to cut, shape and smooth the exteriors of metal components* but no details of these machines are known, except another source which states these contrivances performed *with cutters and saws work usually done elsewhere with grindstones, chisels and files.*[19]

MAKING THE ENFIELD PATTERN 1853 RIFLE-MUSKET

In 1828 Simeon North was granted a contract with the U.S. Ordnance Department to supply five thousand Hall rifles. By 1834, working to a set of gauges supplied by Hall, North was also producing interchangeable rifles and the components of Hall's and North's rifles, produced in different establishments, were interchangeable. This was a landmark achievement in the evolution of precision manufacturing and Smith[20] suggests that no two individuals played a more important role in this development, or as machine-tool innovators, during the early 19th century. In the absence of any knowledge of Hall's machines it is difficult to assess his contribution, but North's milling machine certainly ranks as a major landmark in manufacturing technology.

Whitney's work marks the beginning of the quest for interchangeability in the United States and this is examined in *Harper's Ferry and the new technology*.[21] As the title suggests, this particularly focuses on activities at the U.S. government's armory established at Harpers Ferry in Virginia, but also explores the vital cooperation with the primary armory at Springfield, Massachusetts, which played a major part. However, there seems little to suggest that there was any new approach to making the lock components.

In this same period a set of machines was developed by Thomas Blanchard and are featured in a later study.[22] The first in this series was the *Machine for turning gunstocks* patented in 1819. A gunstock is far from symmetrical and at no point do they have a circular cross-section, the usual outcome of being 'turned' in a lathe. Accounts of their manufacture by hand[23, 24] covered in book 1 reveal the nature of the skills required even in their overall shaping to correct dimensions.

Additional machines were developed by Blanchard for carrying out other operations on the stock, including accurately creating the three-dimensionally complex recess for the lock, and which, perhaps more than any others, illustrate the problems faced by the designers of machine tools which had to replicate and replace the hand-craft skills.

Hounshell suggests that Thomas Blanchard did not draw inspiration from Brunel's block-making machinery because the blocks were not irregularly shaped. He further suggests[25] that Brunel drew inspiration from *as yet unknown American inventions* on the basis that he was living in the United States at the end of the 18th century. These statements have to be challenged. If there was such an important innovation as to inspire Brunel's block-making machines, how could it remain '*as yet unknown*' after the extensive research by numerous people into this subject?

Brunel's ideas were first raised by him when dining in Washington with Alexander Hamilton, Secretary of the Treasury and often referred to as "the father of American manufacturing".[26] In his Report on Manufactures, Hamilton made an extremely prescient comment – *If there be anything in a remark often to be met with, namely, that there is, in the genius of the people of this country, a peculiar aptitude for mechanic improvements, it would operate as a forcible reason for giving opportunities to the exercise of that species of talent, by the propagation of manufactures.*[27] It is suggested that there seems a greater possibility, therefore, that Brunel's ideas may have passed via Hamilton to inspire *as yet unknown American inventions*.

INTRODUCTION

Blanchard himself, in his revised 1819 patent, refers to rotary cutters *as used in the English machines*[28] and it was even noted in Russia that Brunel's machinery might be adapted to gunstocks.[29] Blanchard's patent claimed no general principles but only a systematic arrangement and method of operation[30] – exactly what was to be found in Brunel's machines. Brunel's 'dead-eye' machine, which was not included in his patent, embodied a feature crucial to Blanchard's gunstock lathe – a cam and follower which caused the rotary cutter to rock backwards and forwards as it followed a rotating eccentric cam to cut a discontinuous groove in the circumference of the circular 'dead-eye'. By giving this cam and 'rocking' cutter a longitudinal motion, it would create the basic principle of the Blanchard gunstock lathe. However, this in no way denies the importance of Blanchard's gun-stocking machinery which ranks as another major landmark of 'Yankee ingenuity'.

An official publication of the United States Ordnance Department[31] describes the manufacture of the then current rifle and bayonet. Although of later date, many of the processes referred to, along with illustrations of the machines used, are either identical or very closely similar to those being used twenty years earlier at Enfield and elsewhere. It is on a par with Gamel's work on the operations at the Tula weapon factory in 1826 but has less descriptive detail.

In 1850, another publication gave a detailed description of the manufacture of a musket for the Spanish government, *Memoria de la fabricacion del fusil de nuevo modelo en la Real Manufactura de Armas de Lieja en los anos de 1847 y 1848* [Mémoire on the manufacture of the new model musket in the Royal Manufacture of Weapons in Liège in the years of 1847 and 1848]. It was published in Madrid but, as seen from the title, it describes gunmaking practices at the Royal Manufactory in Liège in the years 1847 and 1848, and the text is accompanied by numerous drawings. While the musket has several unusual features and although numerous machines were employed at various stages in the processes, it is clear that hand-methods were still relied upon to a large extent. Thus, the barrel was still welded by hand, despite the development in England of rolling mills for this purpose, and the machines used for finishing the barrel are reminiscent of those used at Tula. However, it does incorporate a machine for milling the nipple bolster and is the only illustration of a machine of this nature which has been encountered.

Hand-forging was extensively applied to making various lock components, but die-forging appears to have been reserved for their final shaping, and the stock was a hand-crafted product. In this respect, they reflect more the late 18th/early 19th century practices. Nor is there any mention of interchangeability, but that is hardly surprising. That concept was not going to re-emerge in Europe for another few years.

Rosenberg asserts that die-forging techniques, drilling and filing jigs, taps and gauges formed part of *America's most significant contribution to this emerging technological process* and his final comment uses this and other factors to explain why this particular system first emerged in America.[32] It is suggested that it denies the historical evidence, available at the time of Rosenberg's comment, of dies being used in France in the last quarter of the 18th century and in Britain at least by the early 19th century and possibly even earlier, and other tools he mentions also existed previously. However, it has to be conceded that in the days before the internet, becoming aware of such evidence would have been more difficult.

Adam Smith, in his *Wealth of Nations* speaks of the facility of machines to *enable one man to do the work of many*[33] although the nature of the machines he was referring to were probably those in the textile trades. Writing later when even simple machine tools had been refined, in his work *On the economy of machinery and manufactures,* Babbage made pertinent observations:

> *A considerable difference exists between Making and Manufacturing. The Former refers to the production of a Small, the latter to a very large number of individuals* [items]...[34]

and

> *Whenever it is required to produce a great multitude of things, all of exactly the same kind, the proper time has arrived for the construction of tools or machines by which they may be manufactured.*[35]

He refers to what we commonly understand to be machines, with the observation of their ability to make things with *perfect identity,* thereby also alluding to 'mass production' and 'interchangeable manufacture'.[36]

An important aspect of verifying the accuracy of components, and especially those of the lock, in order to achieve interchangeability was the use of gauges. As noted in book 1, Blanc produced a set of gauges for this purpose around 1777 but only sparse mention of them in other gun manufacturing centres at that time has been found. It is ironic that it was the clockmaker, Eli Terry, who made the first practical application of a gauging system in America in the production of his famous wooden clocks as a standardised product with interchangeable parts made by different contractors working to gauges designed and supplied by Terry.[37]

Accounts of the work of Hall and North stress the importance of gauges to check the dimensional accuracy of a variety of components. They were, and still are, an essential part of the 'tooling' used in precision manufacturing. Uselding[38] suggests that in the 19th century, interchangeable manufacturing relied on 'size' accuracy determined by gauges, but that they were not used to check 'position or form'. In this it is suggested he is mistaken since both those attributes are clearly embodied in various gauges used by Blanc in the late eighteenth century and as will be shown later, in those supplied by Ames for the Enfield rifle.

Whilst the Vernier gauge had been invented in 1631,[39] no evidence has been found of its use in gun manufacturing workshops. Maudslay's bench screw micrometer of 1805,[40] built for his own use primarily, was far from portable and it wasn't until 1848 that the first hand-held micrometer was introduced by Jean Laurent Palmer of Paris.[41] The first of this type was put on the market by Brown and Sharpe in America in 1868 and one of these was the first to be imported into Britain at some time post-1868 by Churchill & Co.[42]. These facts make it all the

INTRODUCTION

more interesting that Lewis[43] in his thesis reports that George Lovell was using a micrometer to measure bore gauges at Enfield in 1833. Although to prepare such gauges, accurate to 0.001-inch, a highly accurate measuring device would have been required, the nature of Lovell's micrometer remains unknown.

Nevertheless, it was the supply of an extensive set of gauges and improved Blanchard stock-making machines supplied by the Ames Co. of Chicopee, Massachusetts, and the highly specialised metalworking machines supplied by Robbins & Lawrence of Windsor, Vermont, in combination with British barrel making technology, which became the foundations of the Royal Small Arms Factory at Enfield for the creation of the Pattern 1853 Enfield Rifle in an interchangeable form.

Using this machinery and the extensive set of gauges supplied at the same time enabled the first fully interchangeable firearm outside of America to be produced and is noted in the following comment:

> *Rifle Muskets made at the Royal Small Arms Factory are distinguished by the word* 'Enfield' *engraved on the lock-plate and have their corresponding parts exactly identical in size and interchangeable.*[44]

It is the paucity of technical details in the early accounts of, and the numerous errors in describing the manufacture of the Enfield Pattern 1853 rifle that are the foundations of this study. After the mid-19th century, no other study of this topic has been undertaken. The aim, therefore, is to provide for the first time a complete and detailed analysis of the machines and processes used in the manufacture of the principal components – lock, stock, and barrel – of this rifle and the achievement of interchangeability.

A number of important collections highly relevant to this topic will be examined in detail in their appropriate contexts to enable a greater understanding of the functional characteristics of the machines and the processes of manufacture they were applied to. The machines in their day were 'cutting edge' technology, the equivalent of the CNC machines of today. However, not all the functions are immediately apparent from existing specimens or drawings or gauges, so it has been necessary to create 'sketches' to elucidate certain points.

Chapter 2

Basic Technology of the Enfield Rifle

2.1 Introduction

It seems appropriate at the outset, for the benefit of those who may have no familiarity with firearms, to outline the technology and construction of the rifle under discussion, since its functional characteristics define what was needed for its manufacture by both manual and mechanical means. Much of what has been said of the flintlock mechanism in book 1 also applies here.

Most firearms until the mid-19th century were characterised by three principal components – lock, stock, and barrel. The lock constituted the means of ignition – firing the gun – and was the only true 'mechanism'; the barrel contained the explosive charge and directed the projectile to its target; the stock was used as the mounting for lock and barrel in their correct juxtaposition.

These items were widely different; the lock comprised a number of moving components which had to interact with precision, otherwise the gun might fire accidentally – *go off at half cock*; the barrel was basically an iron tube accurately finished inside and out and strong enough to withstand the internal forces of the explosion of the powder; the stock was an elaborate piece of wood with accurately formed mounting points for the other components which were fitted to it and, importantly, enabled the gun to be aimed and fired.

2.2 Fundamentals of the percussion system of ignition

By the 1830s the 'percussion system' was becoming the prevalent means of providing ignition – 'firing the gun'. This was based on the 1807 British patent of Reverend Alexander Forsyth[1] who, by combining his interests in chemistry and shooting, created a system in which ignition was achieved by using the flash from highly unstable and sensitive chemicals, such as the salts of fulminic acid, gold or mercury fulminate, or a mixture such as potassium chlorate and sulphur, which detonate when struck, hence the term 'percussion ignition'. Detonation is a virtually instantaneous decomposition and may be over in 1/10,000 second; an explosion is a fast-burning decomposition and may last 1/100 second.[2]

Forsyth's basic principle underwent various mechanical refinements in the ensuing years until the 'percussion cap' was invented by Joshua Shaw in America

BASIC TECHNOLOGY OF THE ENFIELD RIFLE

Fig. 2.1. A sectioned barrel with the gunpowder and projectile in place and the cap sitting on the nipple, and showing the channel to the powder charge. (author's collection)

in 1822. It has been suggested that this may have been invented by John Day in England in 1823[3] but Shaw was granted a United States patent on 19th June 1822 and is supported by additional documentary evidence (see Appendix 1). The 'cap' became the most successful way of exploiting Forsyth's invention.

In the percussion system, a cap, generally made of copper and often in the shape of a 'top hat', contained a small amount of 'fulminate' in the crown secured by a drop of shellac varnish. The cap was fitted to the top of a 'nipple', a hollow column screwed into the breech of the barrel. This provided a channel down which the flame from the fulminate was blasted at the instant of detonation and ignited the main charge of gunpowder propellant in the barrel.

Shaw's combination of Forsyth's detonating principle embodied in a copper cap could be considered as the greatest revolution in firearms development since the invention of gunpowder itself, since it led to, and sustains, all firearms to this day which use self-contained metal-cased cartridges which have a percussion 'primer' or 'cap' fitted in their base.

2.3 The lock

The means by which ignition was achieved by means of a 'lock', which might be loosely compared with a common lock in having several interacting components whose movements were driven or controlled by springs. It was the only 'mechanism' in the rifle. Externally the percussion lock is very simple, having

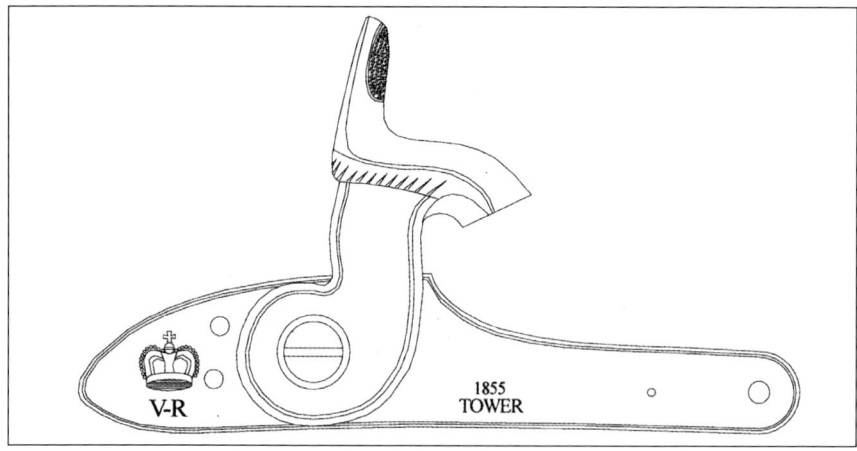

Fig. 2.2. A typical percussion lock as fitted to the Enfield Pattern 1853 rifle.(© P Smithurst)

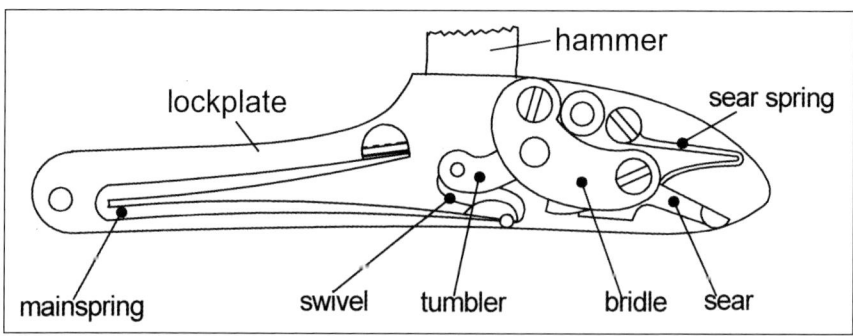

Fig. 2.3. The internal elements of a typical percussion lock as fitted to the Enfield Pattern 1853 rifle. (© P Smithurst)

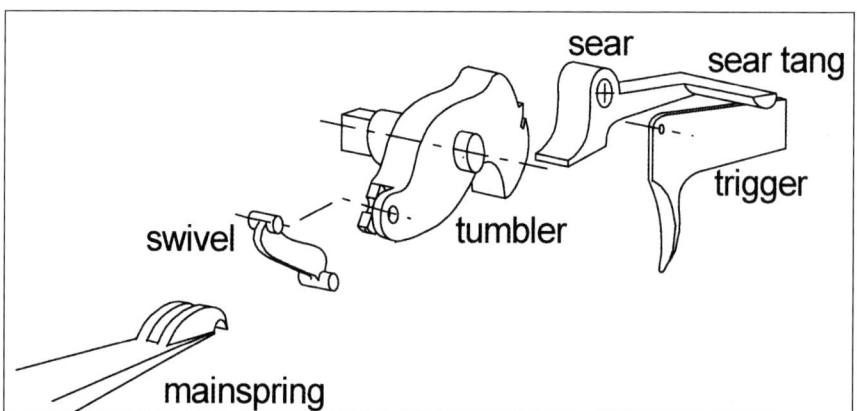

Fig. 2.4. 'Exploded' perspective sketch showing the train of interacting components. (© P. Smithurst)

a hammer that can be drawn back against the mainspring and locked in position until it is released by pulling the trigger.

Internally it has several components which must interact accurately in a 'train' for the lock to work and to do so safely.

2.4 Operation of the lock – 'cocking' and 'firing' the gun

A square hole in the hammer allowed it to be fitted in the correct orientation to a squared shank on the tumbler. As the hammer is drawn back, it causes the tumbler to rotate clockwise as shown in Fig. 2.5. The sear is acted upon by the sear spring so that its thin nose is always pressed against the body of the tumbler which has two notches, or 'bents'. Eventually, as the hammer is drawn further back the sear nose encounters the first 'bent' and drops into place with an audible 'click'. This is the 'half-cock bent' and is undercut so that if the hammer is eased forward again at this point, the nose of the sear is trapped in a slot. Even though the mainspring has been partly compressed by this action, the trapped nose of the sear prevents it being released when the trigger is pressed. This is known as the 'half-cocked' or 'safe' position.

By continuing to draw the hammer back, the second 'bent' is brought into contact with the sear nose. This 'bent' is a simple step cut in the body of the tumbler and the pressure of the sear spring forces the sear nose to fall behind this step. By releasing the hammer at this point, it is held in the 'full-cocked' or firing position by the sear preventing rotation of the tumbler, in the same manner as a ratchet and pawl, and the mainspring is fully compressed. Pressure on the trigger, which engages with the tang of the sear, (see Fig. 2.4) causes the sear to rotate on its axis pin. Because this 'bent' is simply stepped, the sear nose disengages, allowing the tumbler to rotate under the pressure of the mainspring, carrying the hammer with it to strike the percussion cap with great force and detonate it.

The importance of the correct interaction of these lock components, which is dependent upon their accurate formation, cannot be overemphasised and Cotty[4] goes to great lengths to explain the consequences of any errors in their relative proportions and faults in their shapes at their points of interaction. The tumbler is possibly the most important element since it controls the action of the lock

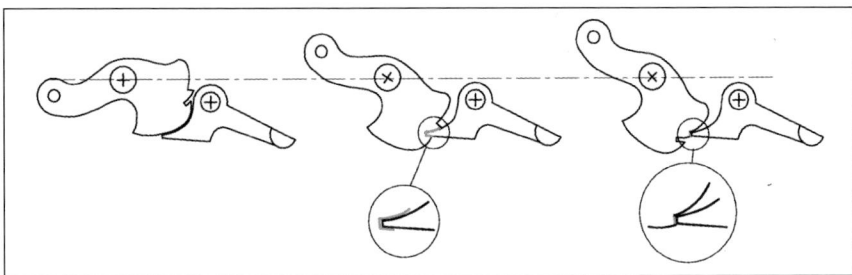

Fig. 2.5. Interactions of tumbler and sear. (© P Smithurst)

and for this reason was often referred to by artillery officers in charge of musket manufacture in late 18th century France as 'the brain of the lock'.[5] Whilst Cotty's comments refer to the flintlock, they are equally applicable to comparable components in the percussion lock. All of the lock components are subject to wear and some, especially the forks at the tip of the mainspring and the swivel they engage with, are subject to shock and therefore need to be made from the highest quality materials and heat treated in ways appropriate to the service they perform.

2.5 The barrel

The barrel of the Enfield rifle actually comprised two elements – the barrel proper which is an iron tube having a cylindrical bore cut with three shallow and wide spiral grooves and tapered on the outside. It is provided with a female thread at the wide breech end to receive a threaded plug, or *breech pin* in the terminology of that time, which is screwed into it. This breech pin was also forged with an integral *tang* in the form of a narrow and relatively thin rearward extension which serves to act as an anchorage point for the breech-end of the barrel.

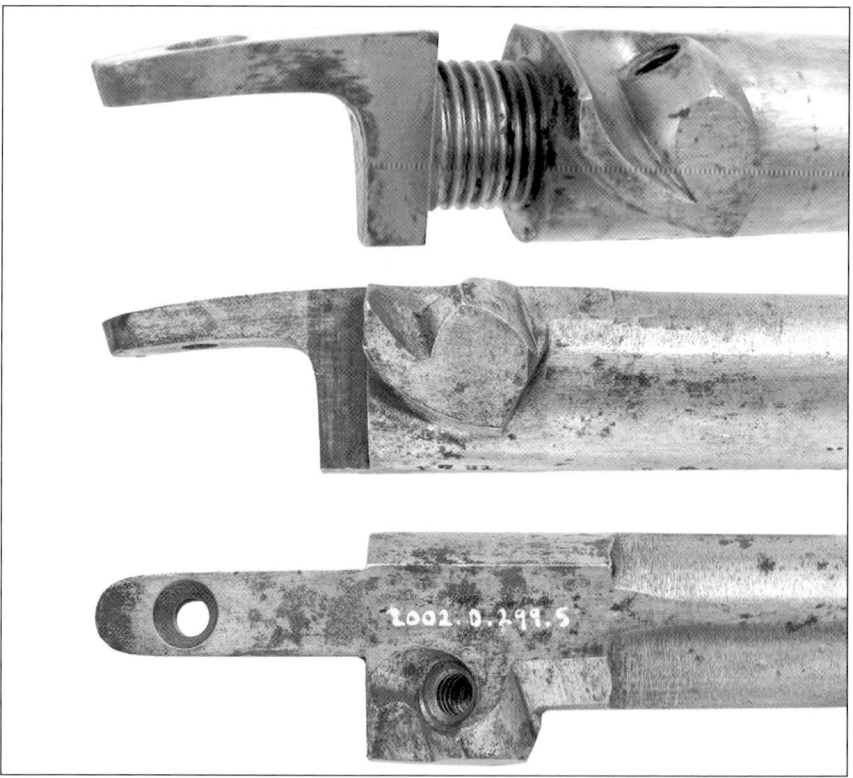

Fig. 2.6. A breech pin fitted to an Enfield barrel. (Courtesy Birmingham Museums Trust)

Above: Fig. 2.7. Enfield rifle front sight. (XII.979, © Royal Armouries)

Right: Fig. 2.8. The Enfield rifle rear sight. (I.L.N. 1855, p. 410)

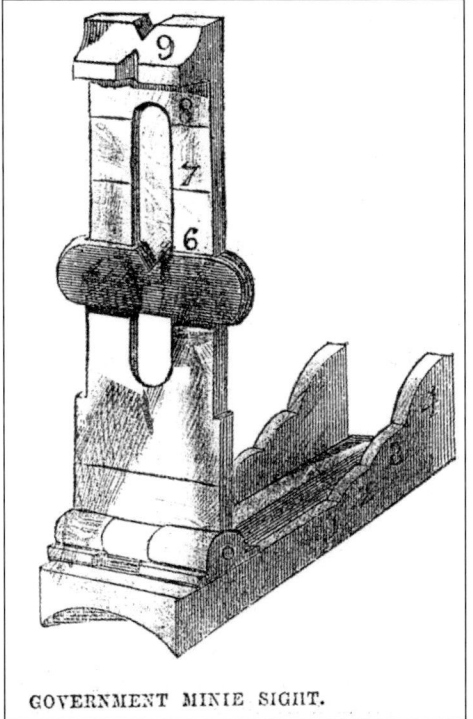

The barrel was fitted with two sights. The front sight was brazed to the barrel as a permanent fixture and in cross-section took the form of an inverted 'V' on a rectangular base which also served as a securing device for the bayonet.

The rear sight was more elaborate in having the facility for adjustment, allowing the barrel to be inclined to the horizontal by pre-determined amounts to accommodate a variety of ranges up to 900 or 1000 yards. It was soft-soldered to the barrel on an accurate centreline and at a predetermined distance from the front sight.

2.6 Rifling

Prior to the development of the rifle most firearms, whether for sporting or military use, were 'smoothbores'; that is to say, the bores of the barrel were plain cylinders. However, it had long been known that 'rifling' – spiral grooves cut into the walls of the bore – substantially increased both accuracy and range of a projectile, both being highly desirable features. Accuracy was improved by causing the projectile to rotate rapidly, giving it gyroscopic stability in flight. Even a modest estimate with the Enfield Pattern 1853 rifle, given a rifling pitch of one turn in 6.6 feet (2 metres), indicates a bullet travelling at 1000 feet (303 metres) per second would acquire a rotation about its axis in its direction of flight approaching 10,000 rpm.

Range was increased by virtue of the bullet filling the bore, as it needed to do to fully engage the shallow rifling grooves, and thereby minimising or eliminating any escape of propellant gases in the gap between the bullet and the walls of the bore, usually referred to as 'windage'. Since range is a function of muzzle energy (kinetic energy of the projectile), which is itself a function of velocity, clearly the greater the work done by the explosive forces on the projectile, the greater its velocity and in consequence, its range. Perhaps the best testimony to the superiority of the rifle over the traditional smooth-bore musket is summed up by the comment of Captain McKerlie of the Royal Engineers following trials of the smooth-bored 1842 Pattern Musket at Chatham in 1846:

> ... as a general rule musketry fire should never be opened beyond 150 yards, and certainly not exceeding 200 yards; at this distance half the number of shots missed the target, 11ft. 6 inches, and at 150 yards a very large number also missed...[6]

2.7 The conundrum of a muzzle-loading rifle

For the virtues of a rifle to be realised in practice, the projectile had obviously to engage with the helical rifling grooves in the bore. For that to happen, the projectile needed to be the same diameter as the *bottom* of the grooves, which is larger than the diameter across the *top* of the ridges between the grooves – the *lands*.

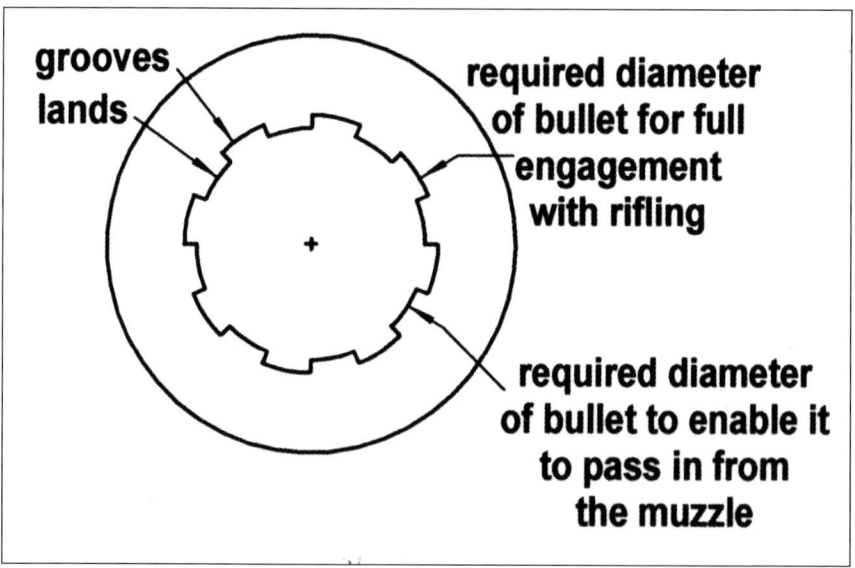

Fig. 2.9. Exaggerated schematic cross-section through a rifled barrel showing the different diameters involved. (© P. Smithurst)

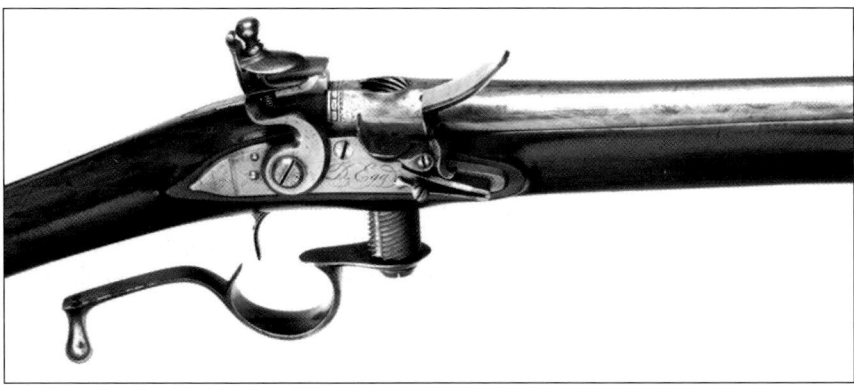

Fig. 2.10. The Ferguson rifle. By unscrewing a plug attached to the trigger guard, access to the breech end of the barrel was provided, whereby a ball and powder could be loaded. Screwing the plug back into place once again sealed the breech, ready for firing. This is Ferguson's own rifle. (XII.11209, © Royal Armouries)

How, then, could a projectile be loaded into a muzzle loading rifle if it needed to engage the rifling grooves but still remain a sliding fit in the bore? Patrick Ferguson provided one solution to that problem in 1776 by designing a rifle which was loaded at the breech. This allowed a projectile to be used which filled the rifling grooves at the outset and which performed excellently:

> *My rifle is in a fair way – by the unanimous suffrages of every officer who has seen it, it has been recommended as superior to any musket rifle or other firearm now in use, & Lord Townsend now talks of having some hundreds made.*[7]

Enough were made to equip a regiment, but wider adoption was mitigated against by costs. The breech-plug, to enable it to open and close the breech in a single revolution, had a seven-start thread and would pose an interesting problem to manufacture even today; the threaded hole into which it fitted closely even more so.

2.8 Solutions to the problem of muzzle-loading rifles

With muzzle-loading rifles, various methods were tried to overcome the inherent problem. For instance, with the rifle developed by Ezekiel Baker in 1800,[8] the projectile – a spherical ball – was wrapped in a thin linen 'patch' lubricated with tallow so that, when loaded, the flexible linen was pressed into the rifling grooves and at the same time gripped the ball with the result that when fired, the ball spun as it travelled along the barrel and jettisoned the patch when it left the muzzle. The Brunswick rifle, copied from its German namesake and adopted by the British army for limited issue in 1837, had two grooves and used a ball with an equatorial

Fig. 2.11. Left: A sectioned Brunswick rifle barrel showing the muzzle with its two notches coinciding with the rifling to guide the loading of the 'belted ball'. (XII.2444, © Royal Armouries). Right: a Brunswick 'belted ball'. (author's collection)

'belt' to match the geometry of the bore and which engaged the rifling in loading – and to ease loading the muzzle had two notches cut to act as a guide.

However, so difficult was it to load a muzzle-loading rifle after a number of rounds had been fired, due to fouling from accumulated powder residues, that in the case of the Baker rifle, the soldiers were issued with mallets to help drive the ball down the bore.[9]

The Brunswick rifle fared no better:

> *The loading of this rifle is so difficult that it is wonderful how the rifle regiments can have continued to use it so long. The force required to ram down the ball being so great as to render a man's hand much too unsteady for accurate shooting.*[10]

In 1826, Captain Henri-Gustave Delvigne of the French army became one of the first to explore ways of overcoming this problem and in 1838 had specimens of his design of rifle made in England. These had a powder chamber at the breech end of the barrel which was of smaller diameter than the main bore of the rifle. This created a shoulder where the two diameters met and against which the easily fitting ball came to rest when loaded. The ball was then rammed hard with an iron ramrod, theoretically causing it to expand radially to fill the rifling grooves.[11]

In 1843 another French officer, Captain Louis-Étienne de Thouvenin, devised a breech plug with a central pillar extending into the bore and against which the ball seated and could be rammed to expand it into the rifling.[12]

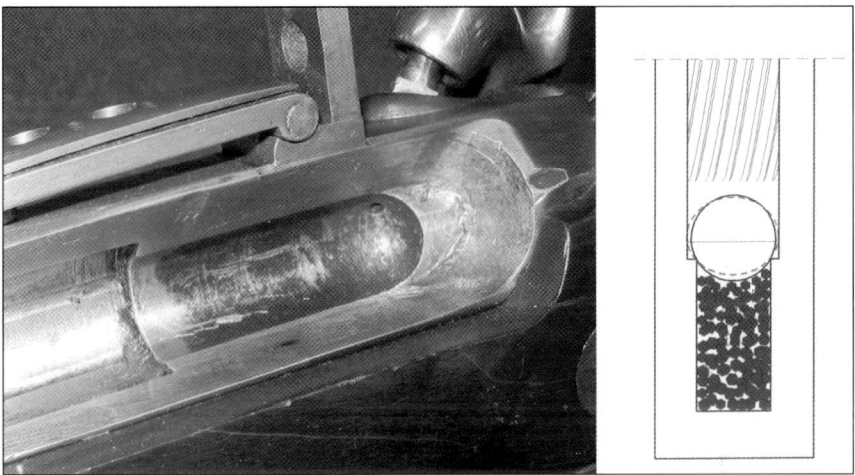

Fig. 2.12. Left: Section through the breech of a Delvigne rifle showing the reduced-bore powder chamber and the shoulder against which the ball was rammed and expanded. (XII.419, © Royal Armouries) Right: sketch to show how Delvigne's system worked.

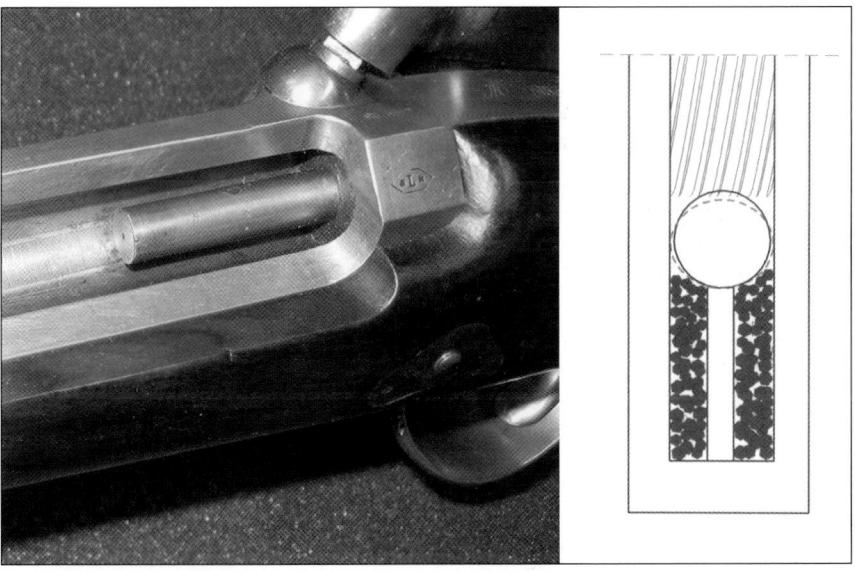

Fig. 2.13. Left: Section of the breech of a Thouvenin rifle showing the pillar against which the ball was rammed and expanded. (XII.2445, © Royal Armouries) Right: sketch to show the principle of Thouvenin's system.

There were various drawbacks with both of these systems, probably the most important one in the author's view being that an initially symmetrical projectile was distorted into an unpredictable asymmetrical shape which would have had a serious effect on its directional stability in flight and, therefore, on accuracy.

2.9 Minié's solution

William Greener, a well-known gunmaker in Birmingham, was also exploring this problem and in 1835[13] he submitted a projectile to the Board of Ordnance in which the force of the explosion itself was used to expand the bullet into the rifling. It consisted of a ball with a cavity into which was fitted a conical plug. At the instant of firing, this plug was driven into the cavity, causing the ball to expand.

The problem here was ensuring that the plug and cavity remained coaxial with the bore in loading and although Greener's idea was dismissed as being 'useless and chimerical'[14] ultimately, in modified form, it was to provide the basis of the system adopted.

In 1842, Delvigne patented another method of expanding the bullet which bears close resemblance to Greener's, except an elongated bullet was used which had a cavity in its base and instead of a conical wedge being used to expand the base, reliance was placed upon the gases from the explosion to cause expansion. It has been noted that the British government were monitoring foreign developments in small arms, especially in France, and the Inspector of Small Arms, George Lovell, had been in regular correspondence with Delvigne from 1838.[15] There is the possibility, therefore, that Lovell had mentioned Greener's 'useless and chimerical' proposal to Delvigne. Be that as it may, the next person to become engaged in the problem was Captain Claude Etienne Minié, and in 1844 he simply took Delvigne's bullet and inserted an iron cup in the base cavity, in effect acting in the same way as Greener's conical plug. However, despite Minié acknowledging Delvigne's work and making a joint statement that they wished the system to be called the 'Delvigne – Minié', the shorter name was adopted and it became the 'Minié', or more often just simply the 'Minnie'.[16]

Greener's name in this connection has been forgotten, despite his contribution being officially recognised in 1856 when:

> *The Lords of the Treasury have, at the recommendation of Lord Panmure, awarded the sum of £1,000 to Mr. Greener, as a public recognition of his priority in bringing before the late Board of Ordnance, in the year 1836, his suggestion for the application of the expansive principle in rifle bullets.*[17]

Fig. 2.14. Greener's expanding ball. (Greener, 1884, p. 124)

BASIC TECHNOLOGY OF THE ENFIELD RIFLE

A claim by a Captain Norton to have invented it in 1828 was summarily dismissed:

> *The committee do not feel they have official grounds or documentary evidence on which to recognise Captain Norton's claim.*[18]

2.10 Minié's system adopted and adapted

In 1851, the British military establishment adopted Minié's projectile and created the Pattern 1851 'Minié' rifle. The original Minié bullet was conoidal and was difficult to load so that it remained coaxial with the bore during loading and when fired. It was very likely to leave the muzzle with an axis of rotation inclined to its axis of symmetry, resulting in inaccuracy.

These and other factors led to the series of experiments in 1852 which resulted in the creation of the Pattern 1853 Enfield Rifle. Along with the rifle was a new

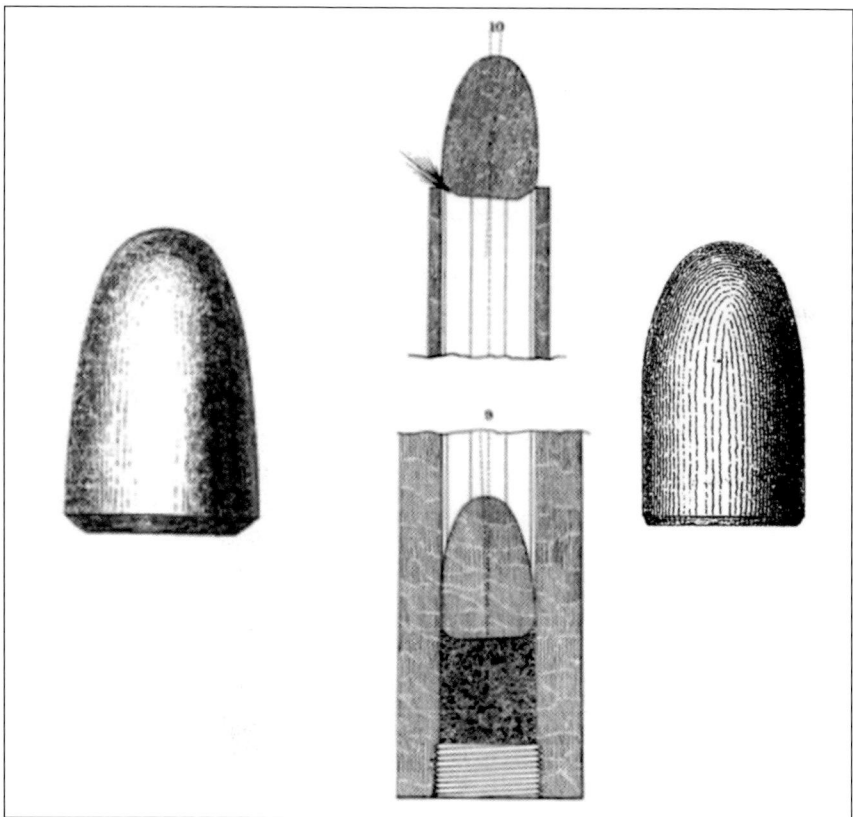

Fig. 2.15. Left: The Minié conical bullet's shape made correct alignment in loading almost impossible. Right: this defect overcome by Pritchett's cylindro-conoidal bullet. (Report, 1852)

bullet. A cylindro-conoidal bullet designed by Pritchett, having a cylindrical base portion, was adopted. Both bullets had hollow conical cavities in the base which, in the Pritchett bullet, was fitted a truncated conical boxwood plug. In principle, on firing this plug was driven into the cavity, causing the 'skirt' to expand and fill the rifling grooves. However, one author in 1858 seriously contested the basic premise in a long article in *The Engineer*,[19] going as far as to suggest the plug served no useful purpose. Nevertheless, the Enfield rifle using Pritchett's bullet was to prove a formidable combination and was used in conflicts across the world.

2.11 The stock

The stock acted as the foundation of the weapon, upon which all the components were mounted. The stock, therefore, had to be made with a similar precision to the components fitted to it so that they were neither too loose nor too tight. The barrel and lock had to be very accurately positioned in relationship to each other so that the hammer would fall centrally on the nipple.

In selecting timber, it had to be free from knots and any sign of incipient cracks, and the grain had to be carefully judged so that any curvature might be utilised rather than create a weakness. The wood had then to be properly seasoned for up to three years so that it would not warp in the extremes of climate encountered in Britain's widespread Empire. It was then sawn into the rough 'blank' shown below and when finished was liberally treated with linseed oil.

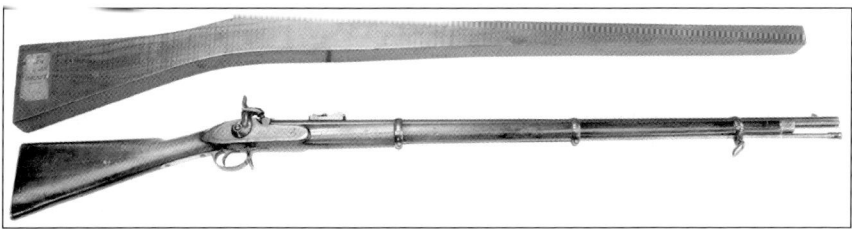

Fig. 2.16. Top; The Italian walnut stock blank. Bottom; the finished rifle. (© Royal Armouries)

Chapter 3

Origins and Procurement

3.1 Origin of the Pattern 1853 rifle

In the early 1850s, the decision to arm the British serviceman with an accurate and long-range rifle in place of the inaccurate and short-range smooth-bore musket had been brought about by ground-breaking developments in the design of projectiles as outlined in Chapter 2. The first step in this process was taken on 13th October 1851, when, after a series of trials, the Committee on Small Arms recorded the specifications for a new Rifled Musket based on Minié's principle:

> Barrel of .702 in. calibre rifled with four grooves of equal depth throughout their length. Minié projectile to weigh, including its iron cup, 680 grains. Charge to be 2½ drachms, and 50 rounds of ammunition to weigh 5 pounds. 10 ounces. The entire musket to work out at 8 ounces less than that in service.[1]

> [Note that avoirdupois weights were used in 19th century English ballistic terminology: 680 grains = 44.2 g.; 2½ drachms = 0.156 ounces = 4.43 g.; 8 ounces. = 226 g.; 5 pounds 10 ounces. = 2.55 kg.]

The Marquess of Anglesey, Master General of Ordnance, with the agreement of the Commander in Chief, the Duke of Wellington, determined that the 'Minié' rifle, designated as the Rifle, Pattern 1851, should become the standard issue weapon to all infantry. However, a variety of problems were soon experienced with the new rifle, the principal one being its weight, but also its large calibre which limited the quantity of ammunition a soldier could carry. As a result, in 1852[2] the new Master General of Ordnance, Viscount Hardinge:

> invited some of the principal gunmakers of England to submit patterns of muskets [rifles] for the use of the army, in the hopes of obtaining a lighter and more efficient arm for the service. The following makers prepared and sent in muskets for trial: Mr. Purdey, Mr. Westley Richards, Mr. Lancaster, Mr. Wilkinson, and Mr. Greener. Mr. Lovell, the Inspector of Small Arms, also prepared a new musket at the Government manufactory.

MAKING THE ENFIELD PATTERN 1853 RIFLE-MUSKET

A series of experiments ensued in which each of the rifles submitted underwent rigorous comparative trials to determine the merits of each with regard to a variety of factors including the requirements of rugged military service. The outcome was that two rifles were made at the Royal Manufactory at Enfield:

> *in which were embodied the improvements and alterations suggested by the experience obtained during the course of the trials with the experimental arms, and which, it was hoped, would possess the necessary requirements for a military weapon.*

> *The musket including bayonet to weigh about 9 pounds 3 ounces*
> *The bore decided upon was .577 inch*
> *The barrel to be, in length, 3 feet 3 inches*
> "   "   *weight, 4 pounds 6 ounces*
> "   *to have 3 grooves.*
> "   *to have a constant spiral of 1 turn in 6 feet 6 inches.*
> "   *to be fastened to the stock by 3 bands.*
> *The ramrod to have a swell near the head.*
> *The bayonet to be fixed by means of a locking ring.*
> *The lock to be made with a swivel.*[3]

[The Imperial weights and measures approximately correspond to: weight, 4.16 kg.; bore size, 14.7 mm; barrel length, 1 metre; barrel weight 2 kg; rifled 1 turn in 2 metres.]

These were significant departures from what had become the traditional design of the musket and its components for British military service. Fig. 3.1 shows the Pattern 1851 rifle along with the flintlock musket of 50 years earlier. In its overall form, the Pattern 1851 Minié rifle owed much to its English forebears, especially in regard to the attachment of the barrel to the stock using cross-pins passing through lugs attached beneath the barrel, and 'ramrod pipes' to hold the ramrod in place.

In contrast the Pattern 1853 rifle owed more to the French Model 1777 musket in the use of iron 'bands' encircling both barrel and stock and securing

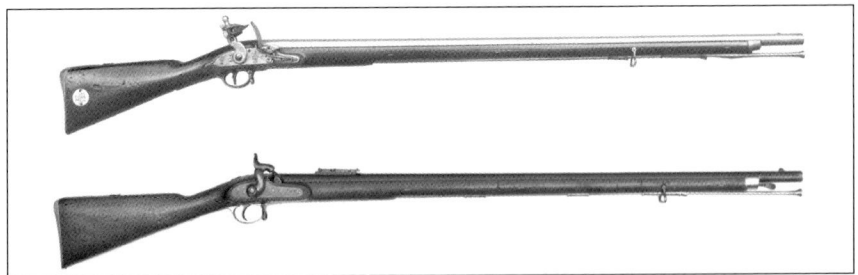

Fig. 3.1. Top; New Land Pattern musket, the last of the flintlock muskets, c 1802 (XII.132). Bottom; 1851 Pattern Minié rifle (XII.1907). (© Royal Armouries)

ORIGINS AND PROCUREMENT

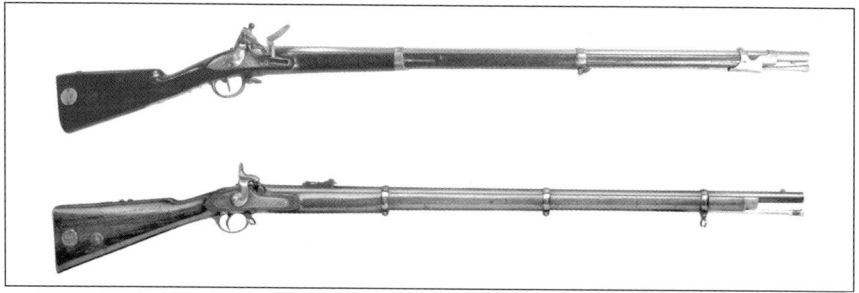

Fig. 3.2. Top; The French Model 1777 musket. (XII.201). Bottom; Pattern 1853 Enfield rifle. (sealed pattern, XII.1918). (© Royal Armouries)

the ramrod, features that were to become widely adopted for military firearms, Fig. 3.2.

At the core of this British military rifle, then, were French innovations in design and more especially, in ballistics through using Pritchett's modification of the Minié bullet.

For any new weapon which was accepted into service, one or more specimens were prepared to which were attached one or more wax seals bearing the arms of the Board of Ordnance or other official bodies and which were referred to as 'sealed patterns'. These sealed patterns then became the reference standards upon which manufacture was based and products assessed, a system initiated in 1631 by Charles I.[4]

3.2 Procurement by contract

From the earliest days of the musket becoming the primary infantry weapon, procurement had been through contracts with the private gun trade which, by the 1850s, was concentrated in Birmingham. Most firearms until the mid-19th century were characterised by three principal components – *lock, stock and barrel* – and their purpose and characteristics have been discussed in Chapter 2. In addition, there were numerous other items such as trigger, trigger guard, buttplate, barrel bands, and various screws which are all grouped together under the term 'gun furniture'. The wide variety of specialised skills used in the production of firearms has been shown in Chapter 1. It is a perfect illustration of Adam Smith's observation on the division of labour:

> *This great increase in the quantity of work, which, in consequence of the division of labour, the same number of people are capable of performing, is owing to three different circumstances; first to the increase in dexterity of every particular workman; secondly to the saving of the time which is commonly lost in passing from one species of work to another and, lastly, to the invention of a great number of machines which facilitate and abridge labour, and enable one man to do the work of many.*[5]

MAKING THE ENFIELD PATTERN 1853 RIFLE-MUSKET

The way the contract system worked was that:

> *the component parts of a musket are in the first instance procured by open contract, and after inspection and approval, placed in the Ordnance Stores. There is a separate contract for each of the separate parts; each is subjected to a strict view or examination by viewers specially appointed for the purpose; accuracy of shape and size is secured by a system of gauges, while other tests are employed to ascertain the strength of the work and the soundness of the material... A portion of the materials so received into store is issued to the Government establishment at Enfield, to be there set up; that is, fitted together and finished into complete muskets. The chief part of the materials is, however, delivered to contractors who contract for the setting-up, and return the arms to the Board of Ordnance finished. In this state it is again subjected to a view.*[6]

['setting-up' was the fitting and assembling of all the components to produce a finished rifle and entailed creating special features on the stock to accept the various components.]

Despite the fact that gauges were used in the inspection process, none, it would appear, were issued to the contractors, nor were they issued a 'pattern' of the item they had contracted to make, Fig. 3.3.

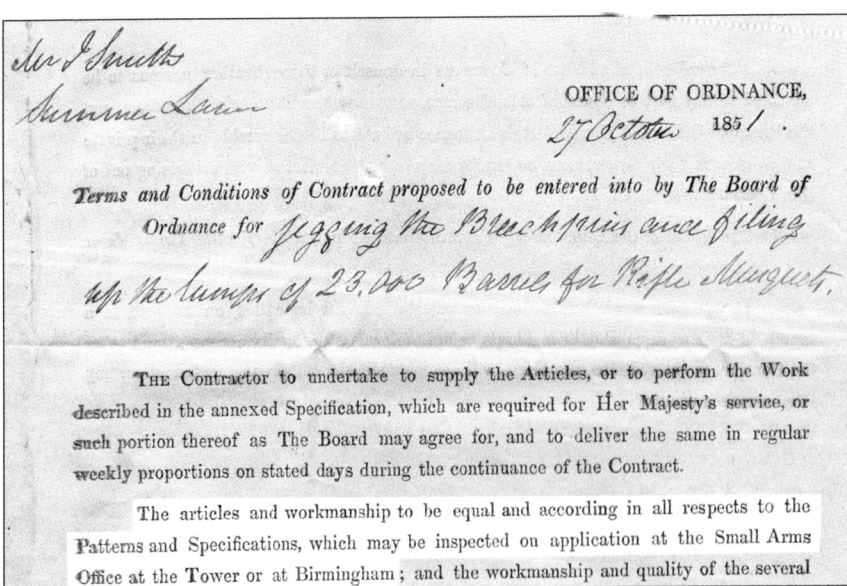

Fig. 3.3. Contract details issued to Joseph Smith in 1851. (Appendix 2; © P. Smithurst)

They had to work from a single 'pattern' weapon deposited at the Ordnance Department office in Birmingham or the Tower of London and obtain their specifications by examining that, or the relevant parts of it. James Gunner, Superintendent at Enfield, in his evidence to the Select Committee of 1854 stated that three sealed patterns were created at Enfield upon the introduction of a new weapon, and one was issued to the Tower and one to the Ordnance Office in Birmingham.[7] This raises an immediate question – were these made using the same filing jigs, gauges etc.? If they were, then were they interchangeable? It seems very unlikely since interchangeability was not actively being sought at that time. In that case, workers in Birmingham and London would be working to two close but different standards.

Gunner also suggested that these pattern weapons should be loaned to individual contractors for a few days to enable them to make their own gauges and that if they worked together, it would only take them two or three days.[8] Considering the number and nature of the gauges later supplied by Ames for the lock, for instance, an allowance of two or three days to make as many sets as there were contractors, each set allowing each component to fit them as accurately as required by the inspection process, this was a very unrealistic time estimate. Gauge making is demanding of both skill and time. For the Board of Ordnance not to have issued patterns or matching gauges to the contractors was a very inefficient way of proceeding.

3.3 Discontent amongst the contractors

The system was almost guaranteed to cause delay and arouse discontent and complaints by the contractors over the strictness of the inspection process or 'viewing', and became a serious bone of contention for those engaged in 'government work':

> *When a man finds his work rejected and incurs a serious loss for some deviation so slight, that, for all practical purposes, it is immaterial, he will naturally hesitate to continue in a business where his week's earnings depend upon such doubtful, and, it may be, capricious decisions.*[9]

When it is realised that the gauging system in use only specified an upper limit on size and the lower acceptable limit was at the discretion of the inspector, such discontent and distrust of the system is understandable. Added to those problems faced by contractors was the fact that while the 1851 Pattern 'Minié' rifle largely copied its forebears, the *rifled* barrel was a new departure and created at least one other step in the manufacturing process. But that was a minor problem in comparison with the Pattern 1853 rifle which was different in almost every respect. Such changes, following rapidly upon one another, meant that contractors were still producing one while having to re-tool to produce the other and for which only a limited amount of their existing tooling could be applied.

3.4 Discontent of the government

Following the issue of a contract for 28,000 of the new 1851 Pattern rifles in February 1852, they were not delivered until November 1853. Also, after the adoption into service of the 1853 Pattern rifle and arms derived from it, in April of that year contracts were entered into for the supply of 2,000 artillery carbines. Only 500 had been completed by January 1854.[10] An order for materials for 20,000 rifles placed in July 1853 was seriously delayed. By this time the situation was critical because the Crimean War created a pressing need for arms. Part of this problem was historical in nature. J. Wood, Secretary to the Board of Ordnance, in a memorandum dated February 1854, suggested that after the peace in 1815:

> *the manufacture of arms for the government ceased and the workmen were dispersed. Little afterwards was done with regard to the provision of arms, until the adoption of the percussion principle, when a re-equipment of the army became necessary.*
>
> *The trade had then fallen into a very disjointed state and there was difficulty in collecting together men capable of making a new arm in a satisfactory manner...*

He went on to comment on:

> *the injury to the service and the high prices which resulted from the organised combinations both of the masters and men in the gun trade.*

Various other deficiencies of the system were cited in a similar vein, but perhaps the most significant was:

> *The rifle musket of 1851 having been superseded in 1853 by another of smaller bore, and somewhat different construction, the Board, in July last, called for tenders for materials for 20,000 muskets of the latter description. The offers received were so unsatisfactory as to price, and evinced so perfect a combination amongst the parties, that they were, after some correspondence, declined...*[11]

That such 'combinations' were very real is evident from a resolution at a meeting of the Setting-Up Contractors at the Stork Hotel in Birmingham on Thursday, March 27th, 1856, chaired by J. Goodman Esq., and sent to Joseph Smith (see Appendix 2, item 9).

In response to this dire need for rifles, on the 31st October 1854 the Board of Ordnance issued a contract for 20,000 rifles to manufacturers in Liège.[12] These rifles are easily recognizable by the date in italics and no place of manufacture on the lockplate and Liège proof marks. On 12th February 1855, Messrs. Fox, Henderson, & Co.'s offer to supply, through 'a manufacturing firm in the United States of America', 25,000 Enfield rifles at £3.50 each, was accepted. This manufacturer was Robbins & Lawrence. Fox, Henderson & Co. had prepared Joseph Paxton's submission of his design for the Great Exhibition building[13] and became the main contractor for the construction of the 'Crystal Palace' and, as such, would have had intimate knowledge of the exhibitors. There was no formal contract between the Board of Ordnance and Fox, Henderson & Co., they had merely accepted an offer; the contract was between Fox, Henderson & Co., and Robbins & Lawrence (see Appendix 8). Whilst it is a deeply complex legal and financial concatenation of events which have been discussed in varying detail from varying perspectives[14, 15] it was the foreclosing of this contract with Robbins & Lawrence by Fox, Henderson & Co. that precipitated the final demise of Robbins & Lawrence in 1856 and ultimately of Fox, Henderson & Co. shortly afterwards. The contract is an event preserved in the rare Pattern 1853 rifles Robbins & Lawrence did succeed in producing, distinguished by being the only 'Enfield' rifles to have the place of manufacture outside of England, WINDSOR, inscribed on the lockplate along with the crown but without the 'V.R.' monogram.

The Board of Ordnance were also possibly stimulated in their dissatisfaction with products obtained through contract by the products exhibited at the Great Exhibition by Samuel Colt of Hartford, Connecticut, and Robbins & Lawrence of Windsor, Vermont. But it was not the firearms themselves which were the cause of this stimulation; it was the way in which they had been manufactured. Both companies exhibited firearms made almost completely by machinery. Some have claimed that Colt's revolvers were interchangeable, but it has been argued,[16] using documentary[17] and material evidence, Fig. 3.5, that at that time such claims are unsustainable. Serial numbers stamped on each component acted as 'assembly marks', traditionally used in non-interchangeable firearms to allow parts to be correctly reunited after further processing.

By contrast, the Robbins & Lawrence rifle locks have all the hallmarks of interchangeability; their lack of such 'assembly marks'.

Fig. 3.4. Robbins & Lawrence contract Pattern 1853 rifle. (XII.1628 © Royal Armouries)

Fig. 3.5. Components of an early production Colt 1851 Navy revolver exhibiting the serial number on every component. (© P. Smithurst)

Fig. 3.6. Lock from a Robbins & Lawrence 'Mississippi' rifle dated 1851 - possibly one of those exhibited at the Great Exhibition. (XII.430, © Royal Armouries)

This latter fact itself was unheard of in English or European gunmaking, other than the work of Honoré Blanc in the late 18th century and at Tula in the 1820s. These factors led the Honourable Board of Ordnance to declare that:

> *Owing to the delays constantly recurring in the fulfilment of contracts for arms, the high price demanded by the contractors, and the inconvenience occasioned to the service by these causes, the Honourable Board of Ordnance, towards the end of the year 1853, considered it advisable, in order to secure a regular supply of them,*

ORIGINS AND PROCUREMENT

> *to take this branch of manufacture into their own hands, and erect a government establishment capable of producing muskets in large numbers, and at a moderate price, by the introduction of machinery into every part of the manufacture where it was applicable.*[18]

There were various outcomes from this. When the Ordnance estimates of £150,000 for the building of a new factory were laid before Parliament, it was opposed on the grounds that:

> *the manufacturers of Birmingham and London were quite capable of supplying the Board with any quantity of arms necessary at a lower rate...*[19]

than the Board could make them, despite the evidence to the contrary and which had brought about this determination by the Board in the first place.

3.5 Select Committee on Small Arms

This disagreement resulted in the appointment of a Parliamentary Select Committee on the 1st March 1854 to:

> *consider the Cheapest, most expeditious, and most efficient mode of providing Small Arms for Her Majesty's Service.*[20]

Whilst the Minutes of Evidence of the Select Committee, which supplied the basis for the final report, make interesting and informative reading, it is only possible to include here a fraction of the various testimonies obtained and fragments of the Committee's findings. There was much, and entirely understandable, subjectivity on the part of the gunmakers themselves in defending their own positions. Objectivity in the support of mechanisation came from those who had extensive experience in that arena, whether in gunmaking or other areas of manufacture.

John Anderson, Inspector of Machinery at Woolwich Arsenal, had been responsible for designing and developing machinery for use in the manufacture of artillery equipment there.[21] The abstracts from his testimony as recorded in the Minutes of Evidence of the Select Committee,[22] are instructive in a variety of ways relating to the gun trade in Birmingham, the operations at Enfield and the use of machinery, of which he was an active and ardent proponent. In regard to the Birmingham gun trade, and especially those engaged in government contract work which he and Lt. Warlow, R.A., visited in March 1853, he comments:

> *We then visited a number of establishments engaged in military musket and bayonet work, all of which, however, are in a low mechanical state, and at least fifty years behind most of the other branches of manufacturing industries we have been examining.*

MAKING THE ENFIELD PATTERN 1853 RIFLE-MUSKET

> *From what I have seen in Birmingham, I fear there will be disappointment if dependence is placed on that quarter for a large supply of arms, there being neither system nor adequate machinery in that business.*

And concerning a visit to Enfield in October 1853, he reported:

> *There is also much to be done at Enfield in improving the gun manufacture; a large number of machines are much wanted, both for musket and rifle making; indeed, nothing less than the introduction of self-acting machinery to produce the several parts composing a gun will give satisfaction.*[23]

By December 1853 he had given a report to the Board of Ordnance which embraced the manufacture of small arms by machinery, extracts of which were included in the minutes of evidence:

> *the musket is eminently fitted to be made by machinery... In considering the musket as an article to be made by machinery, we must not look at it as a whole, but make each of its 57 parts a separate study; by so doing most of the difficulty vanishes.*

He was also of the opinion that what was simply a visual adornment did not improve performance of a firearm and might often impede manufacture:

> *Question. You are of opinion therefore that every part of a musket is so simple as to be capable of being produced by machinery?*
>
> *Answer. Yes. I should mention that there are some of the present parts that have an irregular form, which have nothing to do with the musket, as a musket, neither with its accuracy nor its quality.*[24]

This was a creed Samuel Colt, well versed in the manufacture of firearms by machinery, fully endorsed. In his testimony to the Select Committee on Small Arms, in response to questions from committee members he states:

> *I would simplify the gun very much... These detailed forms are not necessary, and it might be made suitable to the application of machinery by the one simple operation, and yet be made handsome.*[25]

These views are certainly evinced in the lines of his revolvers where any three-dimensional embellishments are eschewed in favour of a simplicity of form that created a streamlined aesthetic which set his revolvers apart from any others.

In his paper to the Institution of Civil Engineers, Colt makes his views on this even clearer:

> *Machinery is now employed by the author, to the extent of about eight-tenths of the whole cost of the construction of these firearms; he was induced gradually to use machinery to so great an extent, by finding that with hand labour it was not possible to obtain that amount of uniformity, or accuracy in the several parts, which is so desirable, and also because he could not otherwise get the number of arms made, at anything like the same cost, as by machinery...*[26]

While his claims, and subsequent claims of interchangeability on his behalf have been argued to be unsustainable based on physical and documentary evidence as noted earlier, it remains an excellent and succinct endorsement of the virtues of a proper, systematic, application of machinery to the manufacture of firearms. Taken with Anderson's earlier observations and the interchangeability achievement of Robbins & Lawrence, it is indicative of a new approach in which products are beginning to be designed for ease of manufacture by machine rather than for an elaborate aesthetic merit.

3.6 Pursuit of mechanisation

The deciding factors in the government's decision to pursue mechanisation, however, are contained in the summaries of the Select Committee with regard to the state of the trades involved in the making of the principal components, lock, stock and barrel, despite the occasional contradictory testimonies:[27]

The lock

Mr. Prosser [civil engineer in Birmingham, who was writing a book on the construction and manufacture of firearms] *said that no lock machinery is employed in this country... but that proper machines would reduce the cost of locks 50 per cent., and that the locks of the Russian musket are made by machinery in a very perfect manner.*

Mr. Lancaster [Charles William Lancaster, a prominent gunmaker who had supplied arms to the Board of Ordnance] *considered that much is to be done with machinery, which would ensure uniformity in the lock, and then but little skilled labour would be required to finish it.*

Mr. Goodman [John Dent Goodman, merchant and partner in a gunmaking business; later became Chairman of the Birmingham Small Arms Company] *considered that, although no part of the lock is at present made by machinery, yet it might be applied to locks with advantage, and that such locks would be better and more uniform than those made by hand labour.*

Mr. Swinburne [John Field Swinburne, contractor for 'setting up' guns for the government] *stated that lock machinery which now exists might be improved with some small degree of advantage, but that there are some lockmakers who now use machinery to a great extent.*

Mr. Nasmyth [James Nasmyth, inventor of the steam hammer who had supplied the Russian arsenal at Tula with steam stamping equipment for use in die-forging] *was of opinion that the more intricate parts of the lock... could be done by special machines.*

The stock
Up to the present time little has been done in this country towards making the stocks of guns by machinery.

Mr. Scott, [William Scott, Birmingham gunmaker, at one time viewer for the Board of Ordnance, and a 'setter-up'] *stated that he had attempted to apply machinery to this purpose but found existing machines and tools to be inadequate.*

Messrs. Wallis and Whitworth [George Wallis, Headmaster, Government School of Art, Birmingham and one of the Commissioners sent to visit the New York Exhibition, and visited the Springfield Armory and the Massachusetts Arms Company; Joseph Whitworth, engineer, machine tool builder and pioneer of standardised screw threads] *gave evidence of the existence of machinery in America that accomplished this object in a very superior manner, and which they had seen in actual operation.*

The barrel
According to Mr. Millward [Charles Millward, gun barrel maker in Birmingham and contractor to the Board of Ordnance] *more machinery has been applied to the barrel than to any other part, and the machinery which he employed saved him 30 per cent. in labour.*

Mr. Prosser *thought the present system of boring a 'bungle', and capable of being improved.*

Mr. Whitworth [Joseph Whitworth, the eminent engineer later experimented with forms of rifling and produced accurate long-range rifles which are highly prized even today] *expressed his opinion that great improvements might be made, and that the barrels might be finished on the outside by self-acting machines, without grinding.*

Colonel Gordon [Lt. Col. Alexander Gordon, member of the Committee on Small-Arms] *said that the system of rifling which is practised in Belgium is more correct, cheaper, and does not injure the health of the workmen so much as ours.*

Mr. Whitworth contemplated a... self-acting... machine that would rifle a number of gun barrels at the same time.

3.7 Committee on Machinery

On the authority of the Select Committee, a Committee on the Machinery of the United States, comprised of Lieutenant-Colonel Robert Burn, R.A., Lieutenant Thomas Picton Warlow, R.A., and John Anderson, Ordnance Inspector of Machinery, was sent to America early in 1854 to gather information regarding the machines used in the manufacture of firearms there.[28] This Committee pre-empted the findings of the Select Committee since they had been authorised to purchase machines for Enfield up to a value of £30,000. However, pending the findings of the Select Committee, this sum was reduced to £10,000 but later increased to £12,500.[29] They were also empowered to employ James Burton, the engineer at Harper's Ferry Armory in Virginia, whose expertise and experience in the operation of a firearms factory equipped for gunmaking by machinery would make him ideal to act as Superintendent / Assistant Engineer to oversee the setting up and commencement of operations at the Enfield factory (see Appendix 2). This in itself was a strong indication of the government's intentions and was reinforced by the delegation's contracting for machinery to be shipped to England.

3.8 Equipping the Enfield factory

On 26th May 1854, a tender from Robbins & Lawrence, supplemented on 24th July, was accepted by the Committee on Machinery to supply Enfield with tooling listed below:[30]

For lockplate	7 milling machines
	2 drilling machines, 5 spindles
	Tapping apparatus
	Apparatus for reaming tumbler hole to size
	1 edging machine
For hammer	2 drilling machines, 4 spindles
	6 milling machines
	1 checking machine for hammer hole and for trimming lockplate

For tumbler	2 double milling machines
	3 milling machines
	2 drilling machines, 4 spindles
	1 grooving machine
	1 squaring machine
	1 screw (hand) machine
For swivel	1 drilling machine, 4 spindles
	1 milling machine
For sear	1 drilling machine, 4 spindles
	2 milling machines
	1 double milling machine
For mainspring	4 milling machines
	1 drilling machine, 4 spindles
For bridle	1 drilling machine
	4 milling machines
For lock screws	2 screw milling machines
	2 thread cutting (hand) machines
	1 slitting machine
	1 pointing machine
	1 clipping machine
For nipple	1 clamp milling machine
	1 chasing machine
	1 squaring machine
	3 drilling machines, 6 spindles
	2 hand machines for finishing thread

On the 18th August, the Committee on Machinery accepted a further tender for:
>Mainspring testing apparatus
>6 milling bridges
>Apparatus for holding and stamping barrels

The total number of machines ordered was therefore 82, but it has been claimed[31] the number was 139 and that an additional 9 woodworking machines were supplied. No source is given for this claim and no evidence has been found to support this larger number. Woodworking machines were indeed supplied, but numbered 12, and were ordered by the Committee on Machinery for use by the Royal Carriage Department at Woolwich, not Enfield.[32]

It is often assumed that the machines alone created an accurate, interchangeable, component. That this is not so is indicated by the filing jigs detailed below supplied by the Ames Manufacturing Company of Chicopee, Massachusetts, as part of their contract of 18th August 1854,[33] and used to hand-finish components to size and form so that they fitted the gauges:

>2 filing jigs for the edges of the tumbler
>2 filing jigs for edges and 1 filing jig for end of tang of sear

1 filing jig for edges of bridle
2 filing jigs for hook and tang of mainspring
2 filing jigs for edges and 1 filing jig for length of mill end of sear spring
2 filing jigs for breech [pin?]
Filing jig for edges of trigger guard
Filing jig for mortise of trigger plate
Filing jig for edges of finger piece and 2 filing jigs for edges of blade

In addition to machines ordered from Robbins & Lawrence, on 17th May 1854, the Ames Company were also contracted to supply a set of 15 gun-stocking machines:[34]

Machine for roughing
Machine for rough turning (with patent rights)
Machine for spotting [machining small flat datum surfaces on the stock]
Machine for sawing breech and muzzle
Machine for bedding the barrel
Machine for planing side and edges with an extra spindle
Machine for bedding breech plate [i.e., butt plate]
Machine for fitting bands (with patent rights)
Machine for turning between the bands
Machine for smooth turning breech (with patent rights)
Machine for smooth turning above the lock (with patent rights)
Machine for bedding the lock [i.e., inletting the lock]
Machine for bedding the guard [trigger guard]
Machine for boring side tang screws [side nails?] and pin holes
Machine for grooving for ramrod

A supplementary order on 18th August 1854 was placed for 8 additional machines:[35]

1 Rough stocking machine
3 Smooth turning machines for the butt behind the lock
1 Smooth turning machine for the stock in front of the lock
2 Machines for edging the lockplate [presumably the same or similar to the edging machine supplied by Robbins & Lawrence]
1 apparatus for testing power [The purpose of this remains unknown but may refer to testing the mainspring or the force required to operate the trigger]
bringing the total number of machines from Ames to 23.

From an earlier comment by the Committee on Machinery,[36] patent rights had been purchased to allow them to make copies of machines should the need arise. A few years later, Greenwood & Batley of Leeds certainly produced stocking machines bearing a close similarity to, if not direct copies of, those supplied by Ames.[37] Details of many of those machines and special tools will be examined and discussed in the appropriate chapters.

3.9 Specifications of the Pattern 1853 rifle

There are no known full written specifications for the manufacture of the Enfield Rifle, and it has already been noted that contractors had to work to pattern arms supplied via the Ordnance Offices at the Tower of London or at Birmingham. A few contemporary factory drawings exist and once formed part of the M.O.D. Pattern Room collections. The Pattern Room was the facility at Enfield for the storage of the Sealed Pattern weapons and accessories and was later expanded to house specimens of the then current firearms in use by various nations, and that function continues through to the present time. These drawings, now lodged with Royal Armouries alongside the remainder of the Pattern Room collections, are probably unique but only one, that of the nipple, or 'cone', is dimensioned.

An unfinished and un-dated drawing shows the rifle from the left side, a detail from which is shown in Fig. 3.7

The interesting feature of this drawing is that it shows the relationship between the main functional elements. Another drawing, Fig. 3.8, is also undated and the size it is possible to reproduce it here reveals little detail but shows the rifle as it appeared in 1853 with the first pattern of barrel bands but with the post-Crimean War ramrod, suggesting a date of ca. 1857 or later.

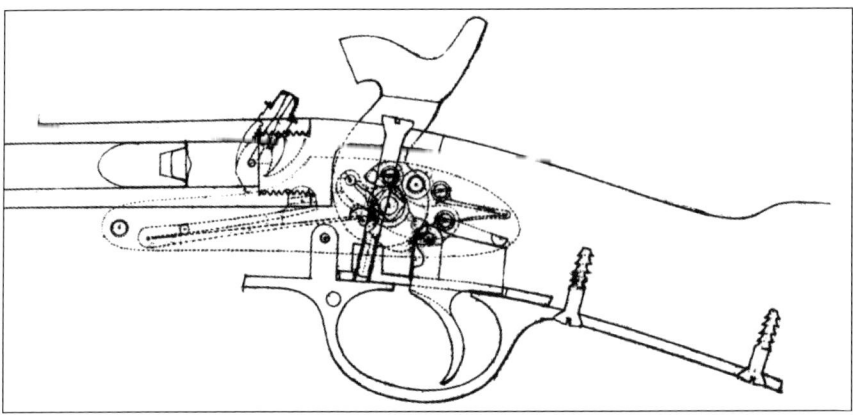

Fig. 3.7. Original Royal Small Arms Factory, Enfield, drawing, No. 458 (part, unfinished), showing the Pattern 1853 Rifle from the left. (© Royal Armouries)

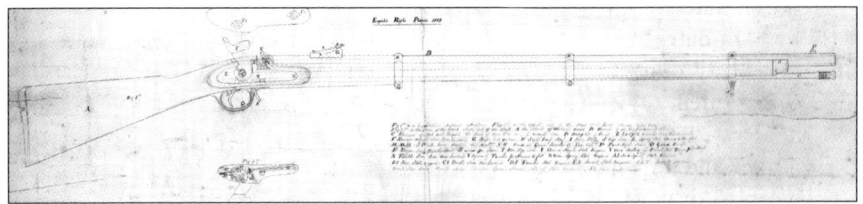

Fig. 3.8. Original but soiled Royal Small Arms Factory, Enfield, drawing, No. 746 of the Pattern 1853 Rifle. (© Royal Armouries)

ORIGINS AND PROCUREMENT

The captions on the drawing, transcribed below, give some understanding of the designated materials:

A. the stock of walnut wood	R. trigger, iron, case hardened
B. barrel of the very best wrought iron	S. breech pin, iron
C. ramrod of steel, well tempered. The head of same CX being of wrought iron	T. rod stop, iron
D. buttplate of brass	U. cone or nipple, steel, tempered
E. lockplate wrought iron case hardened	V. cone seating of barrel
F. hammer wrought iron case hardened	W. trigger plate, brass
G. sight body w[rought] iron	X. tumbler screw, iron, case hardened
H. sight leaf steel	Y. square of tumbler for hammer to fit
I. slide, steel	Z. mainspring, steel, tempered
J. cap, iron [top of sight leaf]	A1. sear spring, steel, tempered
L. spring, steel screws and pin, steel	B1. Sear, steel, tempered
M, M, M. three bands, lower middle and upper, iron	C1. Bridle, iron, case hardened
N, N. band and guard swivels	D1. Tumbler, steel, tempered
O. nose cap brass	E1. Swivel, steel, tempered
P. front sight iron	3, 4, 5, lock screws, turned, sear screw, bridle screw and sear spring screw all of steel, tempered
Q. guard, brass	F1. side cups, brass

These two drawings are discussed in more detail in Chapter 5.

A third drawing, Fig. 3.9, dated 1860 and signed by the draughtsman, Charles Hayes, shows the individual lock components. Details from this will be used in Chapter 4 which deals with the manufacture of the lock

A further un-numbered drawing, Fig. 3.10, has the title "Standard dimensions of nipples with working allowances".

It is annotated "Pitch of thread $18\frac{1}{3}$ to an inch" and the note "Interchangeable Arm" which dates it to c. 1857 or slightly later. It is of particular interest in being the only dimensioned drawing found from this period which provides some quantified tolerances. It also highlights Enfield's, or the Board of Ordnance's – later the War Department's – propensity to use very obscure thread sizes and for which no official reason has been discovered. It may simply have been to discourage 'unofficial' or 'sub-standard' copies being produced and fraudulently offered and which might be injurious to the weapon or the weapon's user.

MAKING THE ENFIELD PATTERN 1853 RIFLE-MUSKET

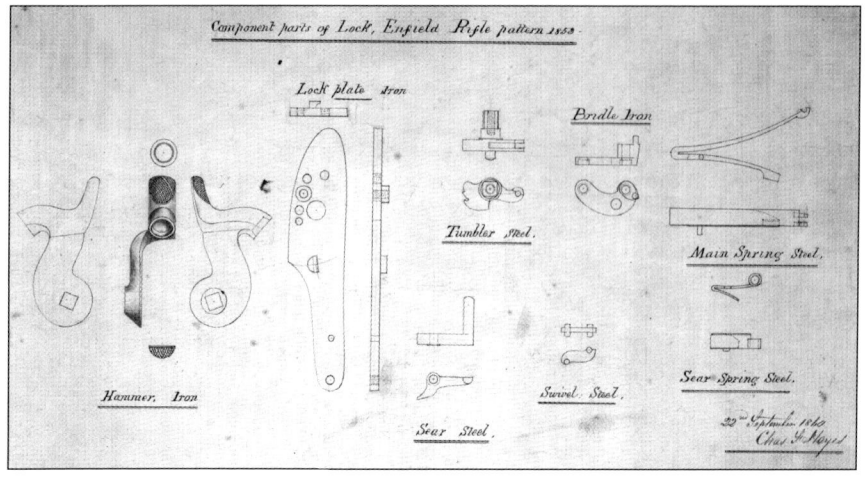

Fig. 3.9. Original Royal Small Arms Factory, Enfield, drawing, No. 749, of lock components. (© Royal Armouries)

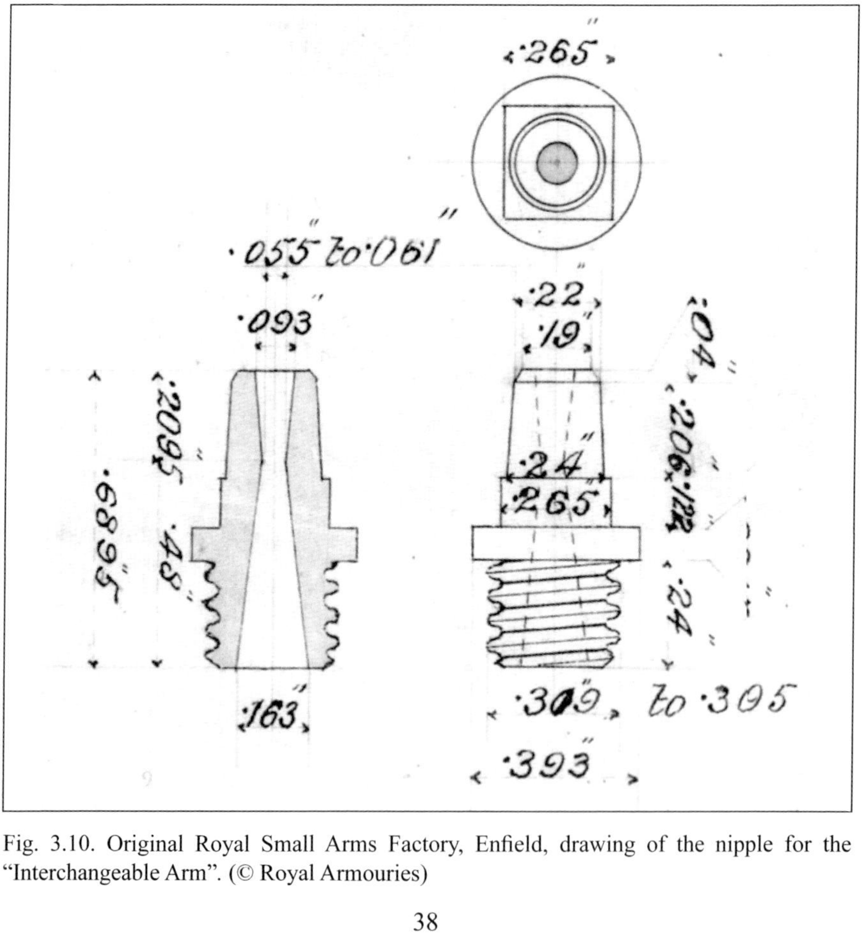

Fig. 3.10. Original Royal Small Arms Factory, Enfield, drawing of the nipple for the "Interchangeable Arm". (© Royal Armouries)

3.10 'Viewing' and inspection procedures – gauges

As already indicated, a key feature of the manufacturing process was inspection, or 'viewing', of a finished item in which gauges played an important role. Few if any gauges in use prior to the Enfield factory being established have survived, or at least, none have been knowingly encountered, but in the pursuit of interchangeability, gauges were vital. The Committee on Machinery ordered a large number and variety of gauges from the Ames Manufacturing Co. which comprised:[38]

For lockplate	1 gauge and plugs for testing the drilling
	1 gauge, receiving for testing the edges
	1 gauge, pattern for testing the position of bolster for mainspring
	1 gauge, grooved, for thickness of plate, bolster, etc.
	1 gauge, plug, for testing the sizes of holes
	2 gauges, plug, for testing the tapping
	1 pattern for filing bolster to height
For hammer	1 gauge, pattern, for testing the punching
	1 gauge, pattern, for testing milling of bolster
	1 gauge, pattern, for testing for straightening
	1 gauge, pattern, for testing edges
	1 gauge, grooved, for testing the finished dimensions
	1 gauge for testing drilling of nose
For tumbler	1 gauge, receiving, for testing the milling, filing, etc.
	1 filing gauge for squares
For sear	1 gauge, receiving, for testing the milling, filing, etc.
	1 gauge, plug, for testing drilling and milling
For bridle	1 gauge, receiving, for testing filing
	1 gauge, pattern and plugs, for testing drilling etc.
For mainspring	1 gauge, receiving, for testing filing, etc.
	1 gauge plate, for levelling bottom edges
For sear spring	1 gauge, receiving, for testing filing, etc.
	1 gauge plate for levelling bottom edges
For lock screws	Gauge plate for testing dimensions, length, cutting threads etc.
For barrel	Gauge for testing counter-boring of breech
	Gauge for testing the tapping of breech
	Gauge, plug, for testing the drilling of cone seat
	Gauge for screw-tapping
	Gauge, receiving, for testing cone seat drilling
	Gauge, receiving, for breech
	Gauge, profile, for testing the underside of tang
	Gauge, profile, for top [of tang?]
	Gauge, plate, for testing the diameter of breech tenon
	Gauge, plate, for testing the barrel at six points
	Gauge, plate, for length and height of stud
	Gauge, plate, for testing the position of stud from breech
	Gauge, nut, for testing threads of breech screw and tap

For butt plate	Gauge, receiving, for testing the filing, etc.
	Gauge, plate, for testing the exterior curves etc.
For trigger guard	Gauge, receiving, for testing the filing
	Gauge, profile, for testing the underside
	Gauge, plate, for thickness, curves etc.
For trigger plate	Gauges, receiving
	Gauges, grooved, for testing thickness etc.
	2 gauges, plugs, for tapping etc.
For trigger	Gauges, receiving, for testing the thickness etc.
For ramrod	Gauges, plates, for testing the diameter and screw
	Gauges, profile, for testing the form of head and swell
For nipple (cone)	Gauges, plates, for testing the exterior and interior diameter
	Gauges and nut for testing screw and tap
Screws	2 gauge plates for testing lengths, diameters, screws etc. of side, tang, breech-plate, guard plate and trigger screws
For bands	3 gauge mandrels for testing interior dimensions
For stock	Pattern for vertical profile of finished stock (brass)
	Gauge, pattern, for testing spotting and angle of butt end
	Gauge, barrel, (solid steel) for testing groove
	Gauge for testing length from breech of barrel of band shoulders
	Gauge, bands etc.
	Gauge for testing position of lock bed
	Gauge for testing the cut for tenon of breech screw [barrel tang]
	Gauge for testing the depth of guard bed
	Gauge pattern for testing the fitting of guard bed
	Pattern for testing the profile from breech of band to breech [butt] plate 4 gauges grooved (16 grooves) for testing the various diameters of stocks
	Gauge pattern for testing margins around lock and sideplate

The Ames company would have needed to have been supplied with a rifle to work from in creating the gauges and that rifle became, in effect, the prototype of the interchangeable series of rifles. If it still exists, its whereabouts is not known. They would also have needed to work in close collaboration with Robbins & Lawrence to ensure that the products from their machines matched Ames' gauges. A cased set of gauges, Fig. 3.11, still survives in Royal Armouries collections, transferred from the M.O.D. Pattern Room when that facility closed.

Despite the large number of gauges, spread over two layers, it does not contain the full set as specified in the Ames contract of 1854 and consists of a mixture of 1857-dated and un-dated gauges, and some marked 'standard', indicating a mixture of sources. The gauges that are dated 1857 may be part of the set supplied by Ames but since they are not named, they continue to remain anonymous.

ORIGINS AND PROCUREMENT

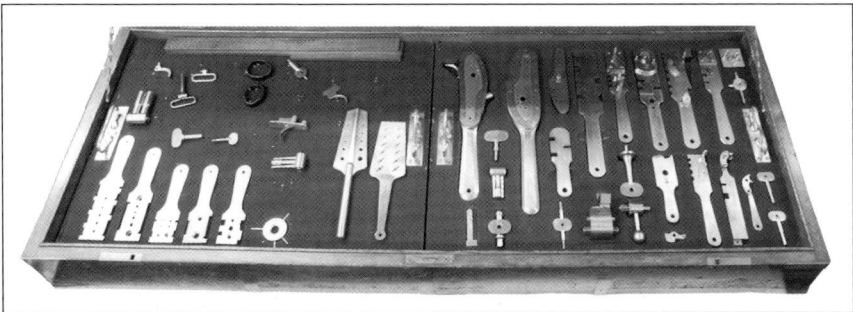

Fig. 3.11. Part of the set of gauges for the Enfield Pattern 1853 rifle. (PR.10142, ex Pattern Room collection, © Royal Armouries)

Nevertheless, the 'set' is a unique and highly important aspect of the manufacture of the Interchangeable Pattern 1853 Rifle.

These gauges only provide an upper limit on size so that over-sized components were automatically rejected. An item that did 'fit' the gauge relied on the judgement of the inspector as to whether that item was of acceptable size or was too loose, and therefore too small, and thus should be rejected. One aspect of 'gauging', therefore, still remained somewhat arbitrary. It might have been expected that in a component such as the lock, where each component had to engage and interact accurately with another, as in, for example, the sear and the tumbler, a lower-limit gauge would have been a more certain way of ensuring correct 'fit' than reliance upon the judgement of the inspector. It could be argued that the concept of 'tolerance' – the variation between maximum and minimum critical dimensions – did not exist at this date, but there is evidence to show that, to some extent, it did. It has already been noted in the nipple drawing, Fig. 3.10. It is clearly expressed in contract documents below, Fig. 3.12, and also in some specifications which will be discussed in Chapter 7 dealing with the finishing of the barrel. It is also clear that accurate linear measuring devices were in use since it has been noted that in 1833 George Lovell was using a micrometer to measure bore gauges at Enfield,[39] but no specific information about this instrument has been found.

> COURSE OF INSPECTION AND PROOF.
>
> ## Barrels.
>
> The Barrel to be sent down for Proof and inspection, percussioned, jointed, and properly breeched; the bore is to be tested with two plugs, viz., one of the diameter of ·572 for the receiving, and one of ·577 for the rejecting plug. They are to be proved with the regulated proof charge of 7 drachms of

Fig. 3.12. Excerpt from a contract document issued to Joseph Smith in connection with the manufacture of the Pattern 1853 Artillery Carbine. (Appendix 2 item 5. © P. Smithurst)

MAKING THE ENFIELD PATTERN 1853 RIFLE-MUSKET

3.11 Operations at Enfield

In 1859 it was reported that 680 machines were in operation at Enfield and the manufacture required 719 distinct operations.[40] It is known from contract details that 104 machines were ordered from Robbins & Lawrence and Ames but, with the demise of Robbins & Lawrence, it is doubtful that all were delivered. Nevertheless, any shortfall had to be obtained from somewhere. This still begs the question of what were these extra 570 machines? Various possible answers present themselves. It is known that during the period January 1854 to March 1859, approximately £71,500 was spent on machinery at Enfield (see Appendix 6). If we take £1 = $5 [derived from the figures quoted in the official document relating to the Fox, Henderson & Co. contract for an advance of £6,000, or $30,000, made to Robbins & Lawrence (Appendix 8)], the cost of machines from the United States, $88,000 in round figures, amounted to £17,600. Therefore, an additional £53,900 was expended on machinery between January 1854 and March 1859. It is also known, for instance, from the details of the American machines listed above, that none were for barrel manufacture. It is therefore significant that a list of 64 machines needed for finishing 250 barrels per day was drawn up by James Burton in a memorandum sent to Captain Dixon, R.A., Inspector of Small Arms at Enfield, on Sept 10th, 1855. This list was augmented in an order placed with Robbins & Lawrence on Nov. 14th, 1855, bringing the total to 70 machines (Appendix 3) but again, it is not known how many, if any, were actually supplied before Robbins & Lawrence ceased trading.

According to the testimony of James Gunner to the Select Committee on the 26th March, 1854, barrel rolling had just been introduced at Enfield "a few days previously".[41] As will be discussed in Chapter 6, three sets of rolls would be needed, one for forming the skelp, one for forming the skelp into a tube and one for welding and tapering the tube. In addition, a die-stamp would be needed for consolidating and forming the welded-on 'lump' that would become the nipple seat and some form of press for straightening the tube after rolling. According to Fitch[42] the daily output of a set of rolls was 200 barrels per day, but that referred to the simpler process starting with a pre-drilled blank tube and only using rolls to lengthen and taper it. If we allow time for bending a flat plate to form a skelp in addition to another set of rolls and welding the tube prior to tapering, that figure might be conservatively divided by 3, indicating another 3 sets of rolls would be required to meet Burton's output of 250 barrels per day.

In December of 1855, a further list indicates £7,250 being expended on machinery but does not indicate numbers, or in most cases, the nature of the machines (See Appendix 4). One exception is 'screw-machines', noted by Thomas as being supplied by Hobbs & Co. It is almost certain that this was the lock making company established by Alfred Hobbs with the prize money he received during the Great Exhibition for 'picking' Bramah's 'unpickable' lock after it had remained inviolate for 60 years.[43] Hobbs branded his locks 'machine made' and used the latest technology in their manufacture. An account of his operations makes the very pertinent observation:

> *The various studs and screws are made... by girls with automaton machines which require little or no skill beyond that of putting in the raw material and taking out the finished products.*[44]

Since the truly 'automatic' screw machine was only invented by Christopher Spencer in 1873,[45] it has to be inferred that Hobbs was using some form of 'turret' or 'capstan' lathe which might be regarded as 'semi-automatic' in requiring a degree of manual operation. Such machines may have been supplied by Robbins & Lawrence since there is no evidence that any similar machines existed in England at that time. Frederick Howe, who was to become one of Robbins & Lawrence's key tool designers, during his apprenticeship encountered a 'turret' or 'capstan' lathe with a multiple tool holder rotating on a vertical, as opposed to a horizontal, axis and this was to become the type favoured by Robbins & Lawrence.[46] Others suggest it to have been invented by Henry Stone, another employee of Robbins & Lawrence and who first produced it commercially in 1854.[47]

On 22nd Oct 1856, an order was placed with Greenwood & Batley by Dixon for:

> *2 lathes for making sight spring screws with cross slide carrying the top two longitudinal slides with levers for milling the wire and threading the screw and 1 cross slide for cutting off the screw*
> (see Appendix 4)

This falls short of a description of the 'vertical-axis-turret' lathe of Robbins & Lawrence but such a machine, similar in action to a lathe, which turned the shanks, threaded them, and then cut them off from the bar stock would be a fast operation. The screws still needed slotting but even so, output probably outpaced the production of the components for which they were needed. It would be logical, in the light of these orders, to infer that screw-making at Enfield used efficient and fast machine methods.

This casts doubt on the accuracy of the only account of screw-making which describes the process as: forging, clamp-milling the head, slotting the head, rounding the shank with a hollow end mill, threading, all in separate machines.[48] Considering that there were 15 different screws used on the rifle, to make them in such a way would have been very inefficient.

Any doubts as to whether Robbins & Lawrence were able to supply some machines before their demise is dispelled by the comment in the report of the American 'Military Commission to Europe' in 1855-56:

> *...the names of Ames, of Chicopee, Massachusetts, and Robbins & Lawrence, of Windsor, Vermont, are accordingly to be read on most of the machines at Enfield.*[49]

MAKING THE ENFIELD PATTERN 1853 RIFLE-MUSKET

The question remains as to which machines *were* supplied? Although a few can be identified, by type more than by name in various accounts since such machines were previously unknown in England, 'screw machines' do not appear unless they are taken to be items in the list of tooling noted earlier (page 45) as being supplied by Robbins & Lawrence:

- 1 screw (hand) machine
- 2 screw milling machines
- 2 thread cutting (hand) machines
- 1 slitting machine
- 1 pointing machine
- 1 clipping machine

The machines were used in making lock screws, but their number argues against them being 'turret' lathes since a single turret lathe could have performed all of those operations except the 'slitting'.

Even if it is assumed that 4 rolling mills and all their ancillary equipment, taken together with the machinery noted above, totalled 100, added to which are the American machines initially ordered, the total still falls short of the 680 noted in '*The Engineer*'. The costs of various machinery – rolling mills etc., Burton's machines for finishing barrels and those contained in Thomas's list, might amount to the sum recorded as being spent on machinery, £25,500, in the financial year 1855-1856 (Return, 1858). The surviving order books of Greenwood & Batley of Leeds, for instance, do indeed show some orders from Enfield for a variety of gunmaking machinery within this period, but only amounting to £2,380 (Appendix 4). There remains, in round figures, £30,100 unaccounted for. If we take the figure of 680 machines being accurate, it can only be assumed that additional machinery was purchased from unspecified sources to possibly eliminate 'bottlenecks' in the production process or to enhance output or both.

There is evidence to support such a view. On 6th October 1858, Dixon wrote to Burton to announce that General Peel had approved Dixon's proposal to increase output. On 3rd December Dixon sent another letter to say that in the preceding two weeks, 1200 arms had been produced (Appendix 3). In the four-year period January 1854–March 1858, *26,739 Musket Rifles (pattern 1853) made by machinery, complete,* were delivered into store from Enfield (Appendix 6, Return, 1858), an average of approximately 6,680 per year. However, in the 12-month period, April 1858 to March 1859, 57,256 rifles were produced at Enfield (Appendix 6, Return, 1859). That almost nine-fold increase in output strongly suggests that much of that additional expense was on machinery for manufacturing the rifle.

In 1860, just over 87,000 rifles were made interchangeably by this machinery at Enfield (Appendix 6, Return, 1860); in the same year it was directed that *the army shall be equipped exclusively with rifles of the interchangeable pattern.*[50]

3.12 Conclusion

Within the period from 1854, when the Select Committee on Small Arms was established to examine the question of procurement, to 1857 when the Royal Small Arms Factory at Enfield commenced production, the processes of both procurement and manufacture of military firearms underwent a revolutionary transformation.

Having all the operations carried out on a single site allowed the process to be systematised and streamlined instead of each component or group of components being manufactured in separate establishments. No longer was there the need for the parts to be gathered into storage at the Tower of London, only to be re-issued to 'setters-up' to assemble the finished weapon. Greater control could be applied to production and greater uniformity achieved through a system of gauging which allowed individual components to be exact replicas of one another, thus leading to interchangeability.

Such a system could only be implemented with great capital outlay beyond the means of individual contractors. In 1856, however, the London Armoury Company was formed as a joint stock company to initially manufacture Adams' revolvers but that was scaled down at some unknown date when a contract for Pattern 1853 Enfield rifles was issued by the War Department.[51, 52] It was acknowledged[53, 54] that these were the only 'contract' rifles accepted as being interchangeable with the Enfield-produced arms and therefore it can be concluded they used similar machines and identical gauges to those at Enfield, requiring significant capital outlay.

In the words of John Anderson in a letter to James Burton on 26th January 1857:

> *It has been uphill work but let us thank God that it is now an accomplished fact, and that success has crowned the efforts that have been made. I will not say how proud I shall be to have a complete Enfield Musket. I look upon it as the most mechanical triumph of the age.* (Appendix 2)

A sealed Pattern 1853 rifle, shown in Figs. 3.13 and 3.14, was prepared and exists in Royal Armouries' collection.

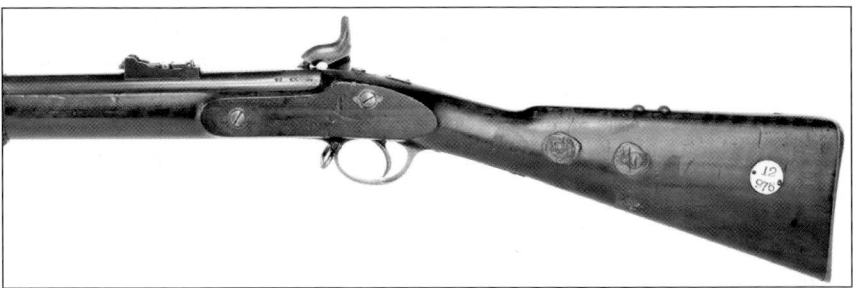

Fig. 3.13. A 'sealed pattern' rifle with the date 1857 and the name Enfield on the lockplate. (XII.976, © Royal Armouries)

MAKING THE ENFIELD PATTERN 1853 RIFLE-MUSKET

Fig. 3.14. Details of the rifle shown in Fig. 3.13. (© Royal Armouries)

The date '1857' and 'Enfield' on the lockplate suggests, even though no documentary evidence has been found, that it marks the beginning of the new era of interchangeable manufacture at Enfield.

Chapter 4

Manufacture of the Lock

4.1 Overview

The lock is the most complex part of the Pattern 1853 Enfield rifle, not just because it was its only mechanism but also because most components, with the exception of the screws, are asymmetrical and incorporate few straight lines as will be seen in the following illustrations. It therefore needed great care in planning and executing manufacture. Mechanisation created serious challenges in requiring machines to replicate hand skills but in describing the processes employed after 1857 the various contemporary accounts provide wholly inadequate detail.

It has therefore been necessary to examine and analyse in detail for the first time a range of unique artefacts and documents to allow an in-depth understanding of this topic to be presented. One of these is a unique factory drawing of the lock components, Fig. 4.1.

Details from this drawing will be used as the manufacture of each component is considered. The correct terminologies of the fully assembled lock components are shown in Fig. 4.2.

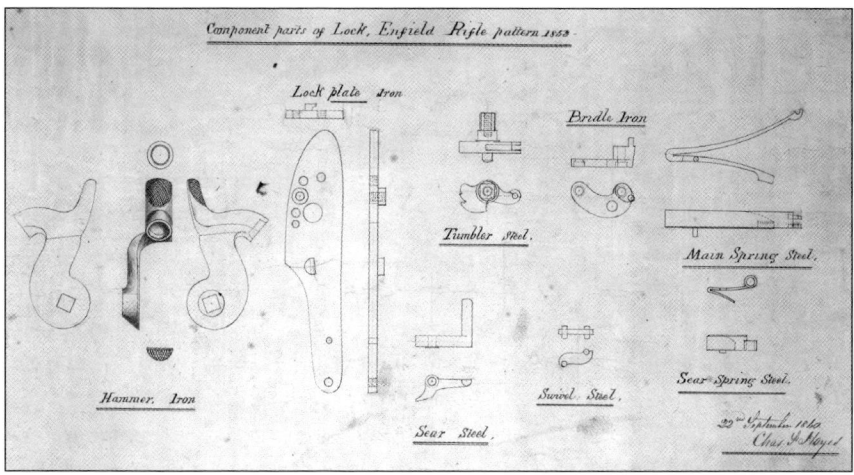

Fig. 4.1. Original Royal Small Arms Factory, Enfield, drawing, No. 749 of lock components. dated September 1860. (© Royal Armouries)

MAKING THE ENFIELD PATTERN 1853 RIFLE-MUSKET

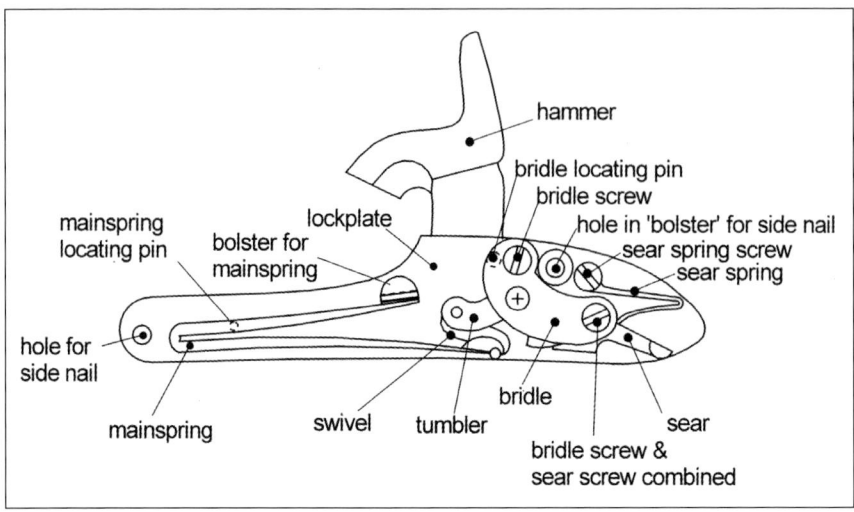

Fig. 4.2. The lock components and terminology. (© P. Smithurst)

Fig. 4.3. The Lock of a Sealed Pattern Enfield Pattern 1853 rifle showing the internal components and their arrangement. (XII.1918 © Royal Armouries;)

Four of the components acted in a 'train' – the mainspring, swivel, tumbler and sear.

It has not been expressed in engineering terms before but for such a mechanism to work effectively and smoothly, a number of criteria had to be met:

1. all of the components were pivoted, therefore the centres of those pivots had to be accurately formed and placed in relation to the interacting elements of the components

MANUFACTURE OF THE LOCK

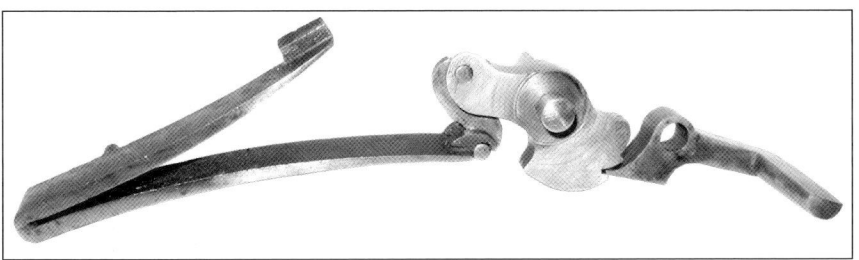

Fig. 4.4. The 'train' of interactions – spring, swivel, tumbler, and sear. (© P. Smithurst)

    2. the centres of the holes into which they fitted had to be accurately located and formed in their mutual positioning of the components on the lockplate
    3. the portions of the components that interacted had to be accurate in terms of form and dimensions of mating parts.

To achieve all of this was difficult enough and the degree of skill needed for what was largely hand-file work prior to mechanisation resulted in extensive specialisation and the use of some 'mechanical aids', such as filing and hand-drilling jigs, as opposed to machine tools. When interchangeability and mechanisation was pursued, the process was elevated to new levels. Not only did the components have to interact correctly with each other in one lock mechanism but had to do so if transferred to another lock mechanism. Details of how lock making was accomplished in England prior to 1857 are not known with any certainty; it can be argued that locks were manufactured using similar technology to that described as being used at the Tula weapon factory since aspects of that technology were imported from Birmingham.[1] However, even that assumption has to be guarded since details given in the Minutes of Evidence of the Select Committee on Small Arms suggest that little machinery was used in lock making.[2] Even with the extensive mechanisation introduced, many of the components were still finished by filing, judging by the number of filing jigs supplied along with the machines noted earlier, and it has to be inferred that in many instances, machines were simply used for roughly forming some components which were finished 'to gauge' with files.

4.2 Forging the Lockplate

The lockplate was the foundation upon which all components were mounted and is therefore an appropriate starting point. The best quality wrought iron was used to provide a close grain structure and good case-hardening properties. It is clear from the only, but very sparse, account which touches upon this that the lockplate was roughly shaped in a 'forging machine' and then 'pressed while still almost white-hot into a die".[3] The 'forging machine' was undoubtedly a small power hammer while the latter implies a 'closed-die' drop-forging process.

MAKING THE ENFIELD PATTERN 1853 RIFLE-MUSKET

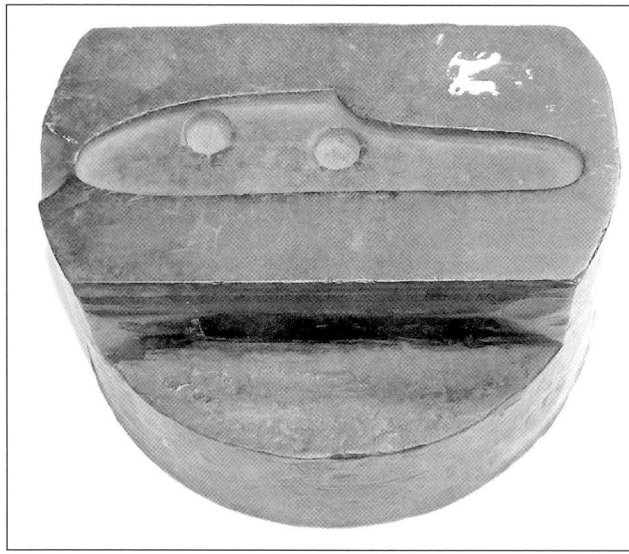

Fig. 4.5. Lockplate forging die. (XVIII.560. © Royal Armouries)

In Royal Armouries' collections is a drop-forging die which has not previously been studied but its two distinctive 'bolsters' and size, which includes a 'machining allowance' on its edges, conform to the Pattern 1853 lockplate and suggest that this die was used to produce that component.

A number of questions raise themselves in this context. The cavity in the die is only slightly deeper than the thickness of the lockplate. It is deduced that a second die, having a matching cavity without the two bolsters, was used in conjunction with the die in Fig. 4.5 to produce a lockplate of sufficient thickness to provide machining allowances on its outer and inner faces. The advantage of a formed top die, as opposed to simply a flat-faced die, is that the thickness of the forging is controlled, provided there is sufficient metal to fill each die, with any excess being squeezed out between the dies to form a 'flash' or 'fin'.

The fin would also act as a cushion, preventing the two dies from making direct and possibly damaging contact, and would then be removed using a trimming die and punch. This was likely to produce some distortion so the lockplate would be returned to the forging die to be straightened again.

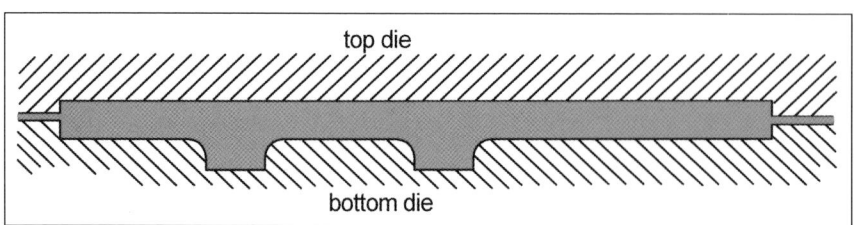

Fig. 4.6. Sketch showing pair of dies in use to forge a lockplate with a thin 'fin' projecting around the edges. (© P. Smithurst)

MANUFACTURE OF THE LOCK

4.3 Machining the Lockplate

The lockplate had to have all the components of the lock accurately positioned and to be given an exact thickness and profile to fit the lock recess in the stock. The forging therefore had to undergo extensive machining. Some machines in the Robbins & Lawrence contract (page 45) for machinery and tooling consisted of:

> *7 Milling machines*
> *2 Drilling machines, 5 spindles*
> *Tapping apparatus*
> *Apparatus for reaming tumbler hole to size*
> *1 Edging machine.*
> *1 Milling tool for milling the mainspring bolster*
> *1 Pattern for filing bolster to height*

It becomes clear, however, that these were to supplement existing tooling.

Machining the outer face

The first operation was to machine the outer face. The various descriptions provide virtually no detail so again it is necessary to apply some deductions. A small clue can be derived from *The Engineer* where it states *the neck is planed to a true face in a small shaping machine* and is *fixed by means of an eccentric handle.*[4] The only meaningful part of this description is the small shaping machine; as for what is meant by the "neck" can only be guessed to be the narrow front portion, but it is a term never before encountered in this context. It also leaves unanswered the question of what the fixture for the lockplate was like. It is unreasonable to consider the lockplate, in whatever fixture was used, to have been clamped on its face since this would have not allowed the planing of the whole face and, had the lockplate been slightly deformed, it could have 'sprung' when released and thereby defeated the object of the exercise. It is far more likely to have been held by its edges in

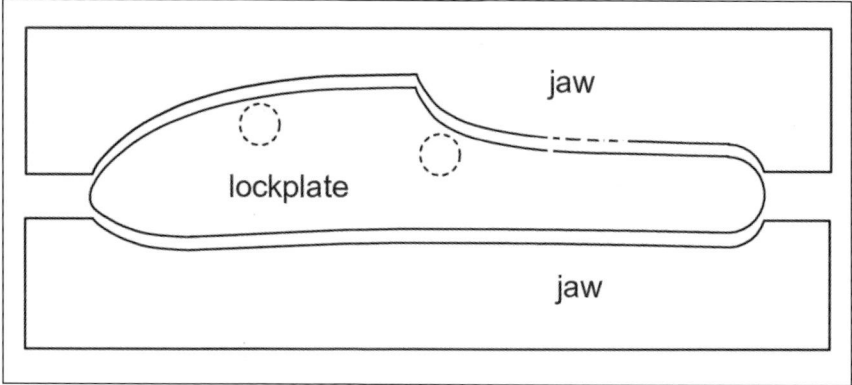

Fig. 4.7. Shaped jaws in a vice to hold the lockplate. (© P. Smithurst)

MAKING THE ENFIELD PATTERN 1853 RIFLE-MUSKET

a vice with suitably shaped jaws, Fig. 4.7, and a fast-acting cam – the "eccentric handle" noted in *The Engineer* - to close the jaws.

A ledge following the profile of the edge of the jaws and equidistant beneath the top face of the jaws would hold the lockplate level and prevent any downward movement from the pressure exerted by the cutter, and the edge of the jaws would give the necessary support during planing.

Machining the inner face

Examination of the inner face of a finished lockplate, Fig. 8, shows the two bolsters, A and B.

Not previously noted elsewhere is that the lower edge of B, referred to as the 'mainspring bolster', which is undercut to provide an abutment for securing the mainspring, has been trimmed at an angle and its edge is tangential to the boss, A. Measurement shows this tangent to be close to 6° to the extremities of the lower edge of the lockplate. This angle would have determined the settings for subsequent machining operations on this face of the forging.

After planing the outside face to produce a flat surface, the brief description simply states:

> *plate is. . . fixed in another machine upon a plate* [fixture] *which is fed under a spindle having keyed upon it a pair of circular cutters placed so far apart as to clear one boss and take a slice off B, and at the same time thinning the plate down from end to end.*[5]

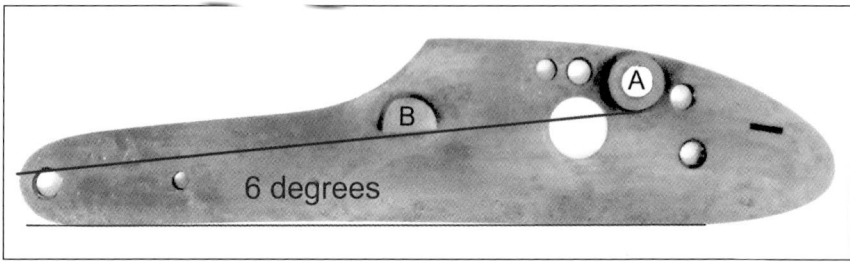

Fig. 4.8. Finished lockplate made by Robbins & Lawrence. (© P. Smithurst)

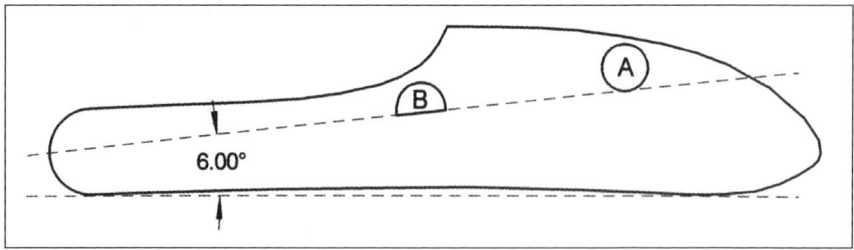

Fig. 4.9. Drawing of the lockplate forging showing, in dotted lines, the alignment of the finished lower edge of bolster, B, with the finished bolster A. (© P. Smithurst)

52

MANUFACTURE OF THE LOCK

The second reference to a 'plate' must refer to a 'fixture'. It would have been necessary for the lockplate in the fixture, or the fixture itself, to be angled at 6°. The above description does not mention how the lockplate was clamped but it would have been necessary to clamp it by its edges, as opposed to its face, to allow clear passage for the milling cutters. This would easily have been achieved using a vice with shaped jaws as shown earlier. The process in question was clearly a horizontal milling operation because the description continues to state the cutter arbour is *horizontal and parallel to the table on which work is mounted.*

This would have left an un-milled portion as shown below in Fig. 4.11.

Using the same fixture in the third operation, the un-machined portion was *principally taken away by another similar machine with three cutters*[6] cutting at a right angle to the first operation.

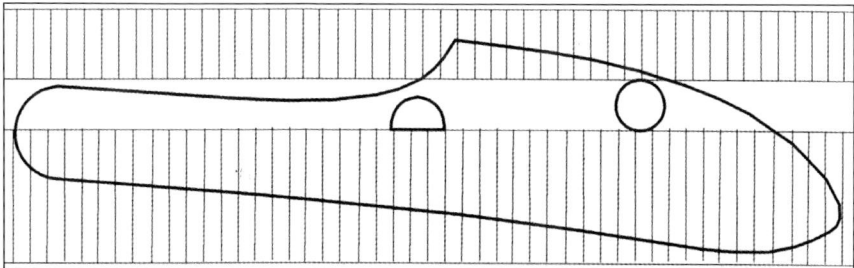

Fig. 4.10. Path of the milling cutters with the lockplate set at 6°. (© P. Smithurst)

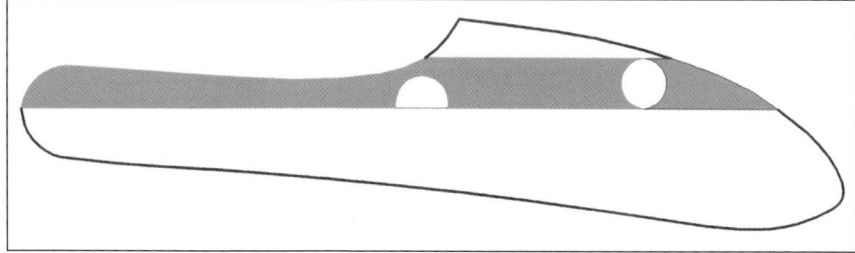

Fig. 4.11. Un-milled portion shown in grey. (© P. Smithurst)

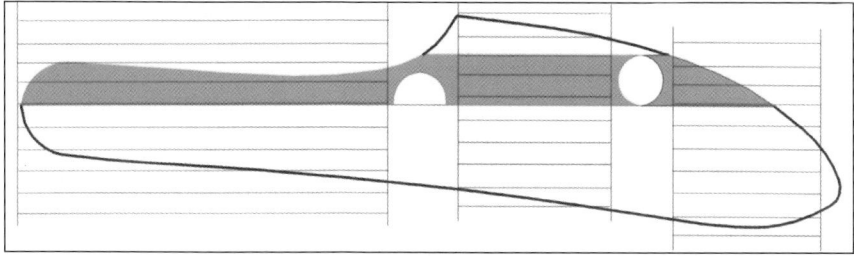

Fig. 4.12. Milling across at 90°. (© P. Smithurst)

MAKING THE ENFIELD PATTERN 1853 RIFLE-MUSKET

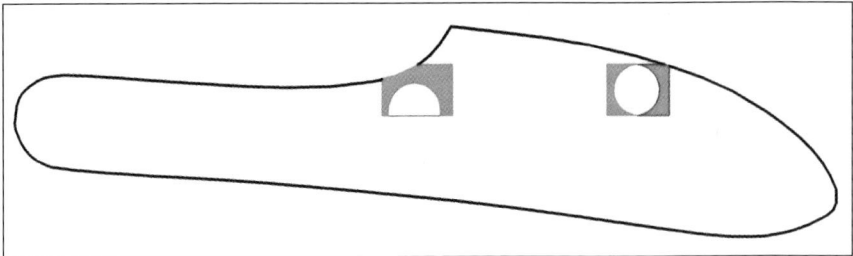

Fig. 4.13. Portions untouched in the milling operations. (© P. Smithurst)

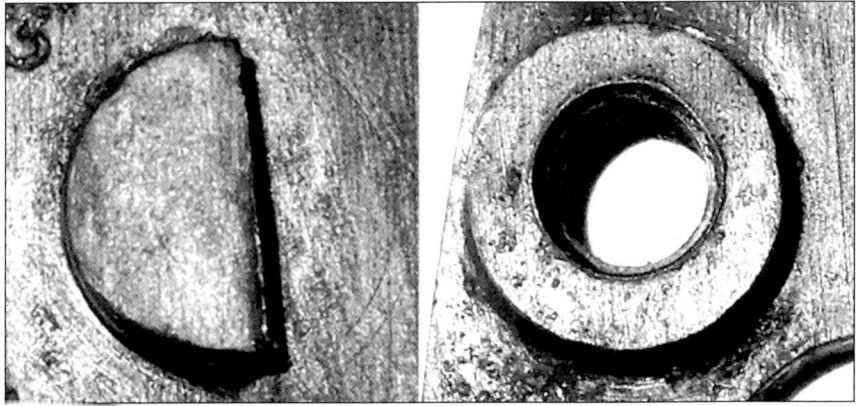

Fig. 4.14. Circular striations visible on the Windsor lockplate. (© P. Smithurst)

The result of these two milling operations would have been two 'islands' untouched by the milling cutters.

These were removed in the fourth operation:

> *the plate is placed under a vertical drill spindle which is fitted with a* [hollow] *cylindrical milled cutter which at once shapes the bosses and takes away what the cross-milling has left: a screw-stop secures the proper depth of cut.*[7]

This is a too simplistic description. Considering the accuracy required in having the hollow end mill concentric with the bolsters, another 'fixture' with the lockplate angled at 6° and able to be moved the precise distance between the centreline of the bolsters while maintaining concentricity would have been necessary. Evidence of this operation is visible as circular striations on the lockplate, Fig. 4.14.

Drilling the lockplate holes

The fifth operation was to drill the eight holes in the lockplate and new information on this topic is presented here. According to a letter from Anderson, Inspector

MANUFACTURE OF THE LOCK

of Machinery at the Royal Gun Factory, Woolwich to Burton, Chief Engineer at Enfield on 26th January 1857, (see Appendix 2) drilling was accomplished using a *box into which you enclose the lockplate while being drilled* and suggests the box was fitted with some form of cover having the requisite holes so that:

> *the whole is moved from time to time by the boy employed so as to allow the drills to enter the holes in the form* [jig] *and in this manner their relative positions is secured.*[8]

The Robbins & Lawrence 5-spindle machines would have been appropriate for drilling the five different sizes of holes in the lockplate. In Royal Armouries, within a collection of gauges believed to have been used by the London Armoury Company, an item has been found which is recorded as a 'dummy lockplate' and simply marked 'standard', Figs. 4.15 & 4.16.

This item exhibits a number of features not previously recorded and all point towards the term 'dummy lockplate' being a misnomer: it is approximately 10 mm. thick, far thicker than any lockplate; its edges are noticeably bevelled towards the inner face; it has a large 'boss' passing through to the inner face where it projects approximately 3 mm; on the inner face a small pin projects. Measurement of these show them to have diameters corresponding to holes on a lockplate for the large tumbler arbour and for the locating pin on the mainspring.

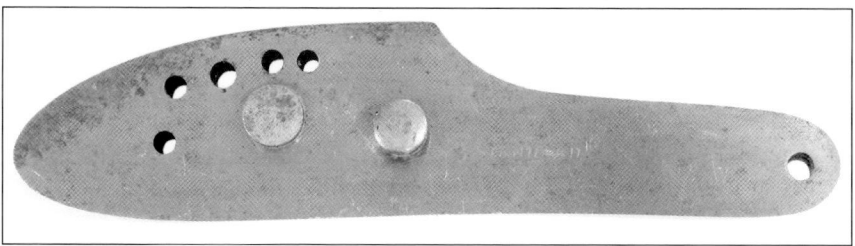

Fig. 4.15. Outer face of 'dummy' Enfield pattern 1853 lockplate. (XIII.949D. © Royal Armouries)

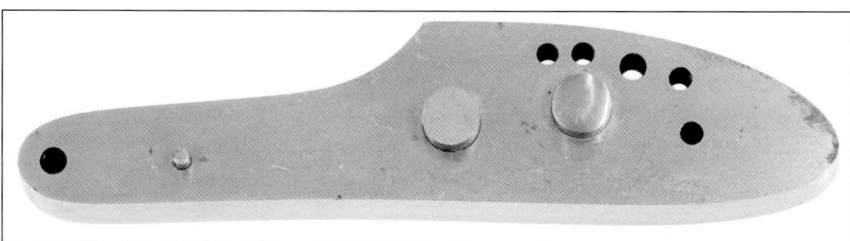

Fig. 4.16. Inner face of 'dummy' Enfield pattern 1853 lockplate. (XIII.949D. © Royal Armouries)

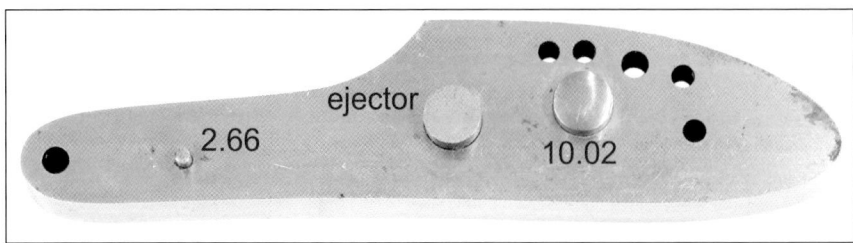

Fig. 4.17. Diameters (mm) of the projecting features and the ejector. (XIII.949D. © Royal Armouries)

A lockplate having these two holes could be fitted to its inner face and be removed by a sliding 'ejector' passing through to the inner face. Any thought that it was used to check the positions of the holes in the finished lockplate by attaching this and inserting the appropriate screws through the holes in this device is countered by two important features:

- Firstly, its thickness prevents the passage of the screws to engage with a finished lockplate fitted to its inner face.
- Secondly, the hole diameters are the same as those in a finished lockplate and therefore represent the *root diameter* of the thread, not the screw *shank diameter*, thus preventing the insertion of the screws.

Any thought that it might be a filing jig to finely finish the lockplate profile is countered by the facts that the inner, smaller, profile of this device does not fit the lockplate gauge and that there would be no need for the holes; such an idea can be dismissed.

It might be considered to be a drilling jig for the remaining lockplate holes, but this idea is countered simply by its not showing the expected wear and tear of factory usage.

The only possibility remaining, therefore, is its being used to check alignment of the other lockplate holes using plug gauges passing through the lockplate into those in the 'device'. The author considers this to be its most likely use since, as will be shown later, the juxtaposition of the four holes associated with the tumbler are crucial, as is their relationship to the hole for the sear axis pin.

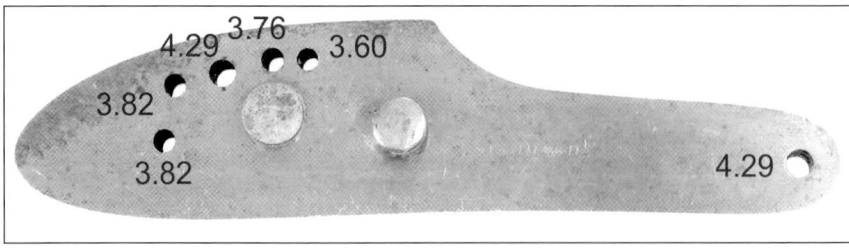

Fig. 4.18. Diameters of the holes. (XIII.949D. © Royal Armouries)

MANUFACTURE OF THE LOCK

After drilling, the appropriate holes were then tapped in the seventh operation *by means of an upright spindle* – no doubt the "tapping apparatus" appearing in the contract. The eighth operation was to form the blind slot in the lockplate for the nib on the sear spring to enter and which prevented rotation of the spring on the lockplate. This was achieved in a simple pressing operation using an appropriate jig to locate the lockplate and a die to form the slot.

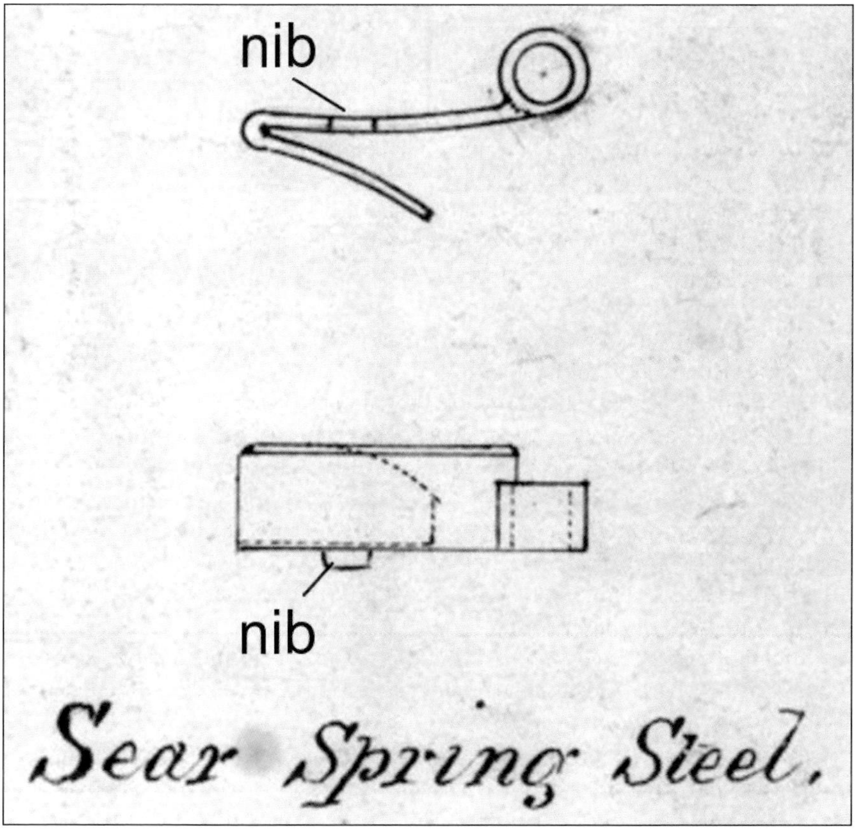

Fig. 4.19. Sear spring drawing with annotation to highlight the 'nib'. (From R.S.A.F. drawing No. 749. © Royal Armouries)

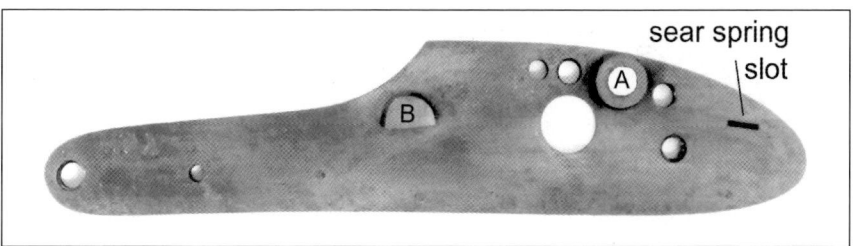

Fig. 4.20. The slot for the sear spring-nib. (© P. Smithurst)

MAKING THE ENFIELD PATTERN 1853 RIFLE-MUSKET

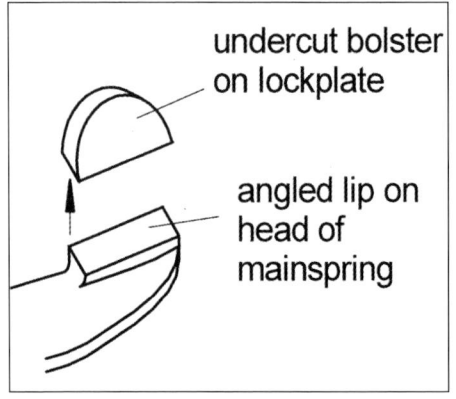

Fig. 4.21. Detail showing bolster, B, and head of mainspring. (© P. Smithurst)

The bolster, B, was then 'undercut' to form the housing for the head of the mainspring, followed by filing to its correct height on the lockplate using the "pattern for filing bolster to height" in the Ames contract.

The large hole for the tumbler arbour was burnished by the insertion of a revolving burnisher which was slightly tapered at the lower end and had a 'flat' on one side with the corners rounded so as not to cut.

Machining the lockplate edges

Although these final operations are described in *the Engineer*, there is much that was left unsaid, making them rather meaningless and that shortfall is rectified here by providing an analysis of how it could be accomplished.

Firstly, it has to be deduced that die-forging produced a lockplate with sufficient uniformity for its edges to act as datum edges for operations up to this point.

Then, as noted in *The Engineer*:

> *The two surfaces, front and back, having been thus formed, it becomes necessary to rough out and finish off the outside edge to an accurate gauge, and this is effected by clipping the plate in an upright position and passing it under a milling-block filed up to the right shape: by this means the top edge is roughed out.*

This can only be a reference to 'form-milling' and such a cutter for milling the front top edge of a lockplate is shown in Fig. 4.22.

Fig. 4.22. A 'form-milling' operation on the front upper edge of a lockplate. (Courtesy American Precision Museum, Windsor, VT.)

MANUFACTURE OF THE LOCK

This 'roughing-out' process was followed by more accurate machining:

The plate is then bolted down on a horizontal bed and acted upon by a pair of horizontal revolving milling tools, which are so placed that one finishes the top and the other the bottom edge of the plate. Forms are placed behind these milling tools, and guide pins are fixed to the frames to which the spindles are fixed; [these are kept up to their work by means of weights], *so that the true shape is imparted to the plate by a simple movement of the bed to which it is fixed.*

This passage undoubtedly refers to the *edging machine* referred to in the contract, but falls short of being an accurate description, especially in stating the cutters were horizontal – they revolved in a horizontal plane about a *vertical* axis. This confusion may have arisen simply from the fact that such a machine had not been previously seen in England.

Fig. 4.23 shows the only Robbins & Lawrence 'edge' or 'profile' milling machine known to survive, which was restored with guidance from the author.

Fig. 4.24 supplies the terms applied to various features of the machine.

Finally, there is the machine illustrated by Fitch, Fig.4.25, which he suggests as being identical to those supplied to Enfield.

The notable feature of the machine in Fig. 4.25 is having what appears to be a weight suspended beneath the machine. A table, weight-driven in one direction, certainly aids understanding of its 'modus operandi'.

Drawings made by Greenwood & Batley of a single-spindle machine are believed to have been derived from the Robbins & Lawrence machines.

Fig. 4.23. The only known surviving Robbins & Lawrence 'edge' or 'profile' miller, recently restored and in working condition. (Courtesy American Precision Museum, Windsor, VT.)

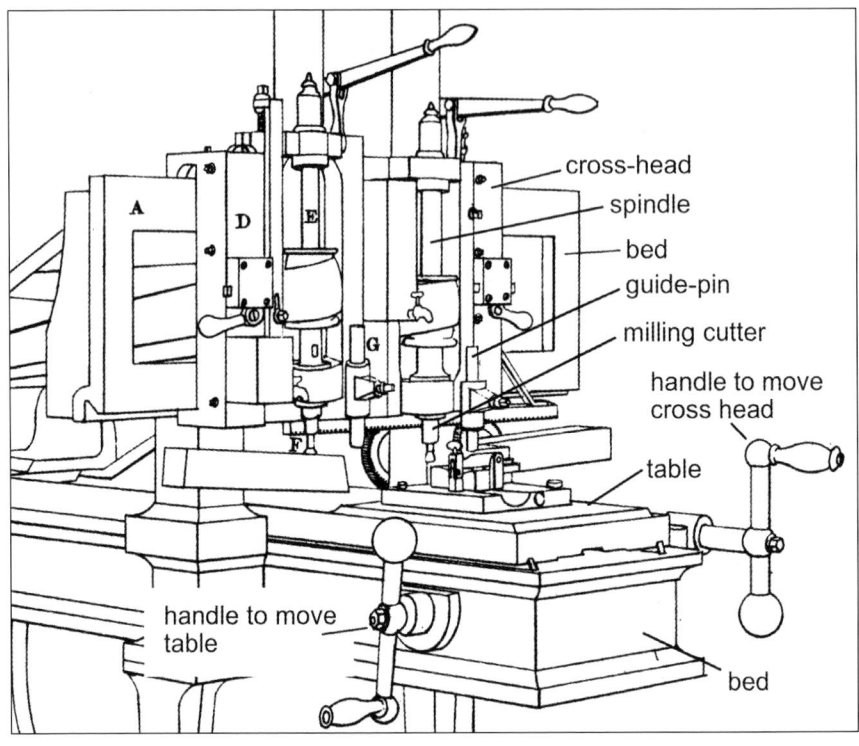

Above: Fig. 4.24. The edge miller, with parts named. (Benton, 1878, Pl. XXII)

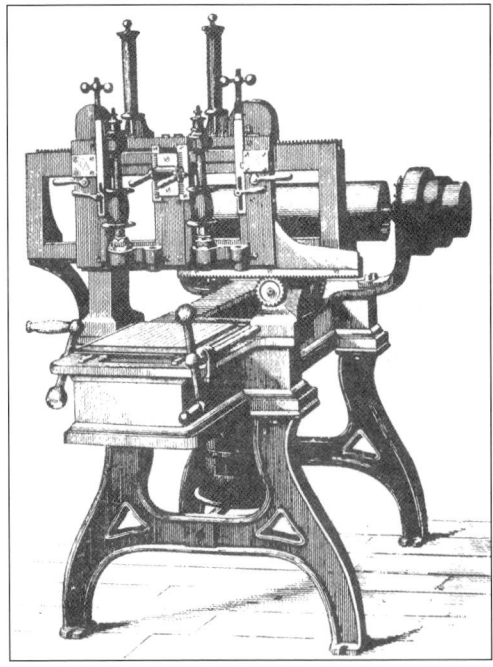

Left: Fig. 4.25. Machine described by Fitch as identical to those supplied to Enfield. (Note the weight suspended between the frames). (Fitch, 1882, p. 29, Fig. 26)

MANUFACTURE OF THE LOCK

This drawing, not previously studied or published, shows the table to have been moved by rack and pinion.

A rack and pinion, as opposed to screw feed, would have been necessary to allow the table to be moved by a falling weight and to follow a contour more easily,

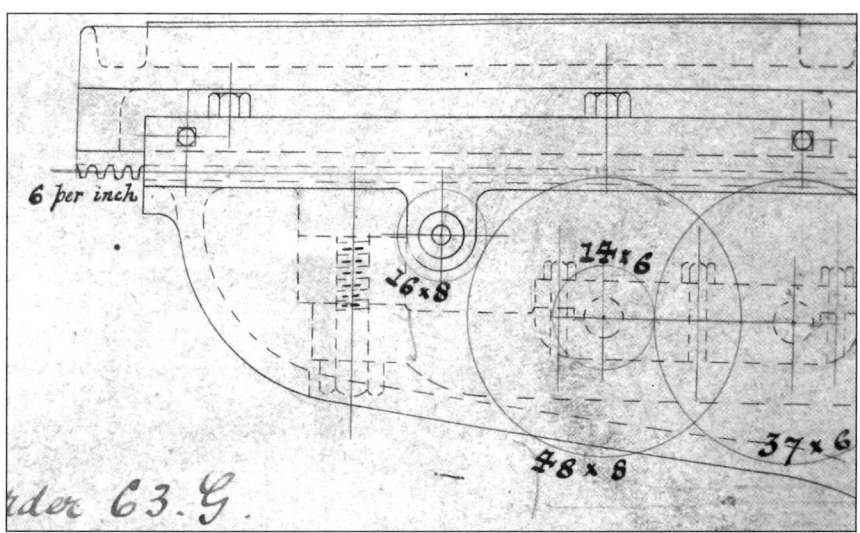

Fig. 4.26. Detail from Greenwood & Batley drawing of 'Edge Milling Machine, Medium Pattern', showing sectional elevation of rack and pinion table feed. (© West Yorkshire Archives, G & B Archive Drawing No. 738)

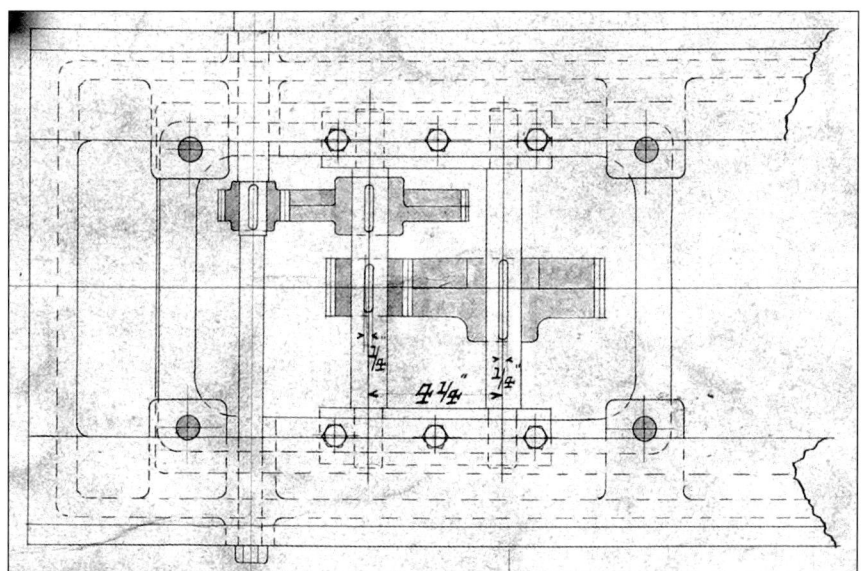

Fig. 4.27. Detail of 'Edge Milling Machine' drawing showing plan view of rack and pinion table feed. (© West Yorkshire Archives, G.& B. Archive Drawing No. 738)

plus the advantage that with a gearing ratio giving an approximately 10:1 reduction of being able to be returned to the starting position, that is hauling the weight up again, relatively easily. The description of the profiling operation in *The Engineer* mentions the use of a *guide pin* which obviously 'follows' a master pattern or template. This *guide pin* can just be seen in Figs. 4.23 and 4.24 fixed at the side of the milling cutter. Again, calling upon the Greenwood & Batley drawing No. 738, this shows the arrangement of spindle and guide pin.

It can be seen in Fig. 4.28 that they are at 4-inch centres and while this is only a single spindle machine, the double-spindle machine was undoubtedly the same since they are of similar proportions. The position and spacing of spindle and guide pin would not have allowed the lockplate, which is approximately 5.4 inches long, and its 'pattern' to be placed cross-wise side by side on the table, nor would it have so easily achieved the desired result. They would have to be placed adjacent longitudinally. As will be shown, this has a bearing on the way the milling operation could be carried out.

There are features of this operation not previously discussed in any of the published accounts but which are presented here. When in use, the table was being continually drawn in one direction by the force applied to it by the weight. This would require that the lateral, manual, movement of the cross head carrying the cutter and the guide pin followed a 'positive' rather than a 'negative', or re-entrant contour. Thus, the 'guide-pin' would be compelled to follow the profile and, therefore, the cutter also. Any, even slight, movement away from the contour would immediately allow the table, acted upon by the weight, to move forward to restore that contact. Studying the lockplate closely reveals various features concerning its contour and Fig. 4.29 shows that it has two 're-entrant' sections.

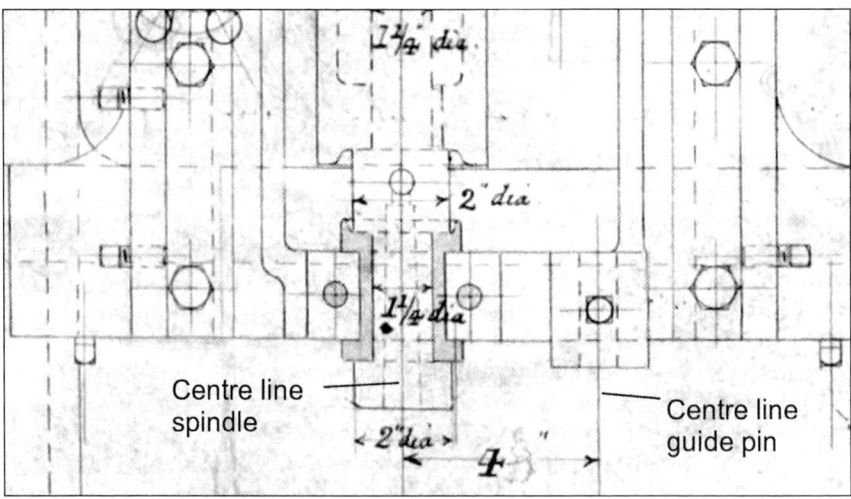

Fig. 4.28. Detail of 'Edge Milling Machine, Medium Pattern' showing the disposition of spindle and guide pin fixture. (© West Yorkshire Archives, G. & B. Archive Drawing No. 738)

MANUFACTURE OF THE LOCK

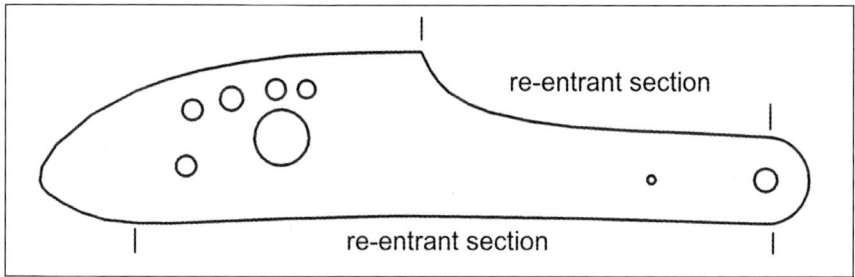

Fig. 4.29. Contours of the Enfield 1853 lockplate. (© P. Smithurst)

Thus, if the lockplate and its template were fitted longitudinally on the table, using the line between the two extremities of the lower edge as a datum parallel to the direction of travel, the guide-pin, following the template while the table was being impelled longitudinally by the weight, would not follow these curves. As soon as the guide pin cleared the rounded end of the lockplate at the top, for instance, the table would immediately be propelled forward by the free-falling weight until contact between guide pin and template, and thus the cutter, was restored at the bottom end of the lockplate, as in Fig. 4.30.

Manual adjustment to maintain contact of the guide-pin with the template during such motion would be impossible.

It is suggested then, that by slightly angling the template and the lockplate by as little as 3°, as in Fig. 4.31, the difficulty of this lower re-entrant section would be overcome and cause the guide-pin to follow a 'positive' contour and transmit this to the cutter and thus, to the lockplate. Using an exact copy of the lockplate, the guide pin and cutter would have to be of the same diameter. At the same setting, the table could be drawn back to its starting position and the second, upper contour, milled by moving the cross-head to the left to allow the table to begin its travel with the guide pin and milling cutter following the profile on the right.

Both these operations would be accomplished with the configuration of template and lockplate suggested in Fig. 4.32, which also shows the paths of guide-pin

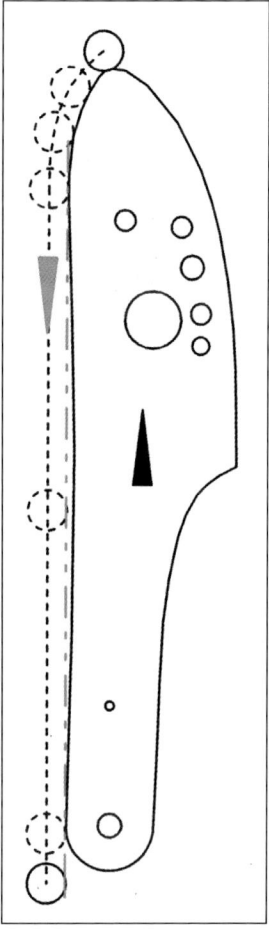

Fig. 4.30. Left. Using the bottom edge extremities to align the lockplate parallel with direction of travel creates a re-entrant section, allowing the guide pin and cutter to by-pass it. (© P. Smithurst)

MAKING THE ENFIELD PATTERN 1853 RIFLE-MUSKET

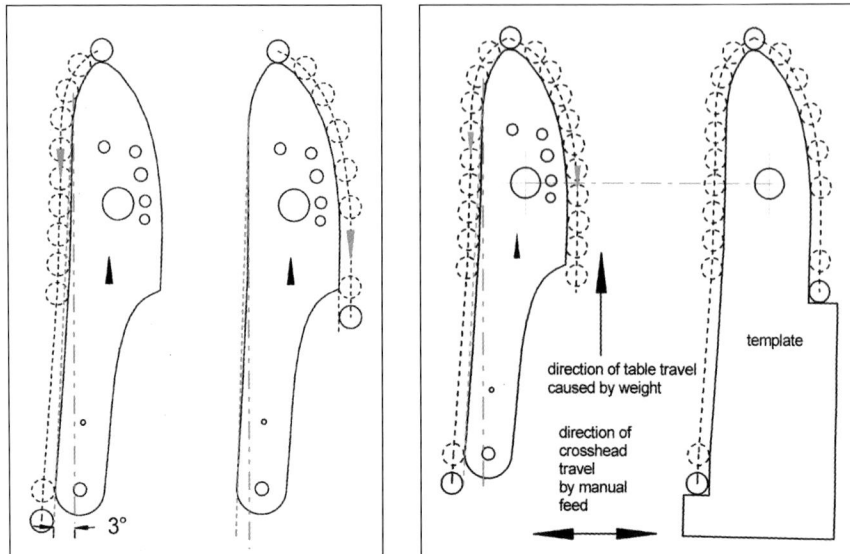

Above left: Fig. 4.31. Positioning of lockplate to achieve the desired result. (© P. Smithurst)

Above right: Fig. 4.32. Template and lockplate configuration to achieve the milling of two edges. (© P. Smithurst)

and milling cutter. Stops would undoubtedly be incorporated into the template to arrest the table when the milling cutter had cleared the lockplate.

The remaining edge and opposite tip of the lockplate could be milled in a similar fashion using a different template and fixture as shown in Fig. 4.33.

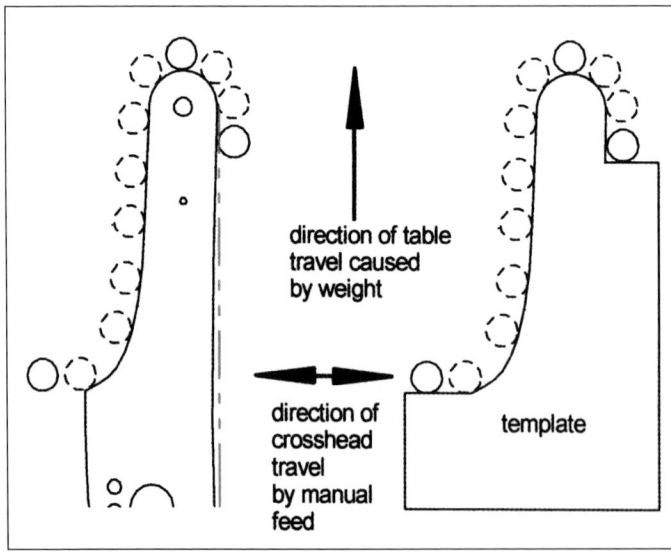

Fig. 4.33. Milling the front profiles. (© P. Smithurst)

MANUFACTURE OF THE LOCK

Of course, all this presumes the accurate positioning of the template and lockplate, mounted in suitable fixtures on the table. This would have been simplified by the fixtures for lockplate and template being an integrated unit, thereby avoiding any displacement of one relative to the other. Work on the lockplate was finished, apart from being polished and case-hardened later.

Gauging the lockplate
Gauges for the lockplate supplied as part of the Ames contract included:

> *1 gauge and plugs for testing the drilling*
> *1 gauge, receiving, for testing the edges*
> *1 gauge pattern for testing the position of the bolster for the Mainspring*
> *1 gauge grooved for thickness of plate, bolster, etc.*
> *1 gauge plug for testing the sizes of holes*
> *2 gauges plug for testing tapping*

The profile gauge is self-explanatory. In Fig. 4.34, the lockplate fits as it should.

However, Fig. 4.35 shows a lockplate made by Robbins & Lawrence as part of a contract of 1855 to supply 25,000 Enfield rifles, and which does not fit.

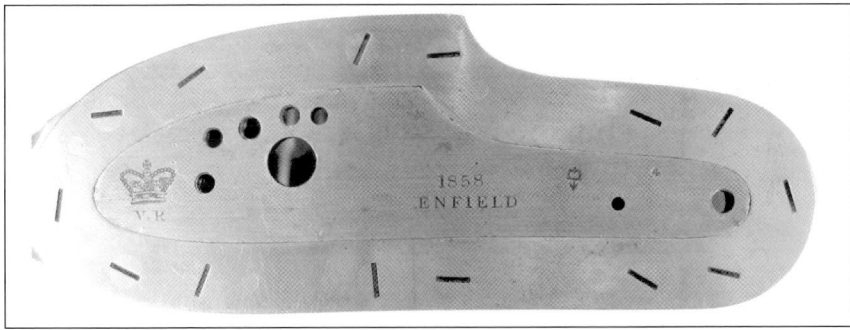

Fig. 4.34. Profile gauge with 1858 dated lockplate fitted. (part of PR.10142. © Royal Armouries)

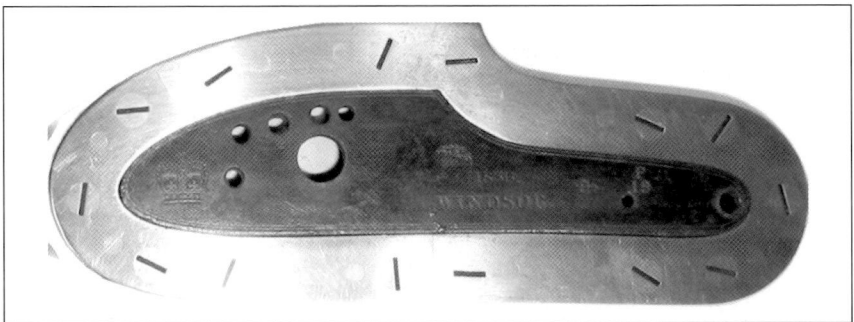

Fig. 4.35. Profile gauge with 1856 dated 'Windsor' lockplate *not* fitting. (© Royal Armouries)

MAKING THE ENFIELD PATTERN 1853 RIFLE-MUSKET

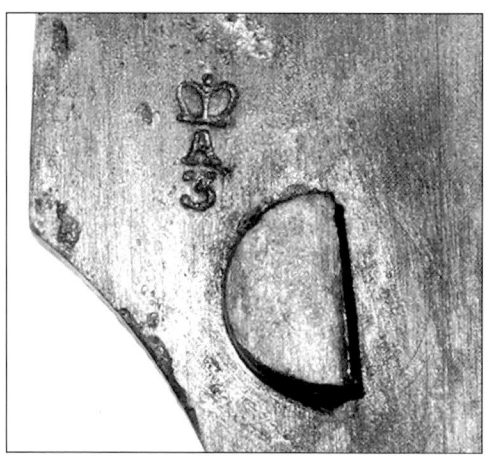

Fig. 4.36. Acceptance mark of the British inspector in America. (© P. Smithurst)

This lockplate has the mark of the British inspector sent to America stamped on the inside, indicating that it was approved.

The gauge shown is a later gauge for the interchangeable rifle and is indicative of the small differences in size which existed between the two classes of arms since the acceptance by the inspector of something considered oversize is untenable and the Robbins & Lawrence arms were ostensibly interchangeable. However, an answer may lie in a very rare publication containing a note relating to the Robbins & Lawrence contract, where it has been suggested, based on the account of Lawrence himself that at the commencement of the Robbins & Lawrence arms contract:

> *the Robbins & Lawrence Company called upon the War Office for sample models [of the] 1853 Enfield rifles but upon disassembling them it was discovered that no two of them were alike. A set of accurate gauges was then requested, and these were also promptly sent. They were made of hard wood – originally accurate possibly to $\frac{1}{64}$ inch – but during their sea voyage they had suffered from dampness and were warped and swollen.*
>
> *With this more or less useless material at hand, Robbins & Lawrence proceeded to make a drawing of the Enfield Rifle as they believed it was intended to be, and from this drawing built an extremely accurate 'master gun'. They then made sets of hardened steel gauges carefully oilstoned after hardening to fit the parts of this master gun; they built sets of accurate jigs and fixtures around the parts which were to be drilled or milled and sets of templates for the profiling and stocking machines were cut from these sample parts.*
> *...they worked very closely to their gauges, making every part like every corresponding part, thus attaining absolute interchangeability in this stand of arms.*[9]

If true, this would account for the disparity between the Enfield gauge and the Windsor lockplate, but it raises other questions. If Robbins & Lawrence had already made gauges, why was another set ordered from Ames, and what were the Ames gauges made to fit? It can only be concluded that they were made to match a 'sample' non-interchangeable weapon different from that supplied to Robbins & Lawrence and which then became the prototype of the interchangeable series.

MANUFACTURE OF THE LOCK

Fig. 4.37. Enfield lockplate plug gauges for the 'plain' holes. (part of PR.10142. © Royal Armouries)

The plain holes in the lockplate were gauged with simple plug gauges, probably used in conjunction with the item shown in Figs. 4.15 – 4.18.

None of the gauges for the threaded holes are in the set of Enfield gauges and an extensive search of Royal Armouries collections has failed to locate them.

4.4 The Hammer

It has already been noted that nearly all the lock components are highly asymmetric with few, if any, straight lines, and the hammer was no exception. It was made of wrought iron and die-forged. Then, as reported in *The Engineer* (referenced earlier):

> *The rough forging is taken and, in the first place, stamped through so as to form the square hole; the presses used for this purpose are very accurately formed, and the hole is left exactly as it comes from the punch, not requiring to have anything done to it afterwards.*

This account also notes that:

> *By this process the hammer is slightly distorted but it is restored to its former position by means of a hand hammer.*

This distortion could not have taken place in the region of the square hole, otherwise correcting it would interfere with the accuracy of that hole. Forming this hole at such an early stage also suggests it may have acted as the datum for future operations, so its orientation in relation to the hammer nose especially was of particular importance. The hammer was arrested in its fall by the nose striking the cap on the nipple; excessive travel could cause the mainspring to strike the bottom of the lock housing in the stock, resulting in splitting at that point.

For finishing the hammer, the tooling contract (page 45) specified:

Robbins & Lawrence:

> *1 drilling machine, 4 spindles*
> *5 milling machines*
> *1 checking machine for hammer hole and for trimming lockplate.*

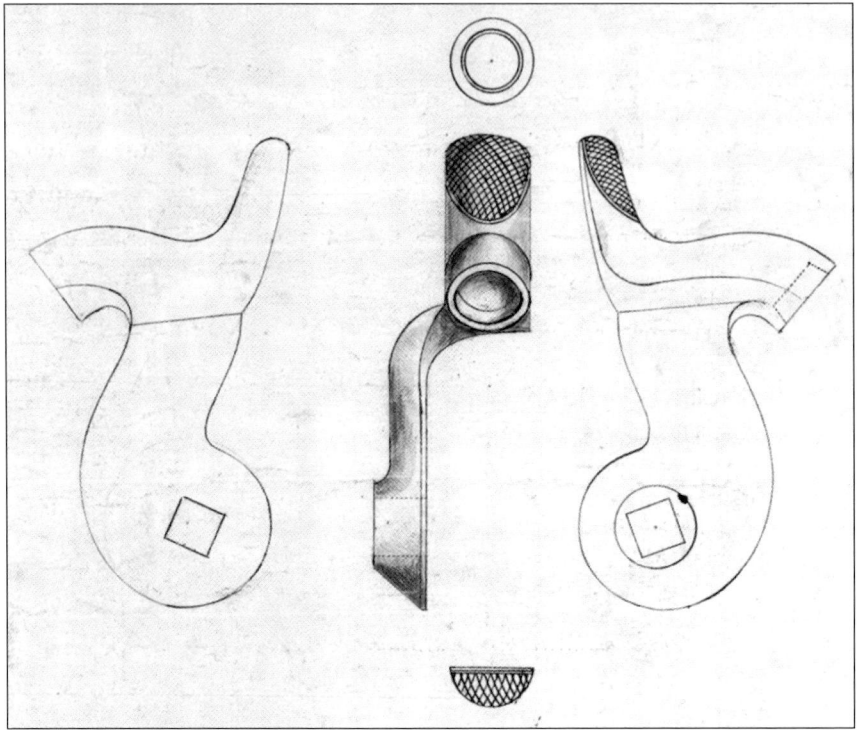

Fig. 4.38. The hammer. (From R.S.A.F. drawing No. 749. © Royal Armouries)

Ames:

> *1 drilling tool for nose*

The description in *The Engineer* of the first of these finishing operations is vague:

> *All the curves required are obtained by a series of machines fitted with circular cutters, and to effect this perfectly it has to be fixed seven different times. As these are all on the principle of those before described, it is unnecessary to say anything more about them.*

Producing these curves accurately undoubtedly required edge-milling operations. The account in *The Engineer* then goes on to state;

> *The eye of the bolster is then rounded by means of a circular cutter on a vertical spindle.*

The 'bolster' was the part of the hammer in which the square hole, the 'eye', was located and this operation was the formation of the circular flat surface around the square hole for the hammer screw to seat on, as shown in Fig. 4.38.

MANUFACTURE OF THE LOCK

The hammer nose was then recessed so that it covered the cap and the nipple when it struck them to prevent any cap fragments striking the face of the shooter. To do this the hammer was secured in a moveable fixture so that the face of the nose was horizontal and the hole 'roughed-out' by means of what needed to be a flat-nosed drill of some form, similar to a 'slot drill' – and then finished with a slightly tapered drill [end mill/slot drill] to make the recess conical. It is possible to draw a comparison with the hand-method used in Liège in the manufacture of the Spanish musket where a hand-powered 'end mill' was used to form the hammer-nose recess using a special fixture, Fig. 4.39.

The outside of the hammer nose was then rounded concentric with the recess by means of what can only have been a hollow end mill. This was fitted with a spring-loaded guide pin which was entered into the recess and allowed the hollow mill to be fed downwards to perform its work on the outside of the hammer nose.

Finally, it is noted:

> *The last process, that of getting the body of the hammer to figure, is more uncommon in its action, and so merits our particular attention.*
>
> *The hammer is fixed in a trough filled with soap and water, and above it revolves a milling cutter, coned to a point; this revolves at a speed equal to 1,300 revolutions* [per minute]; *the spindle of this cutter is moved up and down by means of a lever. Fixed to the same frame is a pointed guide-pin, which travels over a form* [template]

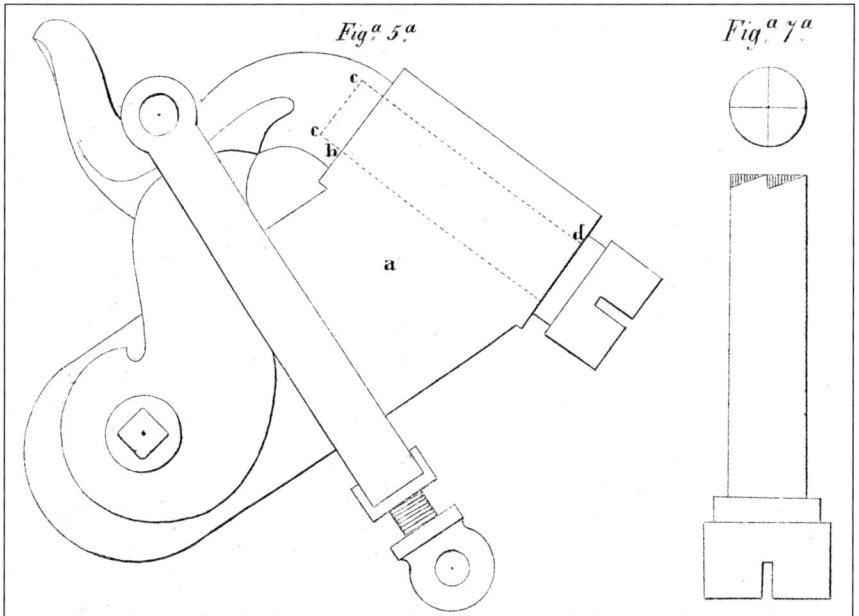

Fig. 4.39. Milling the hammer nose recess at Liège in a special fixture using the manually powered end mill. (*Memoria*, 1850, plate 35)

of the required shape and placed in a similar position; by this means a counterpart is formed to the mould and the hammer is completely shaped.[10]

[The use of soap and water would act as a coolant and lubricant and probably accounts for the term 'suds' being used for 'cutting oil' at the present time.]

This description suggests it only moves on two axes but to copy a 3-dimensional surface the guide and cutter obviously have to move on 3 axes. However, resort can be had again to the unpublished Greenwood & Batley archive drawings. One edge milling machine, Figs. 4.40 and 4.41, has a single lever allowing the cross-head to be moved from side to side and the frame carrying the spindle and guide pin to move up and down.

In this instance, the downward movement of the spindle frame is counterbalanced by a spring, as in a drilling machine, which causes it to rise whenever downward pressure on the lever is removed. The lever operates through a ball joint allowing simultaneous lateral and vertical movement of the cross head and spindle. The third axis is provided by the table movement.

The second Greenwood & Batley machine of this nature, Fig. 4.42, is more conventional in not having the spring-assisted upstroke, but its basic operation is otherwise the same.

On this machine the operating lever is again fitted with a ball joint carried in a fixture on the cross head support. It also passes through a swivelling housing fitted to the vertical frame of the cross-head carrying the spindle and

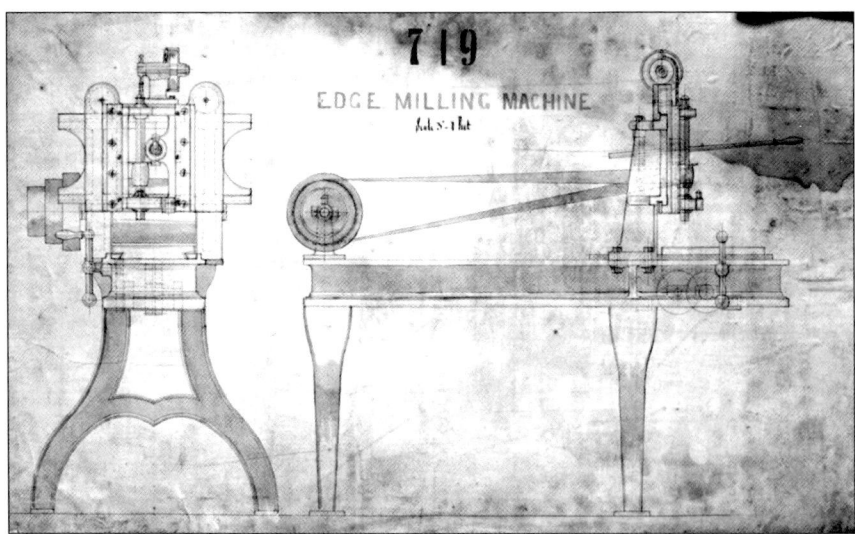

Fig. 4.40. Edge milling machine with lever operated feeds. (© West Yorkshire Archives Greenwood & Batley drawing No. 719)

MANUFACTURE OF THE LOCK

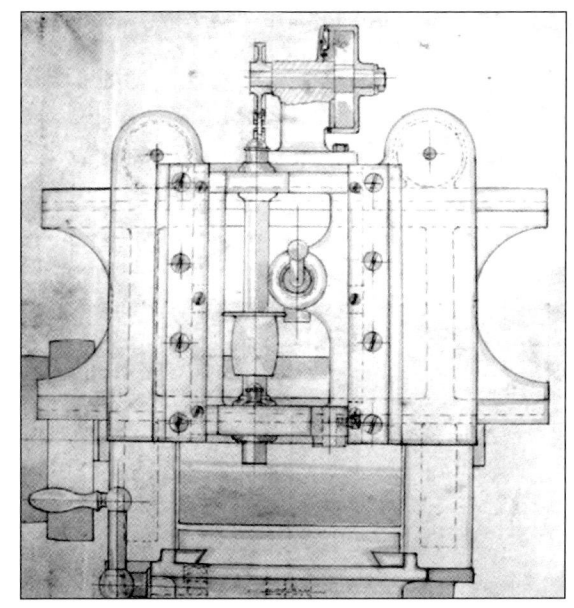

Right: Fig. 4.41. Front view showing cross-head control lever in the centre and the spring-assisted up-stroke at the top. (© West Yorkshire Archives Greenwood & Batley drawing No. 719.)

Below: Fig. 4.42. Sectional side elevation of cross-head of a 'Medium Pattern' edge milling machine with annotations to show the operating system.
(© West Yorkshire Archives Greenwood & Batley drawing No. 738)

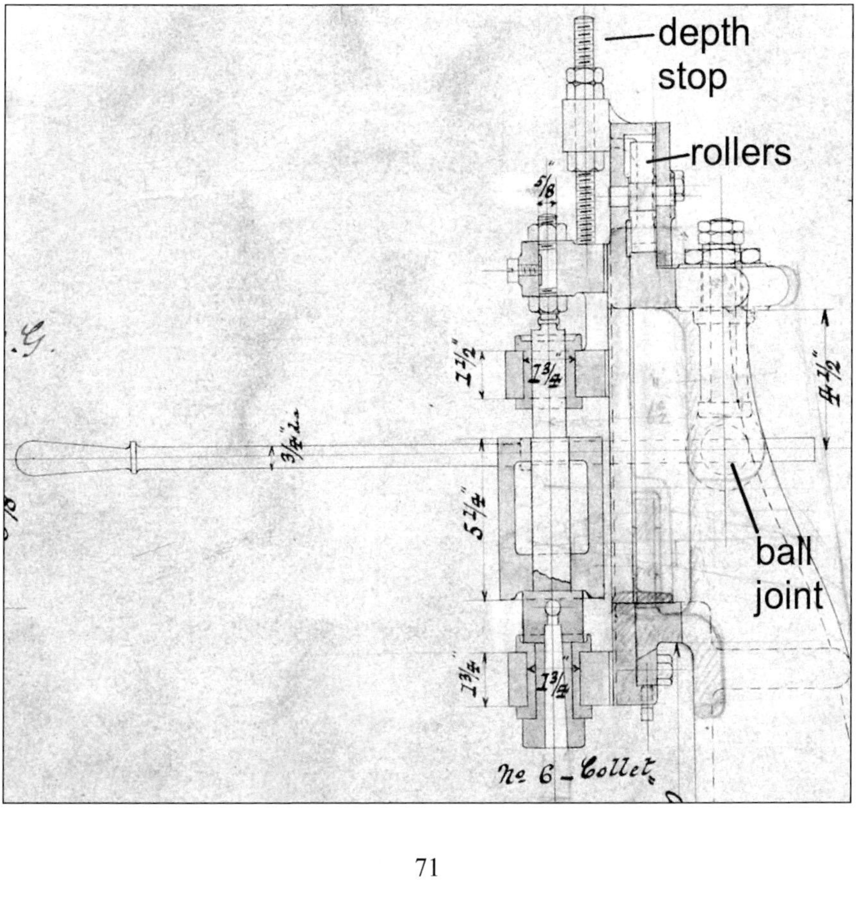

guide-pin. By virtue of these mountings, it can move horizontally and vertically simultaneously, thereby moving the cross head laterally on its mounting and the spindle and guide pin vertically. Where necessary, a depth stop controlled the downward movement.

The traversing of the heavy crosshead of both these machines was facilitated by being mounted on rollers. Either of these machines, accompanied by a high level of manual dexterity, would accomplish the shaping of the hammer but a *pointed cutter* as stated in the account in *The Engineer* would require an almost infinite number of passes. This may have been an instance of poor observation of the cutter if viewed at a distance. It is considered that the task would have been better accomplished with a tapered cutter ending in a small diameter rounded nose. It is stated that thirteen operations were required to finish the hammer.[11]

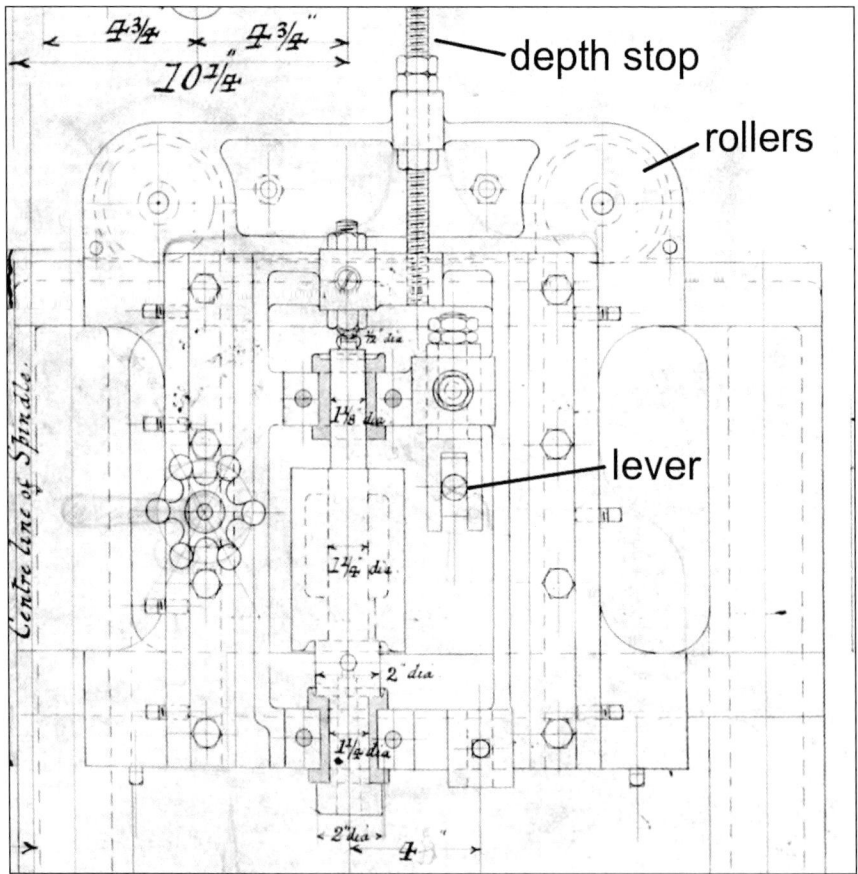

Fig. 4.43. Detail showing a part-sectional front view of the crosshead and spindle housing, annotated to show the lever, rollers and depth stop. (© West Yorkshire Archives Greenwood & Batley drawing No. 738)

MANUFACTURE OF THE LOCK

Gauging the hammer

The contract for gauges specified:

> 1 gauge pattern for testing the punching
> 1 gauge pattern for testing milling of bolster
> 1 gauge pattern for testing for straightening
> 1 gauge pattern for testing the edges
> 1 gauge grooved for testing the finished dimensions
> 1 gauge for testing the drilling of nose

The only three gauges in the set in Royal Armouries collections that can be identified with the hammer are those shown in Figs. 4.44 and 4.45.

The gauge on the left in Fig. 4.44 would fulfil several functions of various Ames gauges; it effectively checks the 'offset' of the head from the main body; the profile; the relationship between the square hole in the body and the hammer nose; and the correct thickness where the square hole, the 'eye', that is, the 'milling of the bolster' (the round projection is a retractable 'ejector' to push the hammer off the square). The gauge on the right in Fig. 4.44 checks the square, the flat portion surrounding the square hole and the recess in the nose. A third gauge in the set, Fig. 4.45, might appear at first sight to be for checking the offset of the recess in the nose from the square hole.

However, its orientation is incorrect, the truncated conical projection on the 'ball' does not fit the hammer nose and to date it has been impossible to determine its function in relation to the Pattern 1853 rifle and it may relate to a different weapon.

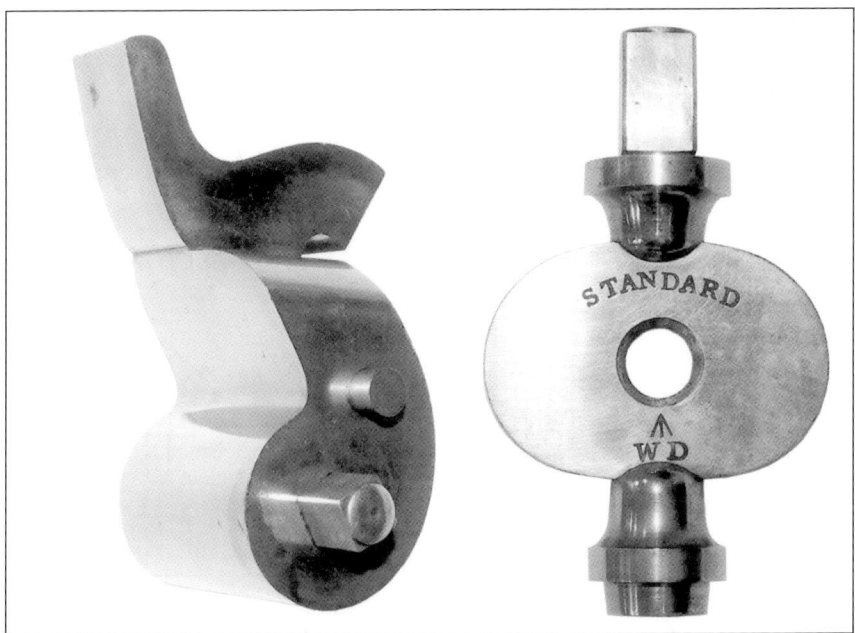

Fig. 4.44. Gauges for hammer. (part of PR.10142. © Royal Armouries)

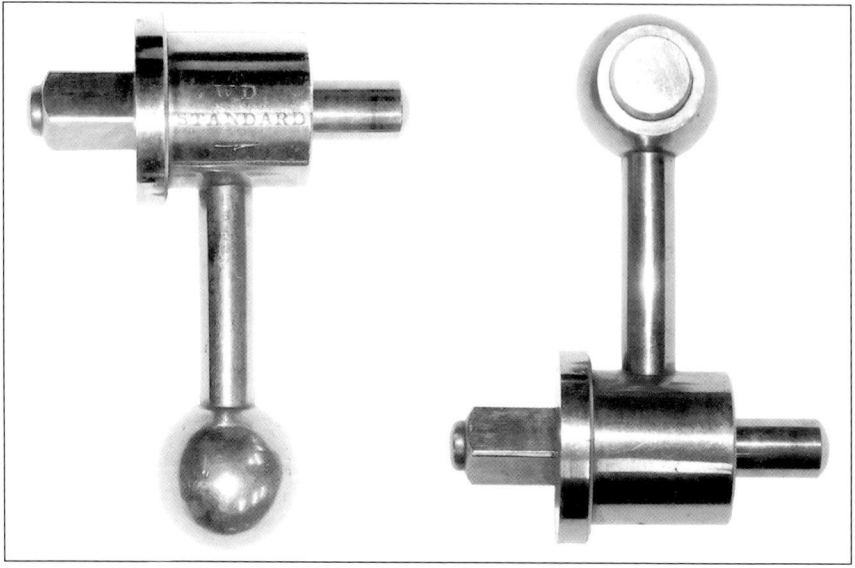

Fig. 4.45. Unknown gauge, front and rear view. (part of PR.10142). © Royal Armouries)

4.5 The Tumbler

This, again, is a component with a highly irregular and complex shape and which needs to be very accurately formed.

The tooling contracts in this instance specified:

From Robbins & Lawrence:

2 double milling machines
3 milling machines
1 drilling machine, 4 spindles
1 grooving machine
1 squaring machine
1 screw (hand) machine

and from the Ames Co.:

2 filing jigs for edges
1 drilling tool for screw-hole
1 tapping tool for screw-hole
1 drilling tool for swivel axle-hole

The tumbler was first forged in dies to give the body its approximate shape. It also had a short stub projecting from what would become its inner face and a long,

MANUFACTURE OF THE LOCK

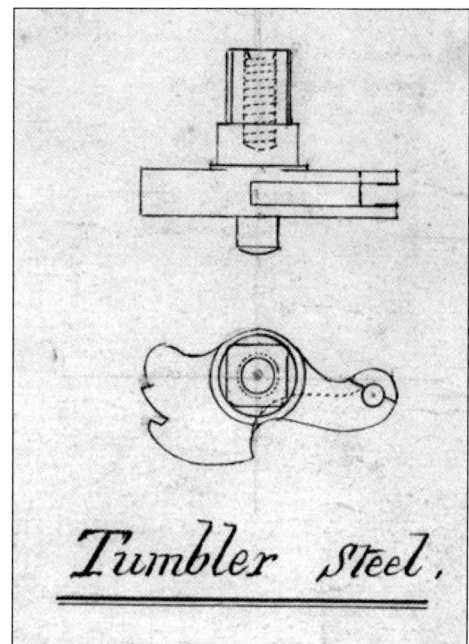

Right: Fig. 4.46. The tumbler, Royal Small Arms Factory drawing No. 749, 1860. (© Royal Armouries)

Below: Fig. 4.47. The tumbler. (author's collection)

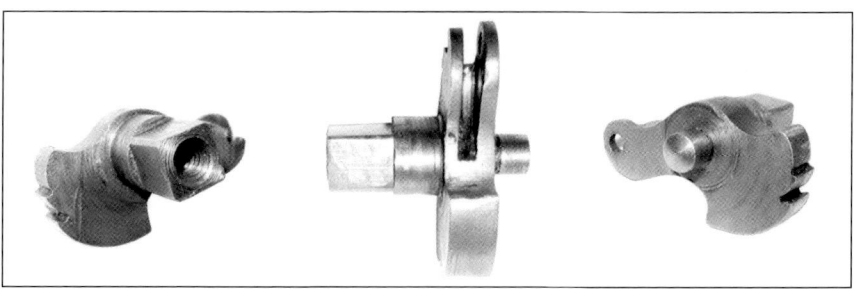

thick, shank projecting from the opposite side and from which the square would be formed.

The accurate formation of the tumbler was the key to the correct functioning of the lock. The first process was to insert the arbours of the tumbler forging into the cavities of two coaxial hollow end mills and revolve it whilst the mills are fed down the arbours onto the faces of the body of the tumbler. This was undoubtedly the *double milling machine* referred to in the contract, and by this means the arbours were rounded and the body thinned down to the correct thickness with its faces square with the arbours. This description is similar to a machine used in Liège in the manufacture of a musket for the Spanish government. The tumbler was rotated, by means of a driving 'dog', h, between two milling cutters, e, e, which 'face' the body to the correct thickness and round the arbours square to the body, Fig. 4.49.

Left: Fig. 4.48. Enfield rifle tumbler forging. (author's collection)

Below: Fig. 4.49. Machine used at Liège for facing the tumbler; the tumbler is rotated by the drive dog, h, between two milling cutters, e, e. (*Memoria*, 1850, plate 33, detail)

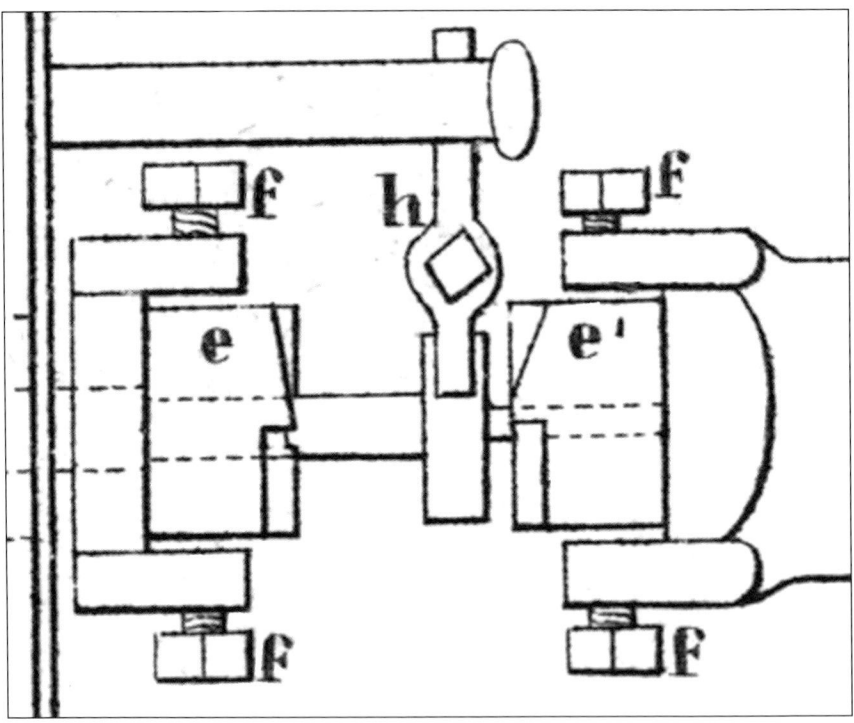

The arbours were then more accurately finished in another operation in which a thin cut was also taken off the face from which the large arbour projected to leave a small shoulder – *anti-friction boss* – around the base of the arbour (*Engineer*, 1859, p. 385).

MANUFACTURE OF THE LOCK

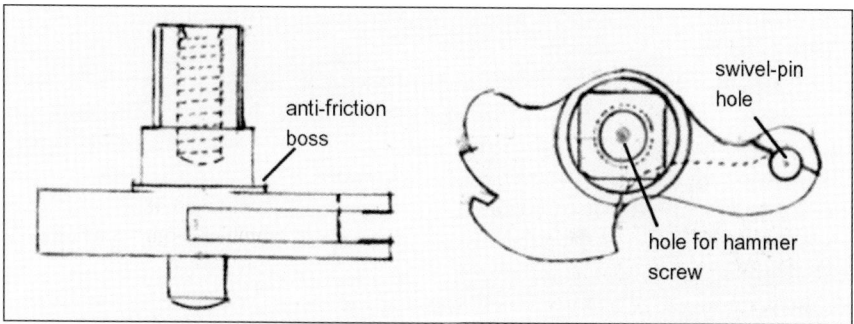

Fig. 4.50. The 'anti-friction boss', swivel pin hole and hole for hammer screw of the Enfield rifle tumbler. (Annotated R.S.A.F. drawing No. 749, 1860. © Royal Armouries)

This 'anti-friction' boss was to ensure the tumbler body did not rub against the inner face of the lockplate. The two arbours were then finished to the correct length and the hole for the 'swivel axle' drilled with the special tool supplied. The hole for the swivel-axle and the large arbour then provided datum points for it to be secured in a fixture for profile milling, requiring four separate operations.

The hole for the screw was drilled in the end of the large arbour and tapped using the 'screw (hand) machine'. The end of the small arbour was rounded with a concave hollow milling cutter and then the slot for the swivel milled with a slitting saw. It will be noticed that this slot follows a double curve (dotted line in Fig. 4.50) so a profile milling machine would be used and, subsequently, part of one flank of the slit, bisecting the hole for the swivel pin, was removed to allow the swivel to be inserted.

The square on the tumbler arbour, onto which the hammer was fitted, was formed by mounting the tumbler in a fixture, the hole for the swivel pin acting as a datum and using a 'gang-milling' operation employing two face-milling cutters mounted horizontally and correctly spaced, creating two 'flats'. By rotating the fixture through 90° enabled the square to be finished. Obviously, the centre of rotation of the fixture would have to coincide with the centres of the arbours and the position of the flats would have to be in correct juxtaposition to the position of the hammer. A similar system was being used in Liège a few years previously, but which relied upon hand-filing the flats in a specially prepared jig, Fig. 4.51.

In this jig two holes the same diameter of the swivel pin and placed at

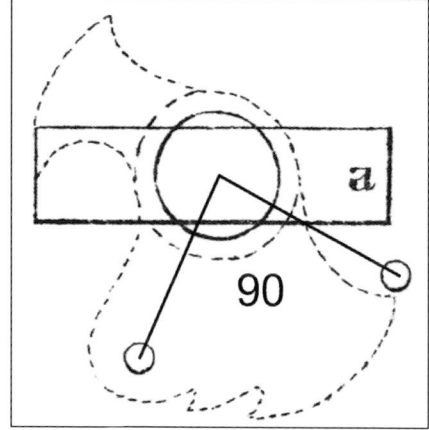

Fig. 4.51. Filing jig used in Liège to form the square on the tumbler shank. (*Memoria*, 1850, Fig. 36 – annotated detail)

MAKING THE ENFIELD PATTERN 1853 RIFLE-MUSKET

90° enable the tumbler to be accurately rotated to allow two opposing flats of the square to be filed in succession.

Finally, the square on the Enfield tumbler arbour was pressed into a die to 'size' it so that it was an exact fit in the square hole on the hammer. This raises the question of the purpose of the *1 filing gauge for squares* ordered from the Ames Company.[12] Since no filing was involved, according to the only account available, it is difficult to envisage its function.

In the fourteenth operation the edge of the tumbler was completed in a copying (edge) milling machine. The last operation would have been the use of the filing jig for finishing to gauge and accurately forming the 'bents', or notches, for the sear to engage with.

Gauging the tumbler
The Ames contract specified –

1 Gauge, receiving, for testing the milling, filing etc.

There are two in the set of surviving Enfield gauges which incorporate profiles and dimensions of various features. Their functions are shown in Figs. 4.52 – 4.56.

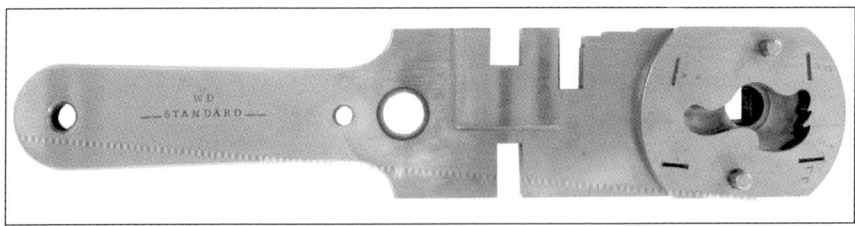

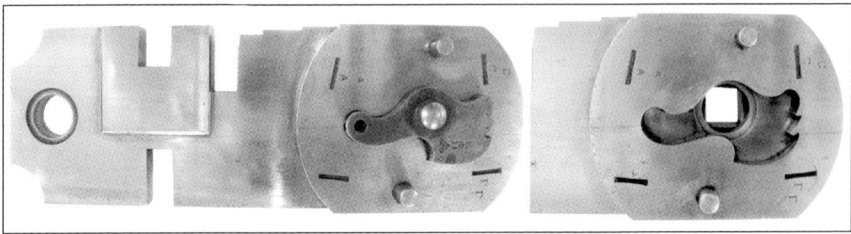

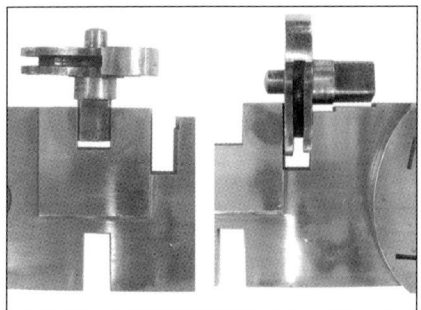

Top: Fig. 4.52. Compound gauge for tumbler. (part of PR.10142 © Royal Armouries)

Above: Fig. 4.53. The gauge in use checking alignment of the square with the tumbler body and body profile. (© Royal Armouries)

Left: Fig. 4.54. The gauge in use checking dimensions of the square and the body/arbour geometry and sizes. (© Royal Armouries)

MANUFACTURE OF THE LOCK

Right: Fig. 4.55. The gauge in use checking diameters of small arbour and large arbour with its anti-friction collar. © Royal Armouries)

Below: Fig. 4.56. The gauge used to check the accuracy of the slot for the swivel and the placement of the swivel-pin hole. (part of PR.10142. © Royal Armouries)

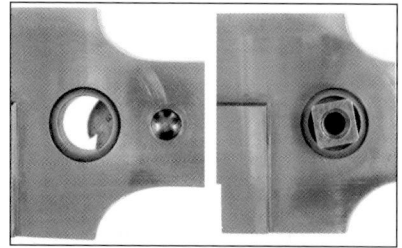

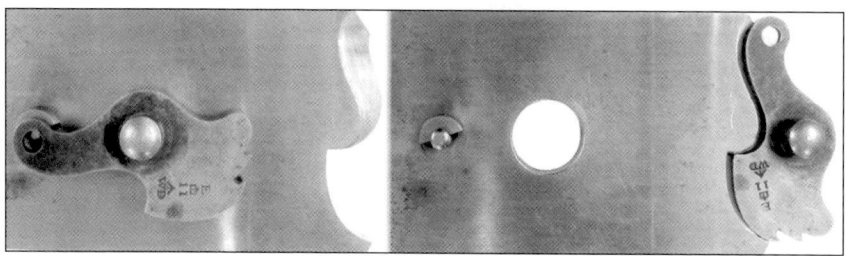

The second gauge is for checking the relationship of the arbour with the swivel axis-pin hole and the geometry of the swivel recess in the body of the tumbler.

4.6 The sear

The sear was probably the least geometrically complex component of the lock but still had to be finished with special care as both the safety and functioning of the weapon depended upon its correct engagement with the tumbler.

It would need to be of high-quality steel, properly heat treated, to withstand the stresses upon it, and any failure in the dimensional

Right: Fig. 4.57. The sear (annotated Royal Small Arms Factory drawing No. 749, © Royal Armouries)

Below: Fig. 4.58. The Enfield sear. (author's collection)

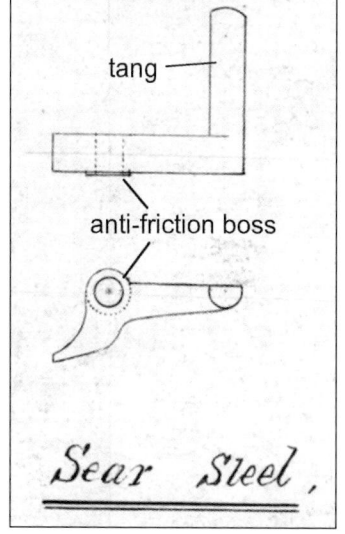

accuracy of the sear nose, where it engaged with the notches on the tumbler, could result in either an unserviceable weapon or the risk of the weapon being fired accidentally.

The tooling contracts specified:

Robbins & Lawrence

> *1 Drilling machine, 4 spindles*
> *2 Milling machines*
> *1 Double milling machine*

Ames[13]

> *1 Drilling tool for axle tube*
> *2 Filing jigs for edges*
> *1 Filing jig for end of tang*

The die-forged sear was pressed through a trimming die to remove any 'fins' and the forging, Fig. 4.59, was then ready for subsequent machining.

Machining commenced with drilling and reaming the hole by which it was suspended in the lock. The faces of the body were then faced parallel and to the correct thickness in a machine similar to that used for the facing of the tumbler, and the body was profile-milled to give it the required form. It would be possible for the flat upper face of the tang to also have been created in this operation since it is co-planar with the top face of the body; its rounded face, which in cross-section forms approximately two-thirds of a circle, *was effected by means of a revolving cutter block.*[14] This is a somewhat inadequate description, and for such a simple operation, it would be machined by being set with the tang upright on a rotating fixture and rotated about the tang axis with its face in contact with a vertical milling cutter, as in Fig. 4.60.

For either of these latter two milling operations, it would have needed to be firmly seated on a suitable fixture using its flat face, which would explain why the

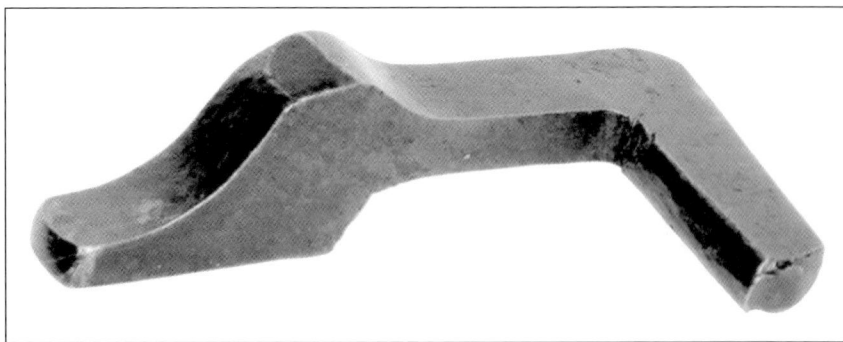

Fig. 4.59. A sear forging with fins trimmed off. (author's collection)

MANUFACTURE OF THE LOCK

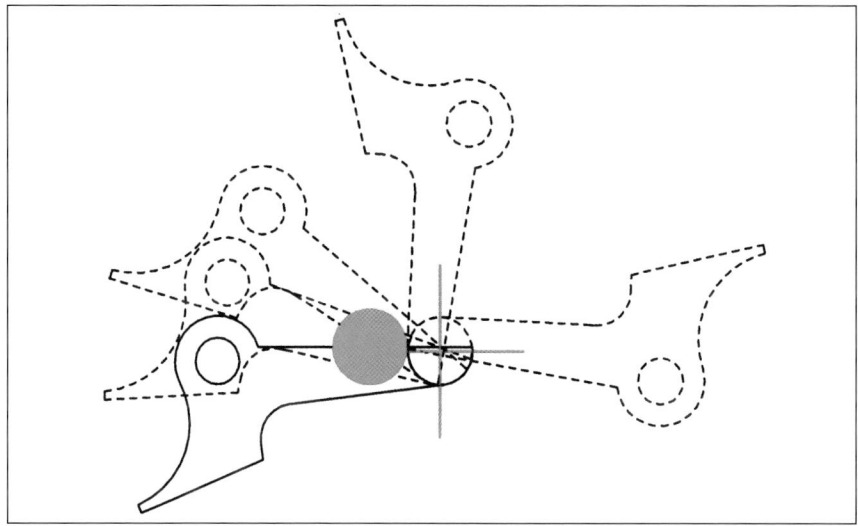

Fig. 4.60. Possible method of milling the sear tang by rotating it about its axis in contact with a vertical rotating milling cutter shown in grey. (© P. Smithurst)

machining of the *anti-friction boss* was left until last. In a similar manner to the tumbler, this was carried out using a single hollow end-mill with a suitable recess and a pilot to centre it in the hole. After this, fine finishing work on the sear was carried out by hand and the use of the filing jigs.

Gauging the sear

The contract for gauges (page 50) from Ames specified:

> *1 Gauge, receiving, for testing filing, milling etc.*
> *1 Gauge plug for testing drilling and milling.*

There is only one gauge is in the surviving set, Fig. 4.61. It has a number of slots to gauge various dimensions and in particular, the sear nose, and the overall profile in relation to the pivot hole, Fig. 4.62.

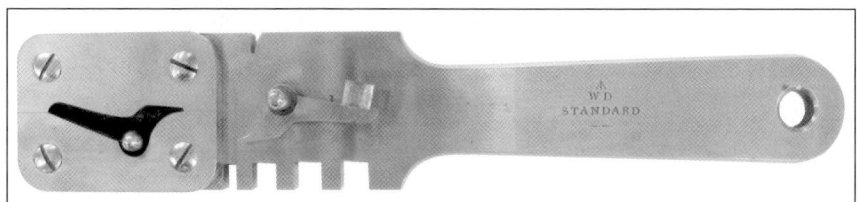

Fig. 4.61. The gauge for the sear. (part of PR.10142. © Royal Armouries)

MAKING THE ENFIELD PATTERN 1853 RIFLE-MUSKET

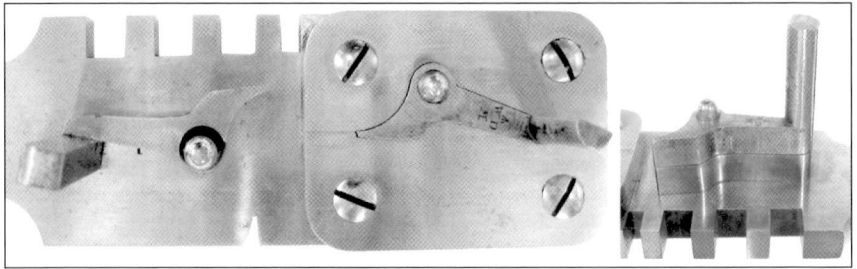

Fig. 4.62. The gauge in use to check the profiles against hole position. (© Royal Armouries)

4.7 The Bridle

The bridle was another complex three-dimensional item with compound curves and few straight lines which is apparent in the three views shown in Fig. 4.63.

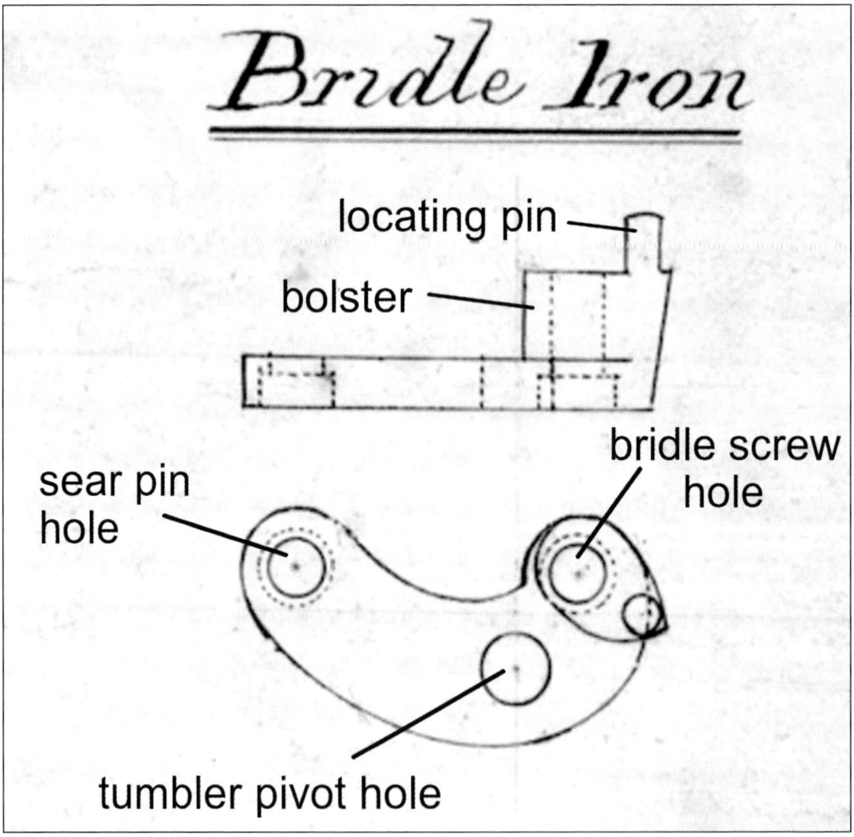

Fig. 4.63. The bridle (annotated Royal Small Arms Factory drawing No. 749, © Royal Armouries)

MANUFACTURE OF THE LOCK

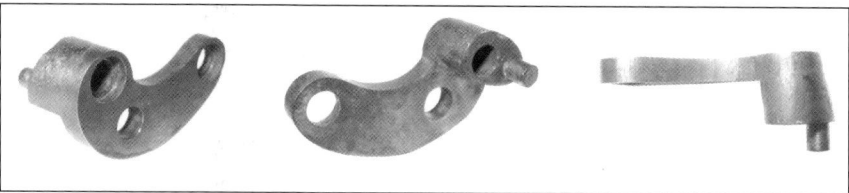

Fig. 4.64. Three views of the Enfield bridle. (author's collection)

The tooling contract specified:

Robbins & Lawrence

> *1 Drilling Machine*
> *4 Milling Machines*

Ames[15]

> *2 drilling tools for holes and milling pivot* [presumably means locating pin]
> *1 filing jig for edges*
> *1 reaming tool for reaming tumbler axis and holes*

Machining commenced with:

> *first, the inside of the bolster. . . is milled after the manner before described, and the same tool faces the upper side and makes it square with the bolster itself.*[16]

This appears meaningless since it would be impossible to mill the *upper* surface in this operation, so it has to mean the *inner* surface from which the bolster projects and would appear to an observer as 'uppermost' when the bridle was inverted and mounted for milling the bolster. The same account goes on to state that the outside face was then roughly milled. After placing in a drilling jig, the holes for the sear pin [screw] and bridle pin [screw] were drilled and in the same machine the locating pin was formed by means of a hollow end mill, described as a 'rose bit'. This indicates that the pin was integral and not inserted. It was then placed in a 'die-clip', [taken to mean a shaped jig/fixture] and the pivot-hole reamered out by hand – so it had obviously been drilled beforehand. It would have been logical to do it in the same drilling jig as the other two holes, thereby ensuring mutual accuracy in their positions. The edges were then rough-milled to shape and, by means of a 'copying machine', the inside of the bolster was finished. This was certainly one of Robbins & Lawrence's 'edging machines' described earlier. The same account states, *The top* [actually the bottom where it seats on the lockplate] *of the bolster was then faced*. Again, this is considered illogical inasmuch as the bolster could not be faced because of the projecting pin and it would be thought to have already been faced in forming the pin. The pin was then reduced to correct length and its tip rounded. The two screw holes were counterbored to accept the heads of the 'pins' [screws]. The top face was milled to the correct thickness, followed by milling the inner curve between the screw holes.

The final operation,

> *which until lately has always been done by hand, the finishing of the outside edge of the bolster and cutting its irregular curve, is effected in a beautiful little copying machine, which is so arranged to take the outline and pitch, at the same time following the irregularities with much greater accuracy than the most experienced workman could do, although the desired effect is so difficult of attainment.*

Considering the curve in question was partly conical with a varying non-circular cross-section, this could only be executed by one of the Robbins & Lawrence 3-dimensional profiling machines similar to that used for forming the curved surface of the hammer and employing a 'pattern' of the bridle. The bridle, being now complete, *as far as it is possible to make it by machine, it is finished by hand*, which is an oblique reference to the Ames filing jig supplied, a further indication that machines often required the support of hand-skills and tools.

Gauging the bridle

The contract (page 50) called for

> *1 gauge, receiving, for testing the filing*
> *1 gauge pattern and plugs for testing, drilling etc.*

Although only two gauges were specified, the second may actually comprise separate plug gauges; the gauges in the surviving cased set and their use is shown in Figs. 4.65 – 4.69.

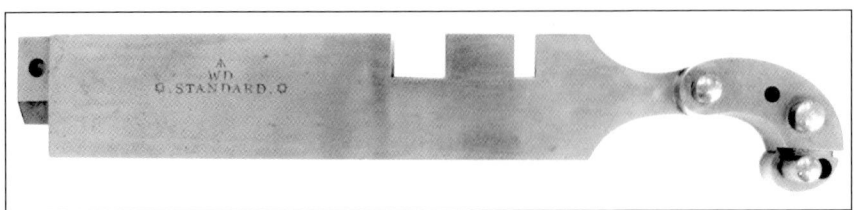

Fig. 4.65. Compound gauge for the bridle. (part of PR.10142. © Royal Armouries)

MANUFACTURE OF THE LOCK

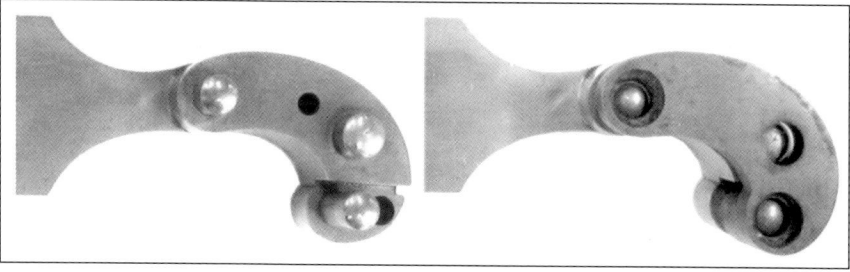

Fig. 4.66. Gauge being used to check the geometry of holes, pin, and profile. (© Royal Armouries)

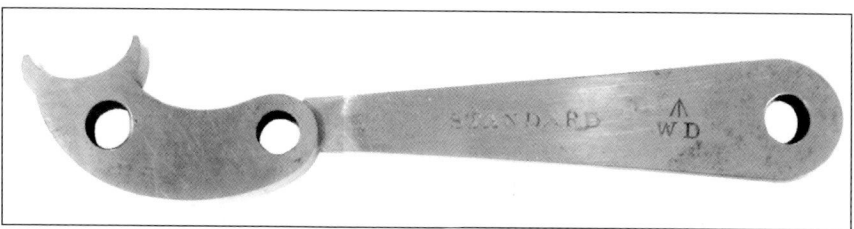

Above: Fig. 4.67. Gauge for checking the inner profile of the bolster in relation to the holes and main profile. (part of PR.10142). © Royal Armouries

Right: Fig. 4.68. Gauge for checking mutual positions of holes and pin, complete with 'ejector' to help remove it. (part of PR.10142. © Royal Armouries)

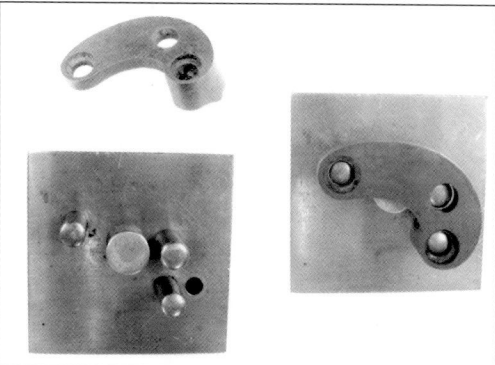

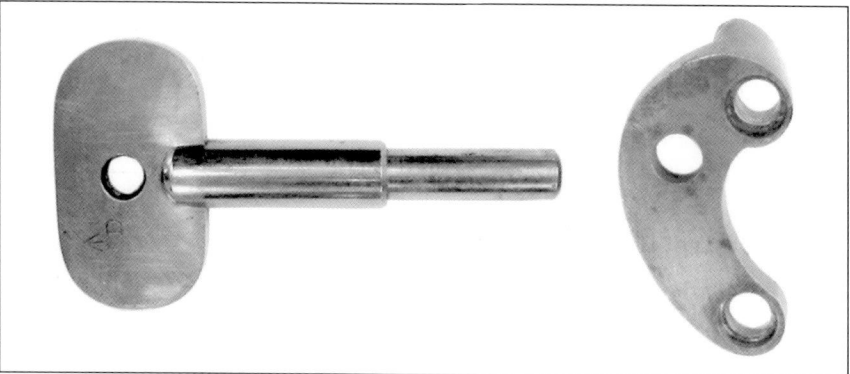

Fig. 4.69. Plug gauge for the tumbler pivot hole. (part of PR.10142. © Royal Armouries)

4.8 The Swivel

The swivel, previously only found on better-quality civilian firearms, acted as the linkage between the mainspring and the tumbler and formed an integral part of the mechanical 'train', previously shown in Fig. 4.4. Compared with the previous so-called 'hook-lock', Fig. 4.72, it offered less friction and therefore a faster lock action with less likelihood of the fall of the hammer being retarded which might have resulted in non-detonation of the percussion cap. In the 'hook-lock',

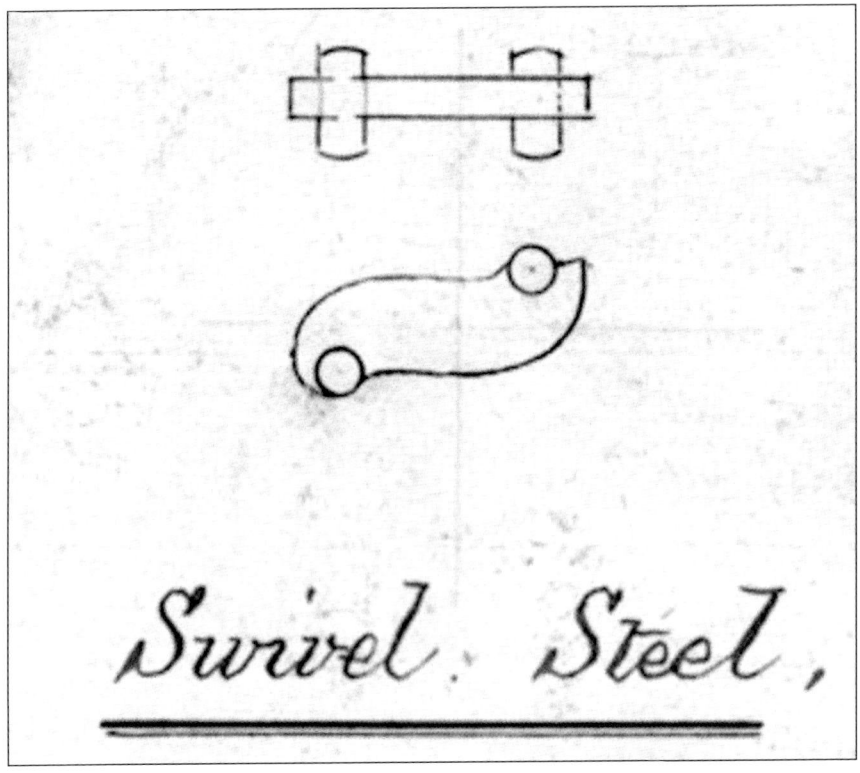

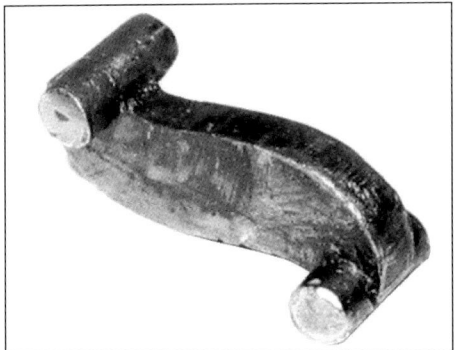

Above: Fig. 4.70. Swivel (annotated Royal Small Arms Factory drawing No. 749 © Royal Armouries)

Left: Fig. 4.71. The Enfield swivel. (author's collection)

MANUFACTURE OF THE LOCK

Fig. 4.72. The traditional 'hook-lock' of the Pattern 1851 Minié rifle. (XII.1907 © Royal Armouries)

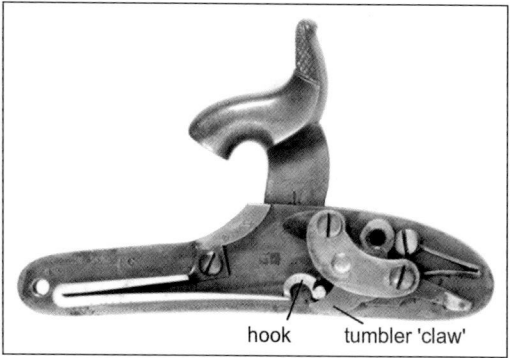

Fig. 4.72, the 'hook' on the end of the mainspring rode up and down the *claw* of the tumbler in action and was a cause of friction and wear on both components. The *hook-lock* was used until the introduction of the Pattern 1853 rifle and can be found in the lock of its predecessor, the Pattern 1851 'Minié' rifle (see Chapter 3). Its replacement with the 'swivel-lock' was just one of the many refinements introduced with the new rifle.

The tooling contract for this consisted simply of a 4-spindle drilling machine[17] but this was obviously to supplement existing machines.

The swivel began its existence as bar of steel $5/16$ (0.312 inches, 7.9 mm) square which was sawn into lengths slightly longer than required. This was secured in a fixture and the portions between where the pivots would be formed was milled away on both sides to leave square bosses at the ends on either side. These bosses were then converted with a hollow end mill into the circular pins or 'axles' and metal removed to the same level as the web between them. The edges were then milled to shape in two consecutive operations and the heads of the pins slightly domed.

Gauging the swivel

No gauges for the swivel were in the Ames contract but one is present in the preserved set of gauges and was used to check form and dimensions.

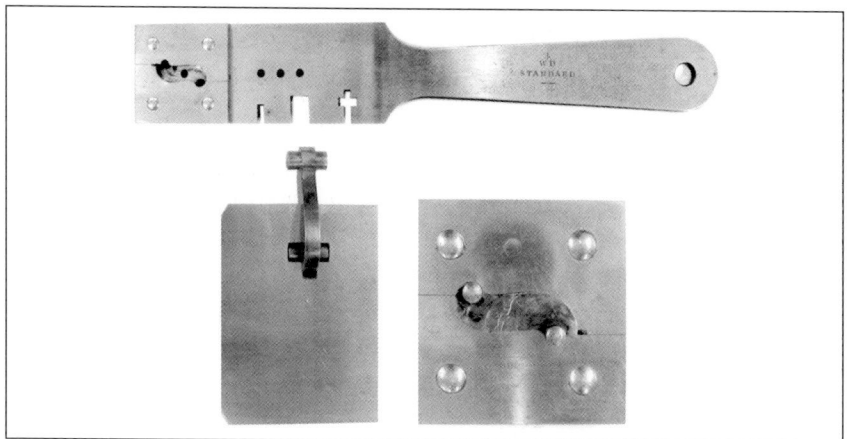

Fig. 4.73. The swivel gauge and details showing the swivel in place on it. (part of PR.10142. © Royal Armouries)

MAKING THE ENFIELD PATTERN 1853 RIFLE-MUSKET

4.9 The Mainspring

The only machines detailed in the contract with Robbins & Lawrence were four milling machines and a 4-spindle drilling machine and some tooling from the Ames Co.:
> *1 gauge plate for levelling the bottom edges.*
> *2 filing jigs for hook and tang.*

Once again, '*The Engineer*' provides a confusing description in which terms like 'lower' and 'top' refer to its position in the various machines, as opposed to its position on the lockplate. The mainspring was first hand-forged to approximate shape and then re-heated to red heat and placed in a bending device in which it was bent to form the 'V' and each arm was formed to correct curvature against fixed templates. The bottom edge of the arm carrying the locating pin was then milled

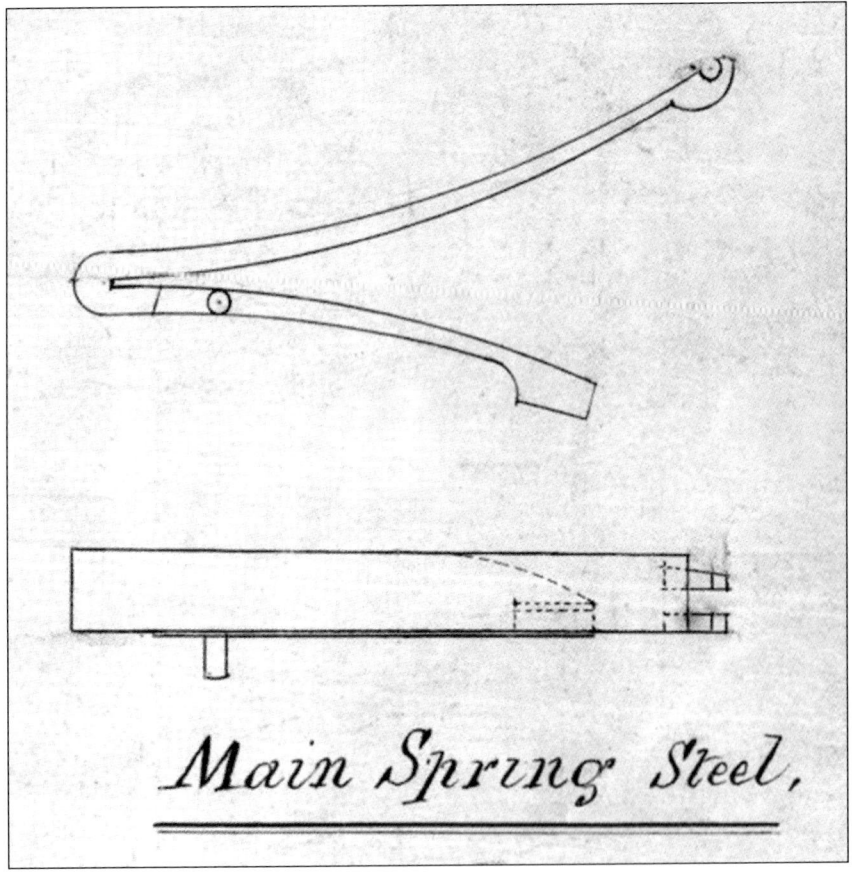

Fig. 4.74. Mainspring. (annotated Royal Small Arms Factory drawing 749 © Royal Armouries)

MANUFACTURE OF THE LOCK

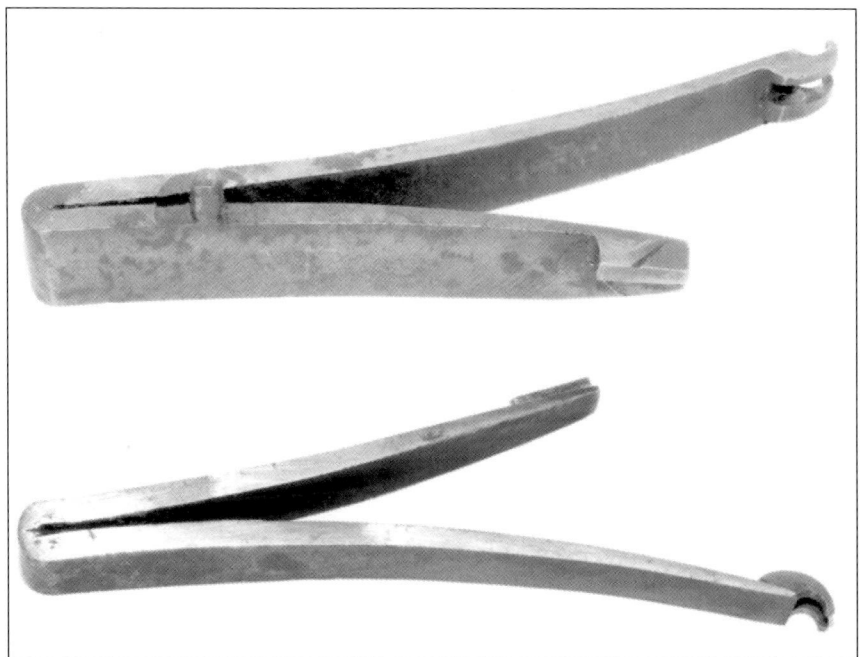

Above: Fig. 4.75. An Enfield Pattern 1853 mainspring (author's collection)

Right: Fig. 4.76. Detail showing circular striations left by hollow milling the locating pin. The other marks on the edges where it meets the lockplate and on the face may be file marks. (© P. Smithurst)

flat, leaving a small block for the locating pin, and the longer arm was milled, removing a little more metal than on the shorter branch so that it did not rub against the lockplate during flexing. This difference can be discerned in the drawing above and provides the equivalent of the *anti-friction bosses* on the tumbler and sear. The locating pin was formed using a hollow end-mill, evidence of which can be seen in the detail in Fig. 4.76.

Some filing was involved and what may be traces of milling or filing on the edge of the spring where it faces the lockplate can be seen in Fig. 4.76.

Gauging the mainspring
Only two gauges were contracted for from Ames:

> *1 gauge, receiving, for testing filing etc.*
> *1 gauge plate for levelling bottom edges*

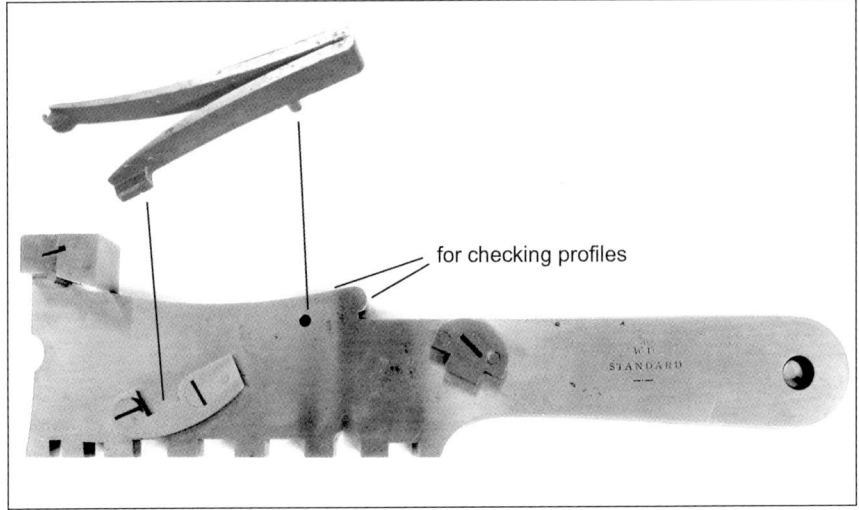

Fig. 4.77. Gauge for mainspring. (part of PR.10142. © Royal Armouries)

The mainspring gauge, Fig. 4.77, in the surviving set is a complex gauge with cut-outs for checking the thicknesses and widths at various points and an arrangement for checking the curvatures on both arms of the spring, the positions of the 'fork', locating pin and the 'tang' by means of which it engages with the undercut bolster on the lockplate. It may well match the Ames 'gauge, receiving' noted above.

It would have been sensible to test the 'strength' of the mainspring and within the list of gauges ordered from Ames is *1 apparatus for testing power* which may have been used for this purpose but is nowhere described.

4.10 Sear spring and its gauging

No machines were contracted for, but the contract for gauges also includes some tooling:

> *1 gauge, receiving, for testing filing etc.*
> *1 gauge plate for levelling bottom edges*
> *2 Filing jigs for edges*
> *1 Filing jig for length of mill end*

The manufacture of the sear spring, considering its overall similarity to the mainspring, used similar methods; bending the piece of steel hot, followed by milling, drilling of the 'eye', then filing to fit the gauge before finally being hardened and tempered. The filing jigs in the contract, as noted above, are a strong indication that much of the work was done by hand, both in forging and filing to finished size and shape. As with the mainspring, the gauge, Fig. 4.79, tests for thickness and widths at various points, profile, and the correct relationship between 'eye' [screw hole] and 'nib'.

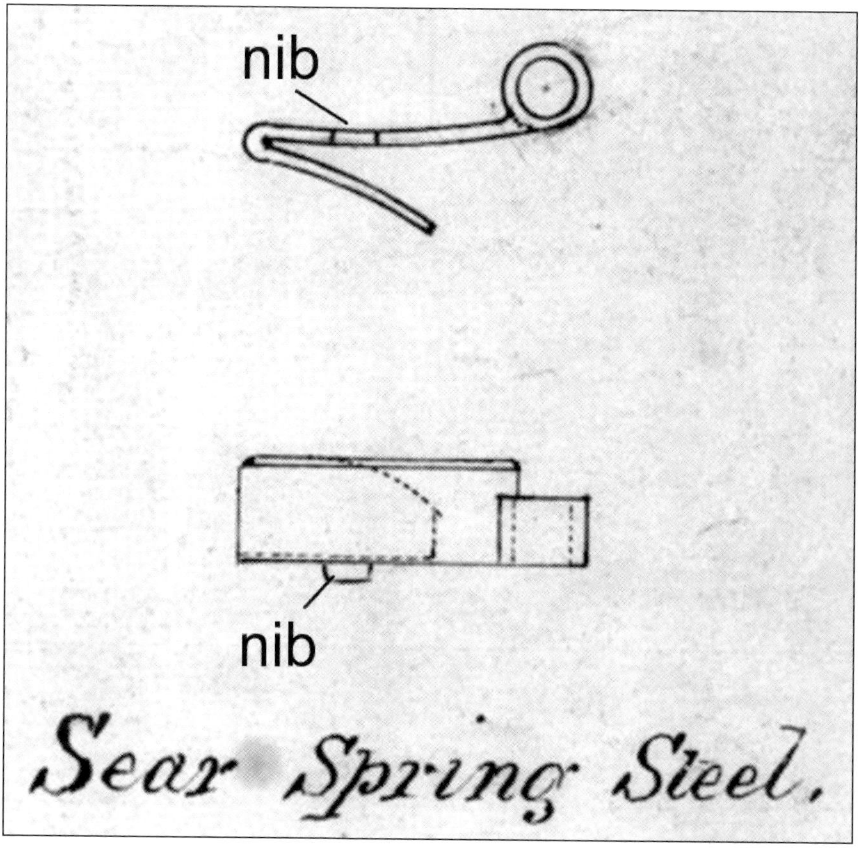

Fig. 4.78. Sear spring. (annotated Royal Small Arms Factory drawing No. 749. © Royal Armouries)

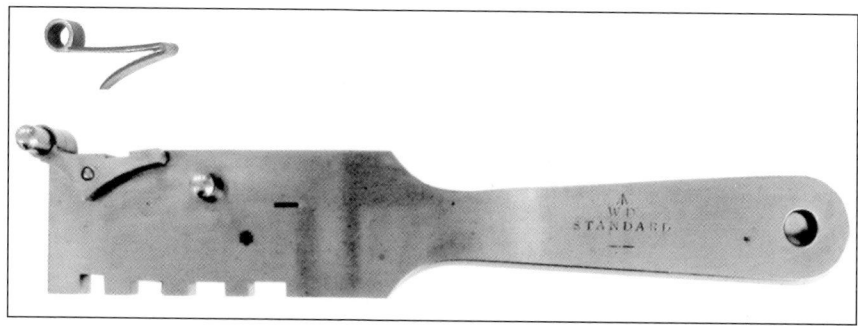

Fig. 4.79. Gauge for the sear spring. (part of PR.10142. © Royal Armouries)

MAKING THE ENFIELD PATTERN 1853 RIFLE-MUSKET

4.11 Screws and their gauging

There are four screws which form part of the lock components: two identical bridle screws, one of which acts as a pivot for the sear; one sear spring screw and one hammer screw. The bridle and sear spring screws have identical threads and only differ in length; the hammer screw has a coarser thread and larger diameter head.

The machines ordered from Robbins & Lawrence for their manufacture comprised:

2 screw milling machines
2 thread cutting (hand) machines
1 slitting machine
1 pointing machine [for rounding tip of screw]

This was later supplemented by:

1 clipping machine [purpose unknown]

The Ames Co. supplied:

Gauge plate for testing dimensions, length, cutting threads etc.

The only mention found in regard to the manufacture of these screws is that:

the tumbler screw is formed in a manner very similar to that before described, only that it is screwed by hand, then hardened and tempered, the sear-spring screw likewise.[18]

The only screw *before described* in '*The Engineer*' was that for the barrel-tang which was forged and the shank rounded by a hollow end-mill – a similar operation to that used on the 'pins' of the mainspring, tumbler, and swivel. The head was reported as being 'clamp-milled', a process using two dies with cavities and 'teeth' formed in them to match the profile of the screw head which were clamped around the rotating workpiece to produce the desired shape. However, there is no mention of a clamp-milling machine in the contract, nor of the cutters for clamp-milling. But the mention of *screw milling machines* raises questions. The context in which they are mentioned suggests that they were used for milling screw threads. However, it is known that the 'thread-milling' machine was first invented, or at least patented, in 1897 by H. Liebert of Rochdale.[19]

It has already been noted in Chapter 3 that 'screw-machines' were supplied by Hobbs & Co. which was almost certainly the lock making company established by Alfred Hobbs, an American who would have had access and knowledge of machine tools used in America. An account of his operations has already been noted earlier.

In the face of that and considering the 'hand-threading' machines mentioned, it can only be concluded that 'screw milling' was simply the use of a hollow end

MANUFACTURE OF THE LOCK

Fig. 4.80. A 'side-nail grinder' for use by hand to form the shanks of screws. (author's collection)

mill to reduce the diameter of a bar of iron to form the shank and underside of the head of the screws, the part-formed screw separated from the bar by the use of a suitable formed parting-off tool to create a slightly 'domed' head, and then a slot cut in the head. The thread could then be cut with the 'hand machine', The absence of a lathe being mentioned is taken to mean that the tooling supplied were special tools to supplement either existing machines at Enfield or which would have been readily available from numerous English makers. This, of course, is all circumstantial but based on experience of engineering workshop practice. However, the nature of the 'clipping machine' mentioned remains a mystery – no such machine is mentioned in any of the contemporary accounts. Also, there is no mention of a 'hollow end mill' in the contract, although this could easily have been produced in the Enfield factory toolroom. Indeed, such tools referred to as 'grinders' for use by hand were supplied to regimental armourers for the same purpose.

No workshop drawings of the screws are known to exist, but their general form is shown by the gauge in Fig. 4.81. Their profiles in the gauge can be clearly seen, the holes being used to check shank/thread sizes plus a larger recess for the head of the hammer screw. This gauge differs from all the others in being the only one not being marked with the War Department mark or designated as 'standard' and was probably made in the factory as a duplicate of, or replacement for, the original gauge supplied by Ames.

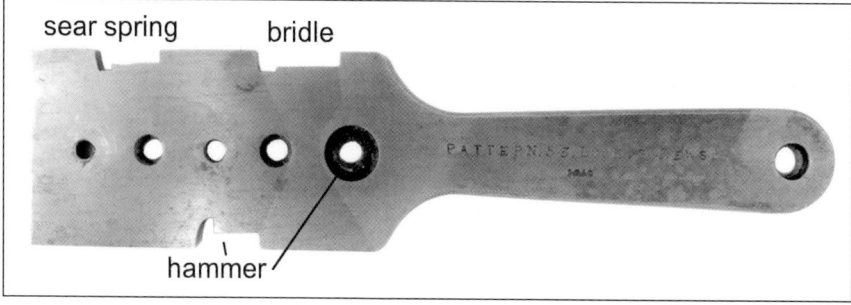

Fig. 4.81. Gauge marked "Pattern 53 lock screws" (part of PR.10142. © Royal Armouries)

4.12 Gauging the assembled lock

A further gauge (Fig. 4.82), not featured in the Ames contract list exists and would have been used for ensuring that the assembled lock would fit the lock recess in the stock.

Fig. 4.83 shows a 'reverse' view of it fitted with the 'working model' lock which is discussed in Chapter 5 and used in the Stocking Department to check the lock recess in the stock.

Being fitted with a trigger, it also allowed the lock to be operated to ensure none of the moving parts would at any time foul the lock recess in the stock. Furthermore, it also allowed the position of the hammer nose in relation to the nipple to be checked and may have functioned as the 'Mainspring testing apparatus' ordered from Robbins & Lawrence or possibly the 'Apparatus for testing power' ordered from Ames since it was listed in the gauges contract.

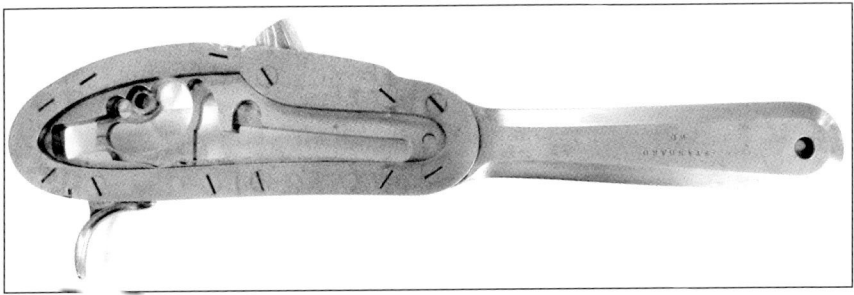

Fig. 4.82. Lock assembly gauge, 'W.D. Standard' (part of PR 10142. © Royal Armouries)

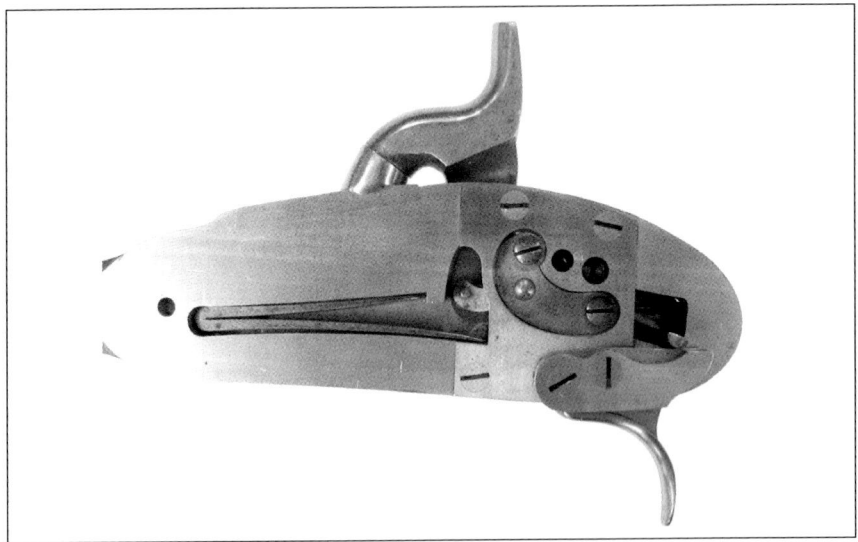

Fig. 4.83. View from inside face of lock assembly gauge fitted with 'working model' of the lock. (part of PR.10142. © Royal Armouries)

MANUFACTURE OF THE LOCK

4.13 Some metrology of Enfield Pattern 1853 Lockplates

No metrology of the lockplate has ever been published and some years ago, through the kindness and cooperation of colleagues at Royal Armouries and of David Eaton, Director, Malcolm Jackson, Metrologist, at the School of Engineering at Sheffield Hallam University, enabled some simple metrology on a group of Enfield lockplates. The focus of the study was to determine any variations in the mutual locations of a specific group of holes, 1, 2, 5 and 6, shown in the drawing below, Fig. 4.84.

The holes relate to the bridle (broken line) which acts as a 'cage' housing the tumbler and the sear and may be more apparent when they are seen in conjunction with these key components, Fig. 4.85.

It is clear that since the bridle is a single entity, the matching holes in the lockplate had to correspond exactly, otherwise the bridle could not be fitted. Whilst there had to be some clearance in the shank diameter of the screws and the holes in the bridle through which they passed, this would have been minimal

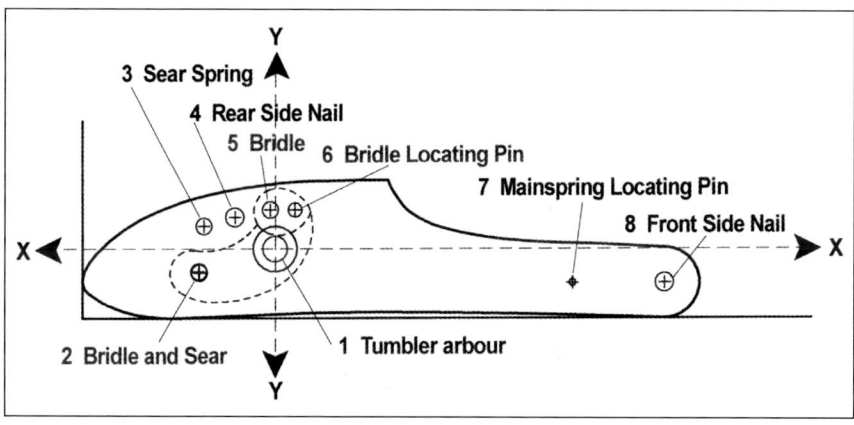

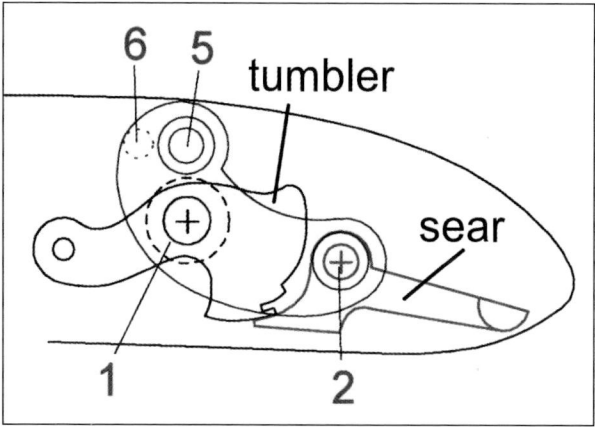

Above: Fig. 4.84. The Enfield Pattern 1853 Lockplate Holes. (© P. Smithurst 2020)

Right: Fig. 4.85. Showing the bridle, tumbler and sear in place. (© P. Smithurst 2020)

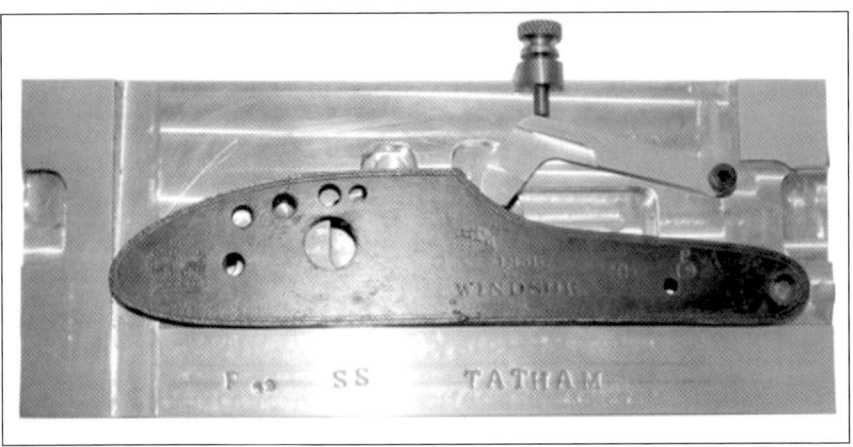

Fig. 4.86. Lockplate held in a jig having an 'L'-shaped lever acted upon by a screw which pressed the lockplate against square shoulders at its left-hand tip and along its bottom edge. (© P. Smithurst 2020)

otherwise the bridle could have moved and caused the tumbler to 'cant over' and 'bind'. It has to be accepted that since all these locks passed inspection and were accepted into service, all were correctly made and assembled to the satisfaction of the inspectors.

This study focussed on the lockplates from two groups of rifles in Royal Armouries' collections; those assembled using contracted locks supplied to the Tower and dated 1855; those supplied through a contract with manufacturers in Liège with locks dated 1856. They were selected because they represented the largest groups with identical dates which it was felt would provide a valid basis for comparison and allow some inferences to be drawn respecting the inspection process.

A special 'jig', Fig. 4.86, was made to hold the lockplates in identical positions during the measuring process and had cavities into which the bolsters of the lockplates fitted to enable them to lay flat.

The shoulders against which the lockplate abutted were square with each other and parallel to the corresponding outside edges. All the holes were measured regarding their positions relative to the large hole for the tumbler arbour as the origin, and centres obtained by averaging diameters using a cylindrical probe to minimise errors in threaded holes. The data, obtained using a Siemens coordinate measuring machine, was never published and the results are presented here for the first time and the tabulated metrology results are shown in Appendix 6.

Tower 1855 Lockplates

Within this group are three lockplates, XII.7235, XII.8999 and XII.9000, manufactured by Joseph Brazier & Son, the most celebrated lock maker at that time.

MANUFACTURE OF THE LOCK

Belgian contract 1856 Lockplates
The same technique applied to the Belgian lockplates, using the same Brazier lockplate as reference, gives similar results. It is clear from this evidence that it would have been impossible to fit the bridle of one lock maker into the lockplate of another.

Damage revealed in the metrology process
During the metrology it was discovered that some holes were noted as 'block':

Acc. No	Hole 1	Hole 2	Hole 3	Hole 4	Hole 5	Hole 6	Hole 7	Hole 8
XII.9074	10.07	4.01	3.96	4.42	4.00	block	2.69	4.43
XII.9076	10.26	4.09	4.04	3.94	4.12	block	2.93	4.43
XII.9078	10.05	4.05	4.04	4.53	4.01	block	2.58	4.47
XII.9079	10.09	3.96	3.90	4.25	3.95	block	2.80	4.44
XII.9088	10.13	4.10	4.11	4.44	4.05	block	2.60	4.39
XII.9090	10.12	4.22	4.22	4.49	4.27	block	2.77	4.44

In these instances, this simply meant that the hole 6, which corresponds with the locating pin on the bridle, was blocked by the remnant of the fractured pin. This could indicate a number of factors. Since the lock was disassembled during the viewing or inspection process, such a breakage must have occurred subsequently. Also, since these rifles never left military service and have been in storage ever since – they have never been in private hands – such a breakage might have resulted from various causes: careless disassembly and/or reassembly by the regimental armourer, which is unlikely.

A more likely cause is that, as noted in the contract document, Fig. 88, the locks were initially examined in 'the filed state', implying that the components had not been case-hardened. Case hardening of very asymmetrical items, such as the bridle, is known to cause some 'warping' when they are quenched. Such warping was known to exist in the manufacture of files, especially 'half-round' files where the larger area of the curved surface allows it to cool faster when quenched, causing the file to bend. File makers counteracted this tendency by judiciously bending the unhardened files so that they straightened when quenched! Moreover, to achieve a reasonable depth of case hardening in the body of the bridle may have resulted in over-hardening of the slender locating pin, making it extra-brittle. Fitting and tightening the screws of a possibly slightly warped hardened bridle to the lockplate might easily have caused this fracture which would have passed un-noticed since the lock had already been inspected.

Questions surrounding inspection of contract locks
The deviation between the holes relating to the bridle alone, and their acceptability, calls into question the nature of both the process of manufacture by private contractors, and inspection. It is known from contractual details, Fig. 4.87, that

each manufacturer, of whatever component of the rifle, could only 'inspect' the patterns at the Ordnance Offices at the Tower of London or at Birmingham and would have to prepare his own jigs and gauges.

Under such circumstances is not really surprising that such variation existed; what is more surprising is that, in many cases, such close correspondence resulted. Another contract document, Fig. 88, implies it was the correct functioning of the lock, its correct overall size and shape and the finishing of the components that were of primary concern, rather than precise component sizes and hole positions.

No reference to the nature of the 'standard jegs [sic] and gauges' is known to exist. However, in view of the fact that these variations in positions of the array of holes associated with the bridle were officially acceptable, it has to be inferred that no gauge could have been used to assess them; it would have been impossible to

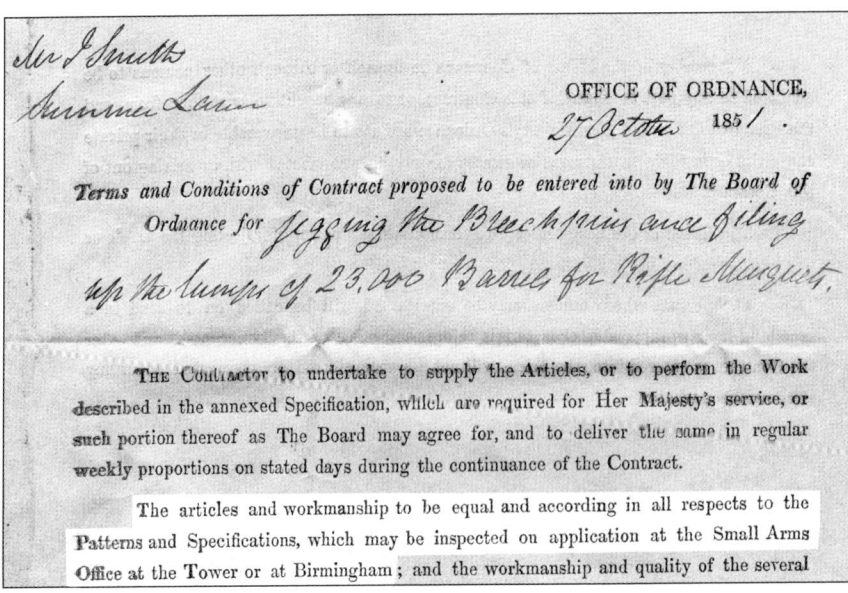

Fig. 4.87. Detail from a contract of 1851. (Appendix 2, item 1. © P. Smithurst 2020)

> **Locks.**
>
> The Locks with Cocks fitted to be sent in for view in the filed state, and to be examined closely to jeg and gauge. The whole of the Locks to be stripped, and the fitting and filing of each part separately examined. The main-springs to draw from 13 lbs. to 15 lbs. at half-cock. The sear spring to weigh from 6 lbs. to 8 lbs. The quality and workmanship to be equal in all respects to the Lock of the pattern carbine agreeably to the standard jegs and gauges.

Fig. 4.88. Abstract from contract document detailing the lock inspection process. (Appendix 2, item 5 © P. Smithurst 2020)

MANUFACTURE OF THE LOCK

construct a gauge that would encompass these variations without risk of sacrificing the functional integrity of the lock. While the lockplate-bridle-tumbler combination had to be an exact fit, the other interacting components would have been made to interact correctly.

Further metrology and interchangeability

A further, smaller programme of metrology was performed recently by Matt Holland at The University of Huddersfield with the intention of establishing some 'benchmarks' for comparison with interchangeable manufacture provided by a Robbins & Lawrence 'Windsor' lockplate and an 1861 Enfield lockplate. The Windsor contract rifles fell within the period when the second pattern barrel bands, held in place by a spring instead of a clamping screw, were in use. Most of these rifles were made under contract in Liège in 1856 and it is therefore likely that it was samples of these which were supplied to Robbins & Lawrence as 'patterns'. It was felt to be instructive, therefore, to compare a Windsor lockplate with these Liège-made rifles and some results are shown below.

In addition to the expected hole centre discrepancies encountered previously, it should be noted that in both cases, the large hole, No 1, in the Windsor lockplate

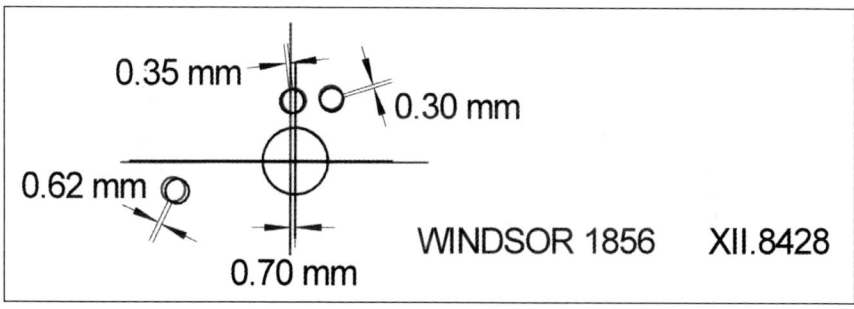

Fig. 4.89. Robbins & Lawrence lockplate with Liège lockplate XII.8428. (© P. Smithurst 2020)

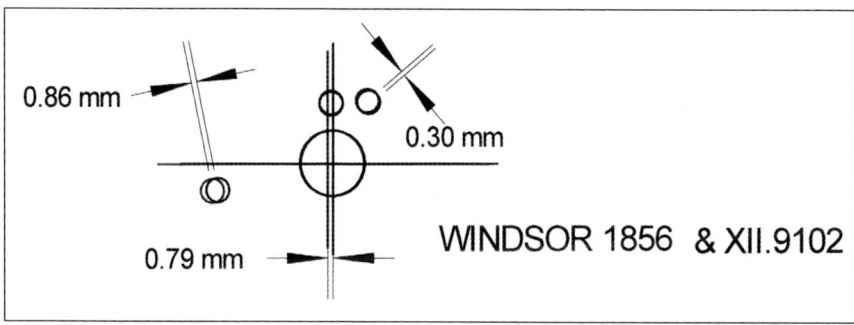

Fig. 4.90. Robbins & Lawrence lockplate with Liège lockplate XII.9102. (© P. Smithurst 2020)

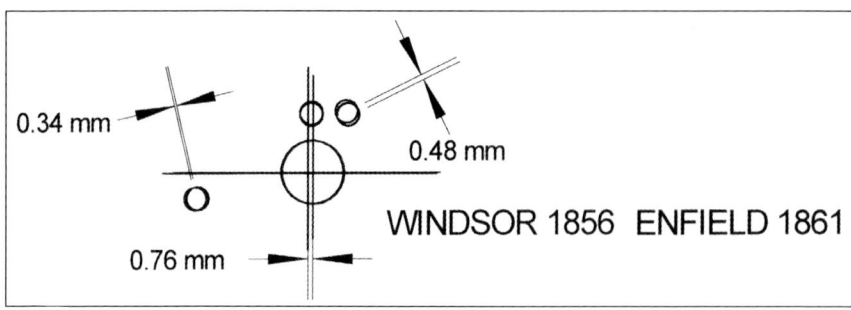

Fig. 4.91. Robbins & Lawrence lockplate with Enfield lockplate of 1861. (© P. Smithurst 2020)

is offset to the left in relation to the Liège-made rifles. This same discrepancy is found in comparing the Windsor lockplate with that of the 1861 Enfield lockplate.

It also should be noted that the lockplate in Fig. 4.91 was from a lock made at Enfield in 1861. This date, in combination with the name 'Enfield', indicates this lock to be one of the 'interchangeable' series which came into production after 1857. It should also be noted that even after 1857, contractors were still used to manufacture short rifles and carbines, and these will have the name 'Tower' and a post-1857 date on the lock and are not interchangeable. Taking the bridle from this 1861-dated lock and applying to it the gauge shows its overall form and the critical three holes and the locating pin to be a perfect fit.

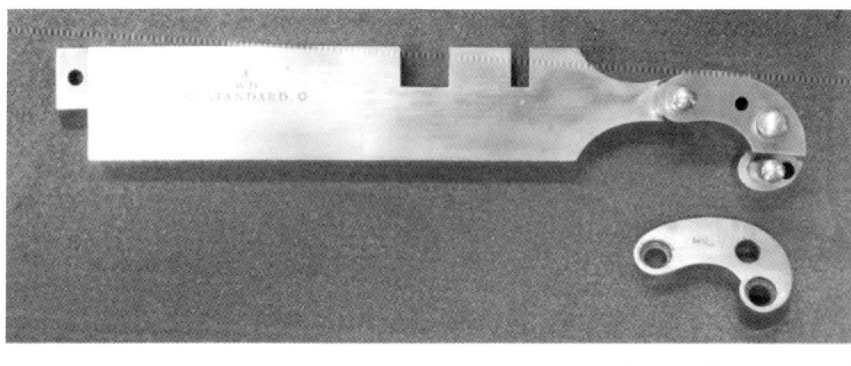

Fig. 4.92. Bridle from an Enfield lock dated 1861 is a perfect fit to the gauge. (The small hole in the gauge is for an 'ejector-pin' in the event of the bridle being a 'tight fit'.) (Gauge part of PR.10142. © Royal Armouries)

MANUFACTURE OF THE LOCK

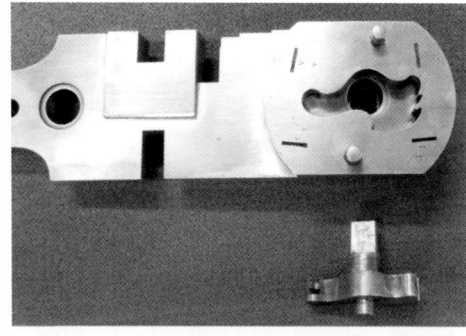

Fig. 4.93. Tumbler from an Enfield lock dated 1861 is a perfect but tight fit in the gauge. (gauge part of PR.10142. © Royal Armouries)

A similar exercise performed with the tumbler, the other item of complex form, from this lock of 1861 was also a perfect fit to the gauge, Fig. 4.93. [The tumbler was a tight fit, so it was not pushed fully into the gauge.]

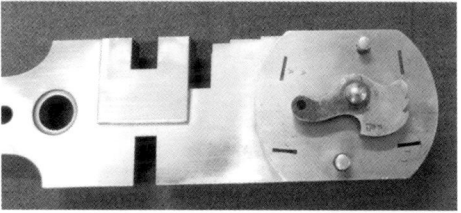

This is taken to be clear indication that interchangeability of these complex components was indeed being achieved. It might be argued that two samples are statistically insignificant. On the other hand, in the absence of interchangeable manufacture, it is suggested that the chance of finding two components at random that were perfect fits to the gauges is vanishingly small. The metrology data of this particular lockplate with regard to the holes for the bridle can therefore be taken to represent the 'standard'. Unfortunately, it has not been possible at this time to further extend this metrology to interchangeable arms.

4.14 Conclusion

In addressing this topic, much new material, ranging from unpublished factory drawings, artefacts, unique official documents, gauges, metrology data and specially prepared drawings, has been utilised and included to provide a more detailed analysis and description of the manufacture of the lock and its attendant procedures than hitherto. The complex geometries of lock components required sophisticated machines for their accurate repetitive production and insights into the operation of these machines, absent in other published accounts, have been provided.

Gauging was the essence of the system but the issue of wooden gauges to Robbins & Lawrence at the time of the contract for rifles is inexplicable and perhaps indicative of the 'establishment's' rudimentary concept of gauging. Furthermore, the contract for gauges from Ames suggests that the 'establishment' was not able to construct the gauges required and, that had gauges such as those supplied by Ames been in use and duplicates issued to contractors, then interchangeability might have been achieved prior to 1857.

There is work still to be done in this field. The metrology study was only able to concentrate on contract-made pre-1857 rifles; only one post-1857 could be studied. The same needs to be applied to more post-1857 rifles to gain some idea of the acceptable tolerances in a gauging system which only set an upper size limit. However, herein lies a problem. It has to be assumed that post-1857 long rifles were mostly, if not entirely, converted to the Snider system, thus transferring interchangeability to the new weapon. This assumption is borne out by evidence presented by Major General William Manley Hall Dixon at a hearing following a writ served on Richard Brown Roden by Thomas Wilson on the 18th May 1876 for non-payment of agreed patent license fees. From 1855 to 1872 Dixon had been Superintendent of the Royal Small Arms Factory at Enfield and was fully and minutely acquainted with the circumstances connected with the adoption of the Snider system into the British Service. He stated all the guns so transformed were new and had not been issued to the army or used, but subsequently, some portion of the Enfields which had been issued were returned to store and converted and the supply of Enfield guns in store had been exhausted in or about the year 1868. So simply searching for Pattern 1853 rifles with post-1857 dated locks inscribed 'Enfield' is likely to prove futile. If further evidence were needed it might be noted that of the 152 'standard issue' rifles in Royal Armouries collections, few have locks dated 1857 and after. Of these, only 8 are marked as 'Enfield' and of these, 5 are sealed patterns or similar, leaving only 3 as a 'standard issue'. It follows that to source post-1857 Enfield locks it is necessary to resort to 1857 or 1858 dated Sniders with Enfield locks.

Chapter 5

Manufacture of the Stock

5.1 Introduction

While the barrel is the key item of any firearm, followed by the 'lock' or other device which comprises the ignition system, it is the gunstock upon which these are mounted in correct juxtaposition, along with other components, that enables the firearm to be used. It is a vital part of that triumvirate, 'lock, stock and barrel'.

The information contained in the few published accounts of manufacture provides incomplete details of the various processes which the stock passes through. However, I have been able to draw upon a collection of stocks in different stages of manufacture plus original factory drawings of some machines, none of which has been previously studied or published. It has thus been possible to arrive at and to present insights and a fuller understanding of the functions of the various machines and their operations. This has been augmented by newly created drawings to elucidate various machine characteristics and processes.

Traditionally, walnut has been the chosen wood but as noted in book 1, other woods were experimented with and at the Russian weapon factory at Tula, elm or birch was employed.

In *Mémoire sur le montage du fusil*,[1] the making of the gunstock for the French 1777 musket at Liège is described and makes it abundantly clear that all the work was carried out by hand. They were supplied with all the components to build a finished musket, and therefore they also acted as 'setters-up' [assemblers] of the musket. Likewise, the gunstock for a Spanish musket manufactured in Liège in the late 1840s relied upon hand skills[2] and it is logical to assume that similar practices were also used in England until c. 1855 since there were no mechanical alternatives.

5.2 The advent of mechanisation – Blanchard's developments in the United States

In the United States, the mechanisation of gunstock manufacture was already underway when Gamel wrote his footnote. In 1819 a patent was granted to Thomas Blanchard for a machine for turning gunstocks.[3] His original patent had four principal claims but was objected to[4] and was re-issued on Jan. 20, 1820 with the claims:

> *The first principle of which I claim to be the inventor consists in the centres of the guide* [model or pattern] *and work* [workpiece] *being placed in a line, and parallel to the centres of a friction wheel and the cutting wheel, which are also placed on a line, by means of which, and a spring bearing against the opposite side of the frame containing the guide* [this is probably meant to be 'friction wheel'] *and cutting wheel, the irregularities of the guide in its revolutions keep the centre of the work at the same distance from the cutters as the centre of the guide is from the exterior of its irregularities, which bear against the friction wheel.*
>
> *The combination and application of the above principles, together with the other parts and movements of the machine (which are not of themselves claimed to be new) to produce the desired effect, constitute the second and last principle on which a patent for this machine is claimed.*

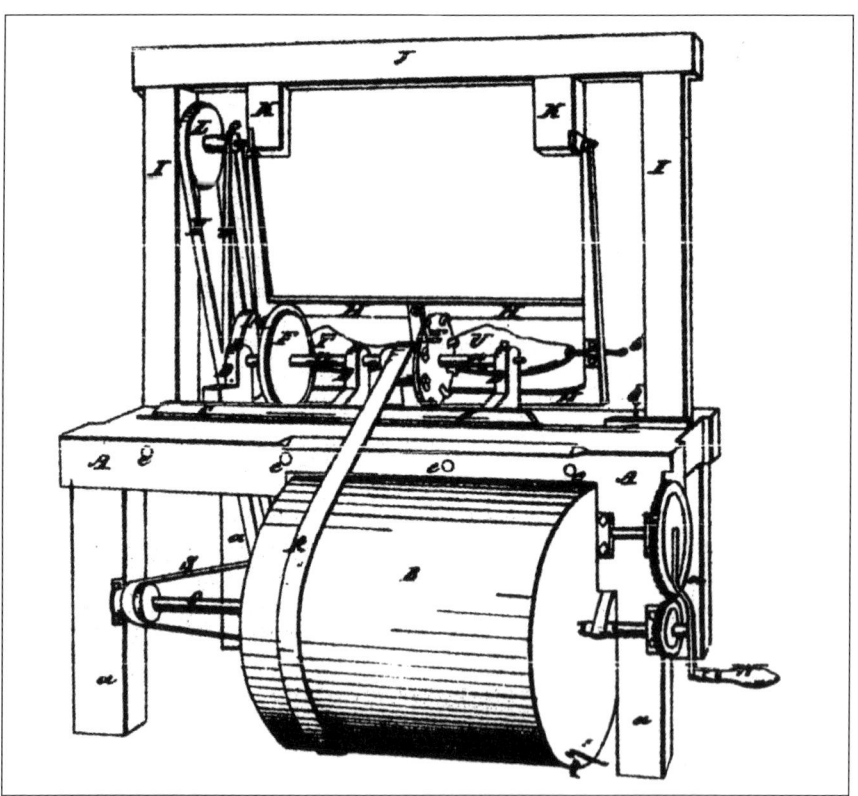

Fig. 5.1. Blanchard's machine as shown in his patents of 1819 and 1820.

MANUFACTURE OF THE STOCK

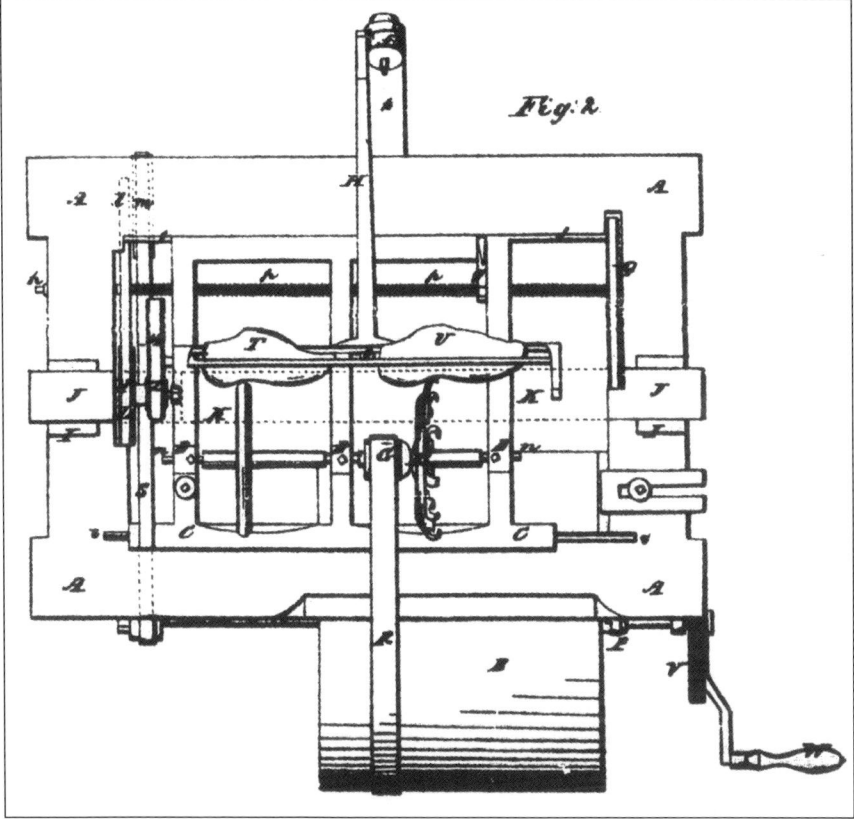

Fig. 5.2. Plan of Blanchard's machine from the patents of 1819 and 1820.

Blanchard's machine differed from Brunel's 'deadeye' machine in giving the 'friction wheel' and 'cutting wheel' the ability to traverse both pattern and workpiece longitudinally as well as operating radially. This placed it in a realm of its own, and it can be regarded as one of the great inventions of the 19th century.

In 1822 a British patent[5] was granted and it is apparent that the applicant must have been a patent agent [while many patents cite the name of the inventor, in some British patents it is often the patent agent whose name appears first] because reading beyond the preliminary statement it is clear that the patent application was lodged by John Parker Boyd of Boston in the United States under the more comprehensive description:

> *Certain Improvements in machinery for shaping or cutting out irregular forms in wood or any other materials or substances which admit of being cut by cutters or tools revolving with a circular motion, whether such motion be continuous or reciprocated.*

Unfortunately, Cooper, in her excellent study of Blanchard, makes no reference to Boyd or any attempt to obtain a U.K. patent, so a question mark must remain over Boyd's role and intentions. An additional point of interest is made by Greenwood who states that the principle of Blanchard's copying lathe was invented by a man named Rigg, near Birmingham, where he used it for making shoe lasts, but offers no date or corroborating evidence.[6]

Regardless of the true provenance of the Buckle / Boyd patent, the drawings which accompany it are clearer and more numerous than those of Blanchard's.

What will be noticed is that in both Blanchard's and Boyd's drawings, the machine is shown set up for making shoe lasts and has the 'model' and the 'workpiece' adjacent on the same spindle centres and carried in a swinging frame, D, in Fig. 5.3. Thus, as the model rotated the workpiece rotated at the same speed, and as they were mounted in a swinging frame the 'friction wheel', which followed the contour of the model, caused the frame, and therefore the cutter wheel, which was coaxial with the friction wheel, to move likewise. In this way the same contour was transferred to the workpiece. Whilst the same arrangement could be used for gunstocks, it would make the machine very long. Instead, it was rearranged so that the model was behind the workpiece, and that the work was carried out separately on the butt and fore-end of the stock, no doubt to avoid any flexing of a long, slender item.

A surviving example of Blanchard's gun-stocking lathe at Springfield Armory, Fig. 5.4, carries a plaque with the date '1822'. This is curious since his machines were reportedly destroyed in a fire on July 1st 1825.[7] It is additionally curious because it is officially recorded that it was in 1821 that Blanchard was paid $288.49 for a 'machine to turn gunstocks' supplied to Springfield Armory.[8]

As shown in Fig. 5.4, the gunstock was fixed at the front of a swinging frame and behind it was fitted the model, usually of bronze or cast iron. The mountings

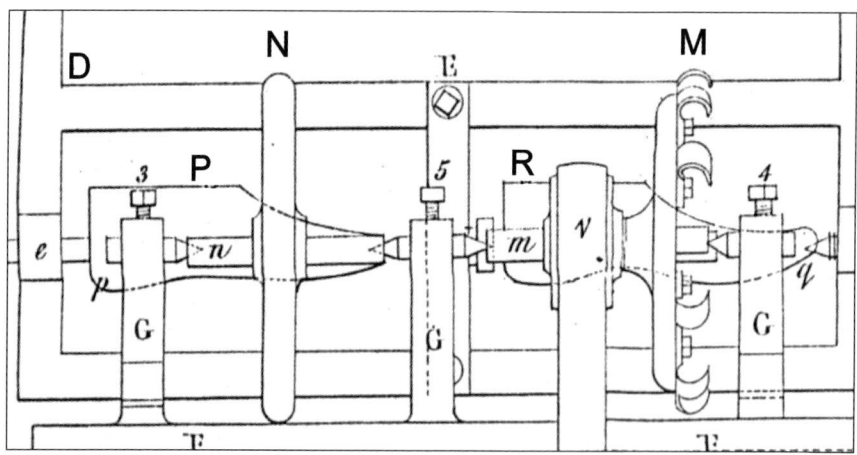

Fig. 5.3. Front elevation from Buckle's patent drawing showing the model, P, the workpiece, R, the friction wheel, N, and the cutting wheel, M, with its 'crooked knives or cutters'. (Buckle 1822)

MANUFACTURE OF THE STOCK

Fig. 5.4. Blanchard's original rebuilt machine now preserved in the U.S. National Armory at Springfield, Massachusetts. (Courtesy U.S. National Parks Service)

for the stock and model had to have parallel axes and the model and stock being worked upon had to be mounted in exactly the same orientation to one another with their rotation synchronised by a train of gears. The friction wheel and the cutting wheel were both eighteen inches in diameter and were mounted with their diameters coplanar in an iron frame which was fed along the bed of the machine by a powered leadscrew. Both the model and stock had to be precisely positioned in relation to each other so that as the model revolved, it caused the friction wheel bearing against it to move radially by following its profile, thus moving the swinging frame, and this movement transferred the profile at that point to the gunstock which was acted upon by the cutter wheel.

Blanchard also created machines for performing other operations on the stock and following a contract with George Bomford, Chief of Ordnance, on 27th July 1828, Blanchard set up his 'production line' which involved 14 stages:[9]

1. *cutting stock blank to thickness and length*
2. *'rough shaping' the stock*
3. *'rough turning' followed by 'fine turning'*
4. *boring for the barrel*
5. *milling the bed for the breech of the barrel and breech pin*
6. *cutting the bed for the tang of the breech plate* [butt plate PGS]
7. *boring the holes for the breech plate* [butt plate] *screws*
8. *for gauging for the barrel*
9. *cutting for the tang of the breech pin*
10. *for forming the concave for the upper band*
11. *for dressing the stock for and between the bands*
12. *forming the bed for the lockplate*
13. *forming the bed for the interior of the lock*
14. *for boring the holes for the side and tang pins*

While these additional machines performed a wide range of operations in preparing the gunstock, they did not carry out all of them, and for that reason Blanchard's process was referred to as 'half-stocking' and the remainder of the work – fitting of the side plate, trigger guard, trigger, ramrod and band springs – was carried out by hand.[10]

None of these additional machines has survived, nor were they patented, so no record exists of what they looked like and how they worked. Even in 1880, it was commented: *Of most of these machines little more than the name remains.*[11]

5.3 Equipping the Enfield factory

By the middle of the 19th century, updated Blanchard machines were being manufactured by the Ames Company of Chicopee Falls, Massachusetts and on 17th May, 1854, they submitted a tender to the Committee on Machinery for the supply of stock making machinery for Enfield which consisted of:

> *Machine for roughing*
> *Machine for rough turning (with patent rights)*
> *Machine for spotting* [machining small flat datum/clamping surfaces on the stock]
> *Machine for sawing breech and muzzle*
> *Machine for bedding the barrel*
> *Machine for planing side and edges with an extra spindle*
> *Machine for bedding breech plate* [i.e., butt plate]
> *Machine for fitting bands (with patent rights)*
> *Machine for turning between the bands*
> *Machine for smooth turning breech (with patent rights)*
> *Machine for smooth turning above the lock (with patent rights)*
> *Machine for bedding the lock* [i.e., inletting the lock]
> *Machine for bedding the guard* [trigger guard]
> *Machine for boring side tang screws and pin holes* [it is unclear what is meant by 'side tang', unless it refers to the cups for the 'side nail' heads]

It will become clear that in the description of the various stages of manufacture additional machines were used and an earlier comment by the Committee on Machinery,[12] states that patent rights had been purchased to allow the War Department to make copies of machines should the need arise. Greenwood & Batley of Leeds certainly produced stocking machinery, some bearing a remarkable similarity to those supplied by Ames. On the 18th August 1854 a further tender submitted by Ames for the following additional machinery was accepted:

> *1 Rough stocking machine*
> *3 Smooth turning machines for the butt behind the lock*
> *1 Smooth turning machine for the stock in front of the lock*

MANUFACTURE OF THE STOCK

and on the same date, a contract for a number of gauges (page 51) for the stock was made:

> *Pattern for vertical profile of finished stock (brass)*
> *Gauge pattern for testing spotting and angle of butt end*
> *Gauge barrel (solid steel) for testing groove*
> *Gauge for testing length from breech of barrel band shoulders*
> *Gauge bands etc.*
> *Gauge for testing position of lock bed*
> *Gauge for testing the cut for tenon of breech screw [barrel tang]*
> *Gauge for testing the depth of guard bed*
> *Gauge pattern for testing the fitting of guard bed*
> *Pattern for testing the profile from breech of band to breech plate*
> *4 gauges grooved (16 grooves) for testing the various diameters of stocks*
> *Gauge pattern for testing margins around lock and side-plate**

[* there was no side plate on the Enfield rifle so what is probably meant is the face on which are fitted the *side-cups* for the heads of the 'side-nails' – lock retaining screws.]

5.4 The stock

As Greenwood noted:

> *To secure success there is one condition which must be rigidly observed throughout this manufacture, namely perfect accuracy in each operation. In a manufacture where twenty operations are built upon one another, each depending for its accuracy on a previous one, it is evident how important this condition is to success.*[13]

The Enfield rifle stock began as a 2.25-inch (≈57mm) thick plank of well-seasoned Italian walnut cut roughly to shape.

Within the box of gauges for the Pattern 1853 rifle in Royal Armouries collections is a folding wooden template, Fig. 5.6.

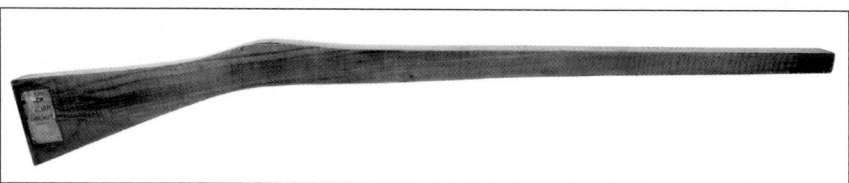

Fig. 5.5. The Italian walnut stock blank. (© Royal Armouries; ex-Pattern Room collection)

MAKING THE ENFIELD PATTERN 1853 RIFLE-MUSKET

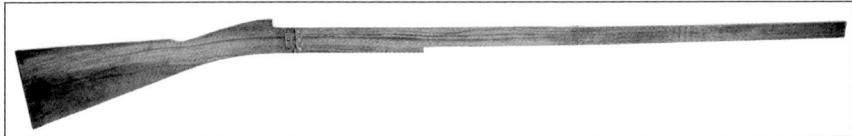

Fig. 5.6. Stock template. (part of PR.10142 © Royal Armouries)

From the inspector's stamp on the 'wrist' – the narrow portion between the butt and where the lock is mounted – it was clearly an 'official' item prepared at Enfield.

It also carries additional markings stamped into the wood relating to a variety of arms within the Pattern 1853 series:

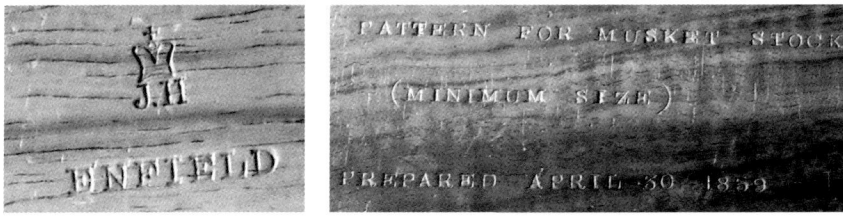

Above left: Fig. 5.6A Butt marking 1. Inspector's stamp on the 'wrist'.

Above right: Fig. 5.6B Butt marking 2. *Pattern for musket stock (minimum size) Prepared April 30 1859*

Fig. 5.6C Butt marking 3. *Width of stock in the rough 2 ¼ inches.*

Fig. 5.6D Fore-end markings for lengths of 1st class musket and 2nd and 3rd class carbines complete with spelling error. (part of PR.10142 © Royal Armouries)

MANUFACTURE OF THE STOCK

As indicated by the butt stamping, it was intended as a 'pattern' for rifle-muskets, and the fore-end clearly matches the rifle-musket length as indicated by the stamping at the tip, but it also carries two additional markings relating to the lengths of 2nd and 3rd class carbine stocks. First-class arms were either new or as good as new, and were issued to regular troops. If a first-class arm had been repaired to any great extent it was relegated to 'second-class' and the stock stamped with the number '2' in place of '1' and issued to volunteers or militia groups. Further repairs were recorded by additional number '2''s stamped on the stock. If yet further damage resulted in a weapon being considered 'unfit for service', they were relegated to third-class and used for recruit drilling.[14]

The 'pattern' is of flimsy construction, made up of three pieces of thin wood hinged together, one of which has come adrift, and is not really fit for the purpose of factory production use. After all, brass stock gauges had already been supplied by Ames and its wooden construction is certainly curious when it is considered that regimental armourers, as part of their 'Armourer's Forge' equipment, were being supplied with 'Gauges, metal, for stocking' in sets of 5, Fig. 5.7.

If the 'pattern' was simply used for checking the length of repaired arms, that is 2nd or 3rd class arms, being marked also for a 1st class arm is inconsistent.

Considering its nature suggests that it would have lent itself to being carried when visiting timber suppliers. Its form and markings would enable different sizes of timber to be selected for the three different lengths of arms, but that does not explain why 2nd and 3rd class arms were included.

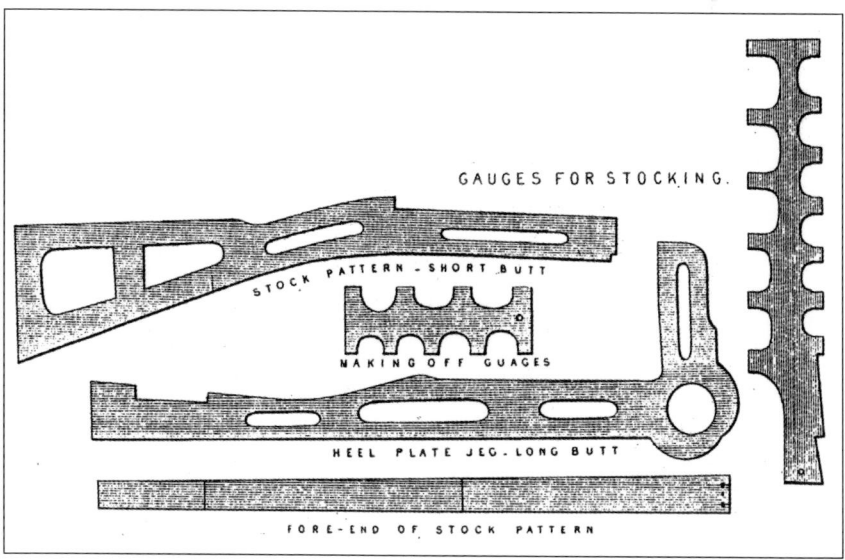

Fig. 5.7. Gauges, metal, for stocking, set of 5. (Petrie, 1865, p. 84 and plate XVI)

MAKING THE ENFIELD PATTERN 1853 RIFLE-MUSKET

Greenwood[15] itemises the various operations as:

1. Slabbing.
2. Marking centres.
3. Fore-end rough turned.
4. Butt rough turned.
5. Spotting – 5 flat places cut on the right-hand side of the stock and 2 on the left.
6. Bedding the barrel – hollow channel cut in the fore-end and the recess for the tang of the breech pin [screw].
7. Hand-finishing the recess for the breech tang.
8. Stock sawn to exact length at butt and muzzle and the butt shaped to the form of the butt plate by a revolving cutter.
9. The two opposing flat faces are planed where the lock is to fit; recesses for the side cups are formed; the upper edges of barrel channel are profiled; upper and lower edges of the butt planed.
10. Tang of the butt plate is let in; the three holes for screws are bored and the two large holes are tapped.
11. The corner under the tang of the butt plate is rounded by hand.
12. Lock bedded.
13. The end of the curved recess for the cone seat (nipple bolster) where it joints against the lockplate is squared by hand.
14. The rigger guard is bedded and the screw holes drilled, and the recess for the trigger plate is cut and the stop for the ramrod let in.
15. The stock is cut under the bands from a copy, and the nose cap let on.
16. The stock is cut between the bands.
17. The arris at the extreme muzzle end of the stock under the flange of the nose cap is taken off by hand.
18. The butt end of the stock is finish-turned in a copying lathe.
19. The fore end of the stock between the lock and the first band is finish-turned in a copying lathe.
20. The groove is cut for the ramrod.
21. The recess to receive the ramrod spring is cut out and the transverse pin hole for fixing the spring is bored.
22. The hole for the ramrod is bored in continuation of the groove.
23. The holes for fixing the lock plate are bored, and also the screw hole for the tang of the breech screw, the screw hole for the nose cap and the pin hole to fix one end of the trigger guard.

Slabbing

In the first operation, 'slabbing', the stock blank was placed on its right side and sawn along just above what would be the centre-line of the barrel to the point where the breech abutted against the stock. Whilst the process was discussed in

MANUFACTURE OF THE STOCK

The Engineer,[16] it is not illustrated. Other accounts do not even mention the process. The appropriate specimens of stocks held at Birmingham Museum were secured to a rack, making it impossible to obtain full-length clear photographs so the sketches below, Figs. 5.8–5.10, illustrate this process.

Care was needed to ensure that what would become the square face of the breech-end of the barrel channel was not undercut, and 'stops' would have been used to limit the travel of the table on which the blank was mounted.

This saw-cut was then cut into obliquely across at the breech, separating the surplus piece of timber from the stock.

A portion of one of the stocks in the Birmingham Museum collections which has undergone the 'slabbing' operation is shown in Fig. 5.11. What is apparent on the specimen, though not so easily seen in the photograph, are the angular lines left by the saw teeth which appear parallel with the oblique cut, suggesting the use of a bandsaw. However the description of the process[17] states a circular saw was used,

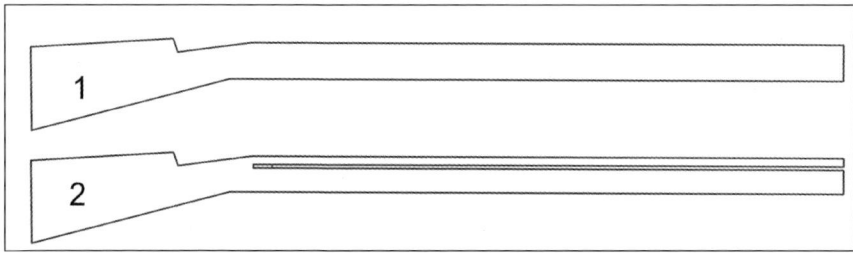

Fig. 5.8. Slabbing. 1, the blank; 2, the first sawcut. (© P. Smithurst)

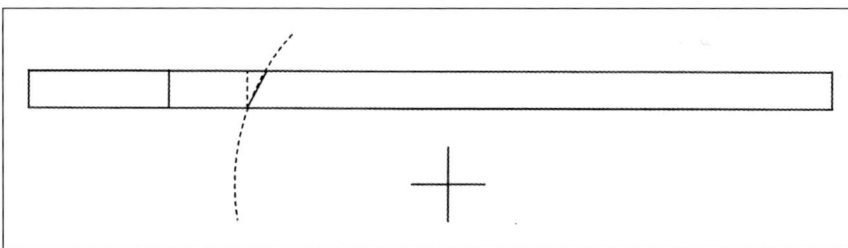

Fig. 5.9. Extent of the saw-cut, avoiding undercutting the breech face. (© P. Smithurst)

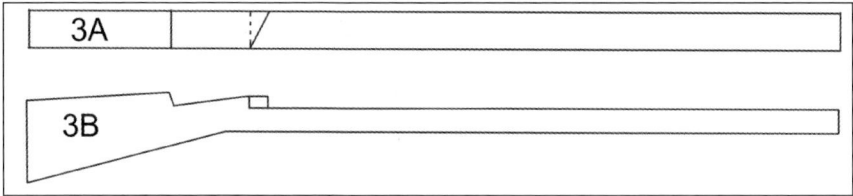

Fig. 5.10. Slabbing; the final stage, making the oblique cut, 3A and 3B, to remove the excess timber strip. (© P. Smithurst)

Fig. 5.11. Top view of portion of a stock which has undergone slabbing. (1885S00607.00001 courtesy Birmingham Museums Trust)

although bandsaws were available as shown by a patent granted to Robert Eadon of Sheffield in 1856 for the manufacture of 'cast-steel' bandsaw blades.[18]

The flat faces thus formed on the top of the fore-end of the stock and the breech were then used as reference faces to mount it on another fixture on the machine and, according to one article[19] the excess material at the butt and muzzle was removed, but another[20] notes that only the muzzle end was trimmed at this stage and goes on to state the butt and muzzle were trimmed at the eighth operation.

Centering

For this operation the stock was mounted on its side on a flat iron plate, the recently sawn surfaces being pressed up against a shoulder on the plate. At the butt end, a slide fitted with four projecting points was placed against the butt and given a blow with a hammer, driving the points into the face of the butt.

At the muzzle end, an accurately placed conical drill bored a centre hole and this was visible on a stock which had been 'centred'.

Unfortunately, because the stock was fixed in a rack, it was not possible to record the 'centering' on the butt. These centre holes matched centre points for mounting the stock on the machines which performed the rough-turning operations.

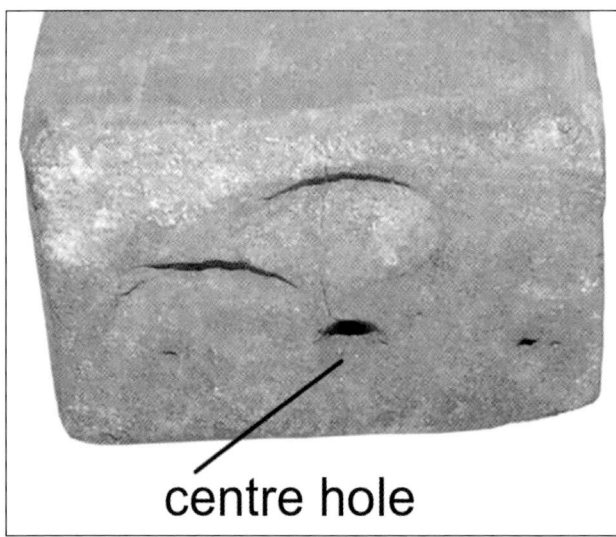

Fig. 5.12. A centre-hole in the bruised muzzle-end of a stock blank after 'slabbing'. (1885S 00607.00002 courtesy Birmingham Museums Trust)

MANUFACTURE OF THE STOCK

Rough-turning the fore end

This operation is discussed in *The Engineer*[21] but again is not illustrated, and the description of the machine is somewhat convoluted and incomplete. As described, the stock was passed through a hollow bearing and the centre-hole at the muzzle was located on a centre point on the machine. The prepared flat surface of the stock fore end rested against a straight, flat iron bar between two bearings. What is not stated, but of great importance, is that this iron bar could be no wider than that of the stock after being turned. It might therefore have been stepped to match the steps in the stock. The portion of the stock where the lock would be fitted was contained within the hollow bearing and fixed by a clamping screw.

From the description it is clear that the stock blank was mounted above the pattern. These needed to be in exact mutual alignment, and also to rotate synchronously for the form of the iron pattern to be transferred to the stock. A schematic conception of this arrangement is shown in Fig. 5.13.

The next part of the description in *The Engineer* offers nothing that gives a real understanding of how this machine worked. It simply states that the follower wheel and cutter *vibrate on levers hung on centres near the ground line*. It implies a system as shown in Fig. 5.14, which clearly would not have worked since the cutter wheel would never have been in contact with the stock.

A simple expedient for creating a working system would have been to use James Watt's 'parallel motion' in which, by means of a system of linkages, he was able to maintain the vertical motion of a piston rod of a steam engine whilst its point of connection on the end of a beam moved in an arc.

Thus, in the turning machine, this could have been achieved by having the follower wheel and cutting wheel on separate levers, the lower ends of which were mounted at two points on the machine vertically separated by the same distance between centres of workpiece and pattern. These two levers would then have had a linkage, of equal length to the distance between the lower mounting points, connecting them at, say,

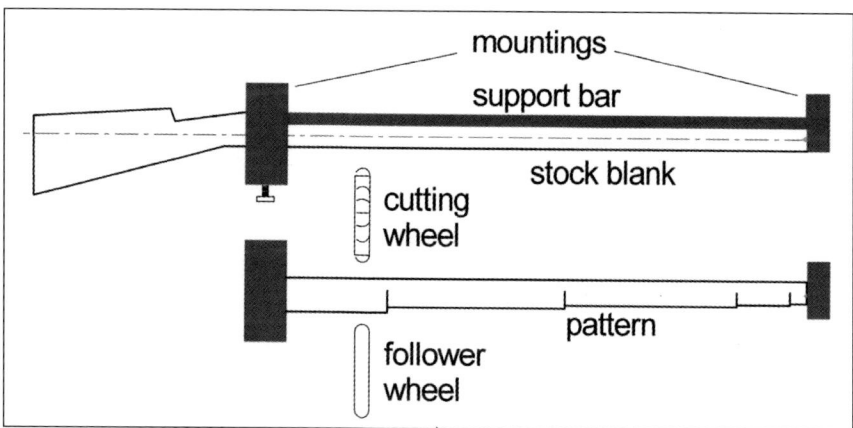

Fig. 5.13. Arrangement of stock and pattern in the turning machine showing the mountings and support bar in blue. (© P. Smithurst)

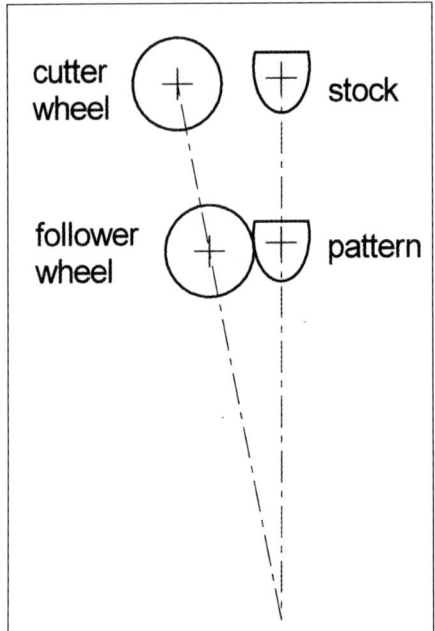 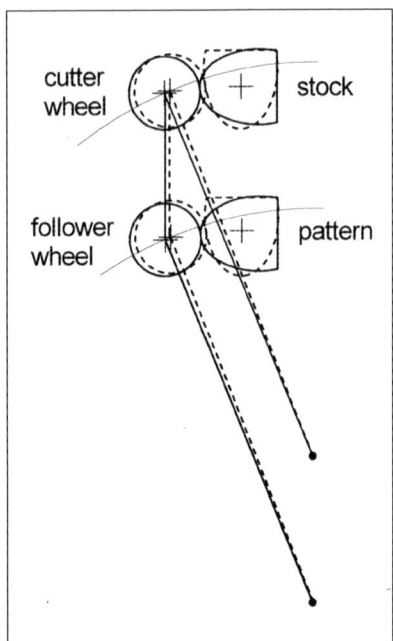

Above left: Fig. 5.14. Interpretation of the system as described in '*The Engineer*'. (© P. Smithurst)

Above left: Fig. 5.15. An arrangement using Watt's parallel motion which would have met the needs of the stock-turning machine. (© P. Smithurst)

the centres of the follower and cutter wheels, or at other points, so that levers and linkages formed a parallelogram, Fig, 5.15. Thus, as shown, the revolving of the lower pattern in contact with the follower wheel would cause the cutter wheel to move in unison and transfer the same profile of the pattern to the stock.

Some means of keeping the follower in contact with the pattern would have been needed and on the Enfield machine this was accomplished by a treadle system operated by the machinist. The machine at Enfield also incorporated an improvement by James Burton and utilised two followers and two cutting wheels, allowing two portions of the fore-end to be machined at the same time thus halving the time required. The machine was 'self-acting', the carriage which carried the stock and pattern being traversed past the cutting and follower wheels by a leadscrew.

An American publication of later date covering the manufacture of a similar component provides an even less informative description but does illustrate a machine closely similar to that used at Enfield for the butt, not the fore-end, Fig. 5.16. There is insufficient detail to suggest that a 'Watt's parallel motion' system was used, but it would be difficult to find a simpler alternative.

The curved rim of the follower wheel meant that the steps where the bands would ultimately fit exhibited themselves as smooth transitions at this stage but it

MANUFACTURE OF THE STOCK

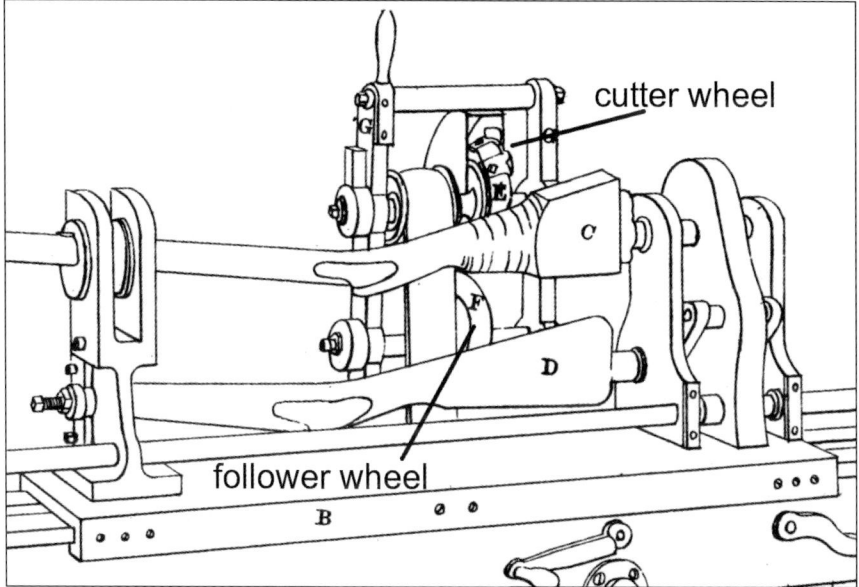

Fig. 5.16. Butt turning machine used in America is of similar arrangement in having pattern below the workpiece. (Benton, 1878, p.110 and plate XV)

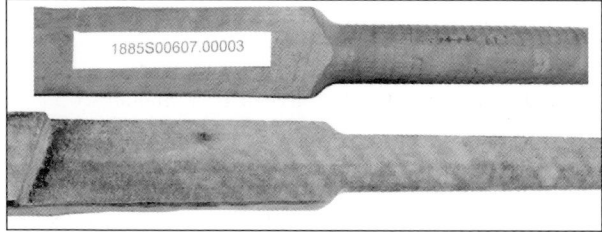

Fig. 5.17. Top and underside views of a stock with a 'rough-turned' fore-end. (1885S00607.00003 courtesy Birmingham Museums Trust)

is not known whether the transitions on the pattern were also smooth or had square steps. In the latter, the steps would be gradually 'turned over' at the corners, making them useless when it came to 'squaring' these steps on the stock later, so patterns with smooth transitions would have been a better option. The specimen in the Birmingham Museum collections, Fig. 5.17, shows a portion of a 'rough-turned' fore-end terminating at the position where the lower band would ultimately be fitted.

Rough-turning the butt

The only description of this process as carried out at Enfield[22] is minimal and again lacks illustrations. It is stated that the machine was similar to that for turning the fore-end and that the cutter-wheel had:

> twelve hooked cutters of three different forms - - ; the first two are for getting rid of the extra timber, and the last for taking a chisel cut. These tool discs make on an average 3,000 revolutions per minute.

The specimens in Birmingham Museum's collections provides examples of this process, although, for reasons already noted, it was not possible to obtain a full-length photograph.

The 'rough-turning' of the butt also included the remainder of the stock up to the position of the lower band, the termination point from 'rough-turning' of the fore-end. Apart from anything else, this is evident on the specimen shown by the change in the marks left by the cutter at this junction.

This evidence presented by the Birmingham Museum specimens parallels Benton's description, *removing the surplus stock* [timber] *from the lower band to the heel of the butt*[23] and he illustrates the machine, referred to earlier, for this purpose, Fig. 5.20.

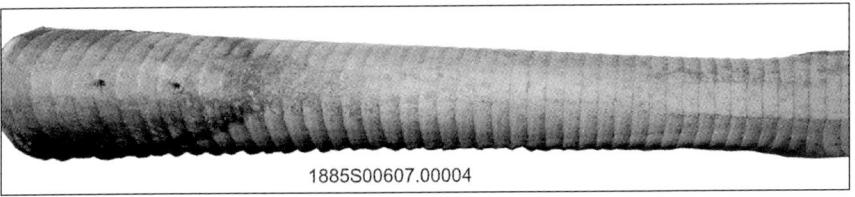

Fig. 5.18. A 'rough-turned' butt. The curved form of the cutter is evident in the turning marks. (1885S00607.00004 courtesy Birmingham Museums Trust)

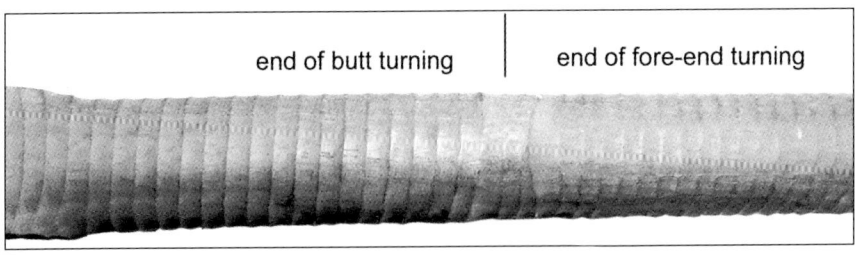

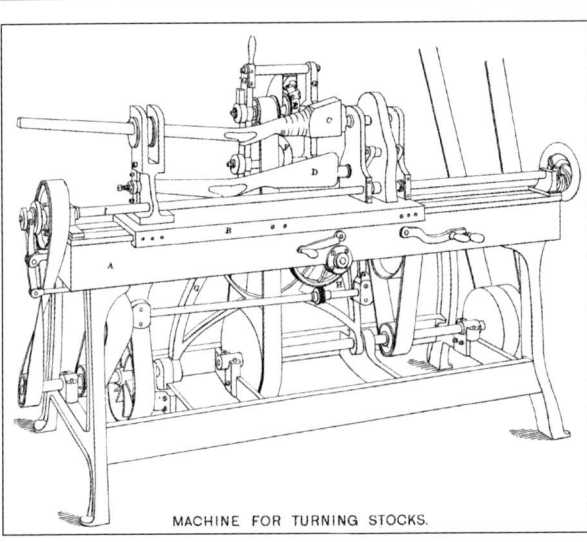

Above: Fig. 5.19. 'Rough-turning' the butt extended to the position of the lower band where 'rough-turning' of the fore-end terminated. (1885S607.0004 courtesy Birmingham Museums Trust)

Left: Fig. 5.20. 'Rough-turning' the butt from lower band to heel. (Benton, 1878, plate XV)

MANUFACTURE OF THE STOCK

Spotting

Examination of one of the Birmingham Museum specimens shows that the purpose of this process was to create seven flat surfaces at 90⁰ to the flat surface on the top of the fore-end created in the slabbing operation. Greenwood notes that two were on each side of the butt and the muzzle; one was where the lock was to be fitted and two more along the right-hand side of the fore-end.[24] Though not always mentioned, these 'spots' became datum planes for future operations.

Both machine and process are described with minimal detail in *The Engineer*:

> *the stock is held down by two spring clips upon a plate the width of the flat surface prepared by the slabbing machine; this plate is weighted and fitted with guides so that it can be worked up and down vertically by means of a treddle* [sic]. - - - *when the stock is fitted and clipped in position and taken as high as it can come, that by application of the foot it can be brought down; the machine then being supplied with seven stout saws on five spindles, two placed the right distance apart on those situated at the extreme ends of the stock, it will be seen that, by lowering the table, these saws will form the seven required spots...*

There are a number of issues with these statements. The first is that if the stock was mounted on *a plate the width of the flat surface prepared by the slabbing machine*, there would be no clearance for the saw blades since after rough-turning, the width of the stock was both reduced and non-uniform. The stock would also need to be correctly aligned, and the same centre-marks used in the rough-turning process would also have served for this purpose. The pairs of saws used for spotting the butt and at the muzzle-end would have to be of sufficient diameter for the spindle on which they were mounted to clear the end face of butt and fore-end and produce a spot of required size.

Photographing the spotting on the butt was made difficult by the required specimens being fixed in a rack, Fig. 5.22.

On the other hand, the three separate saws in the central portion would need to be of the same diameter as the spot width and would need to be flat on their end face without any protrusions, in the same manner as an end-mill or facing cutter.

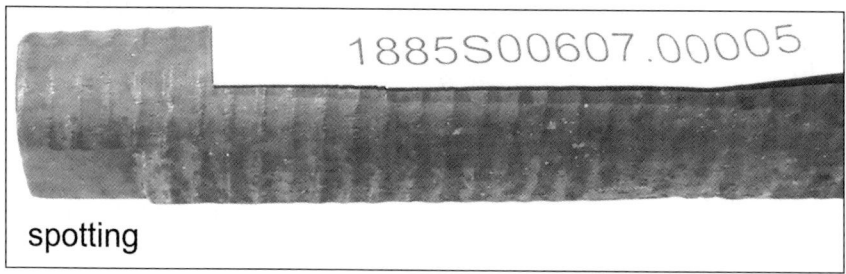

Fig. 5.21. The 'spottng' at the muzzle. (1885S00607.0005 courtesy Birmingham Museums Trust)

MAKING THE ENFIELD PATTERN 1853 RIFLE-MUSKET

Left: Fig. 5.22. The 'spot' on one face of the butt. (1885S00607.00005 courtesy Birmingham Museums Trust)

Below: Fig. 5.23. The 'spotting' at the lock. (1885S00607.00005, courtesy Birmingham Museums Trust)

Bottom: Fig. 5.24. The 'spotting' at the lower band. (1885S00607.00005, courtesy Birmingham Museums Trust)

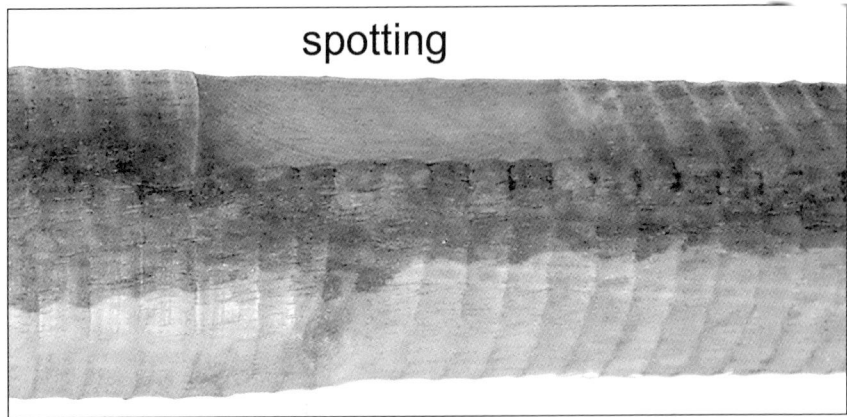

The features of a machine that could accomplish these operations, based on the description available and the evidence presented by the stocks themselves, is shown in Fig. 5.25.

The stock was mounted upside-down in the spotting machine, and for simplicity is been shown as viewed from the rear. The stock was lowered from its raised position on its mounting plate past the rotating cutters which removed the timber in

MANUFACTURE OF THE STOCK

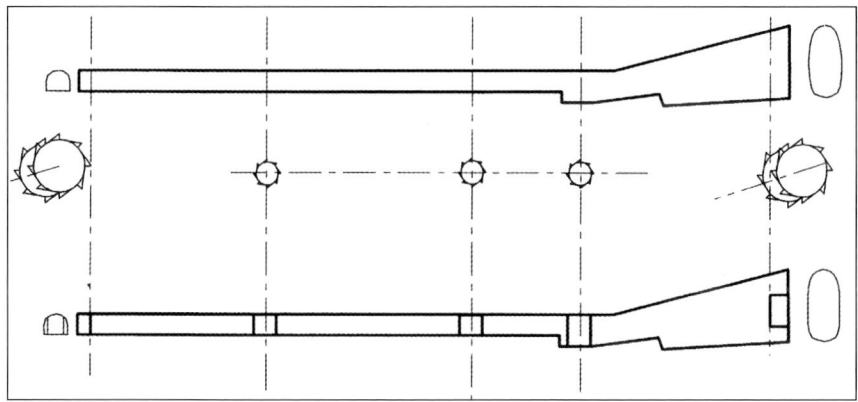

Fig. 5.25. Schematic Interpretation of 'spotting' as described in '*The Engineer*'. (© P. Smithurst)

the required places. Not mentioned in any account is that there were two primary criteria in spotting; the first being that all the spot faces were at predetermined distances and parallel to the centre line of the stock, and the second being the need for precise adjustment of the settings of the saws in the machine. These spots were of importance in the operations which followed.

Bedding the barrel

Barrel-bedding consisted of forming a tapered groove in the fore-end of the stock, and recesses at the breech-end into which the tang and heel of the breech pin fitted.

Fig. 5.26. Barrel-bedding machine used at Enfield. (*Engineer*, 1859, p.258)

MAKING THE ENFIELD PATTERN 1853 RIFLE-MUSKET

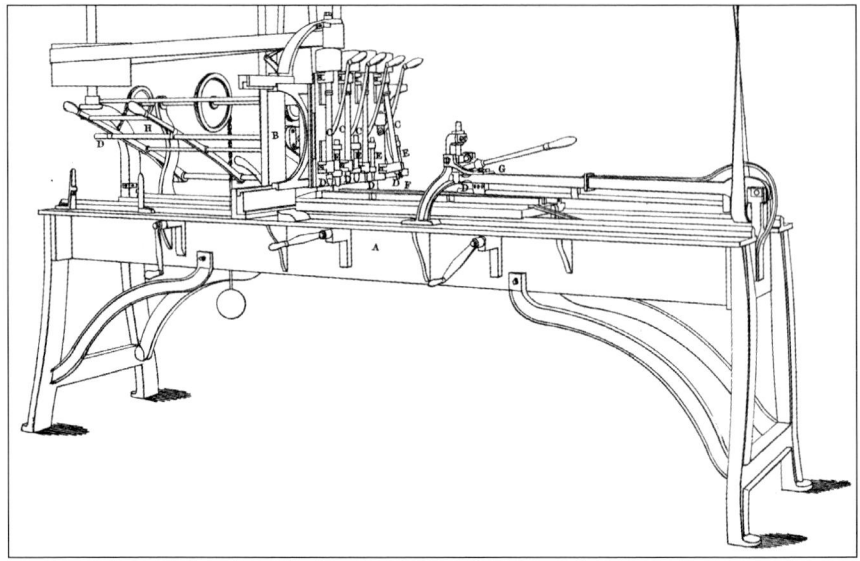

Fig. 5.27. Barrel-bedding machine used at Springfield in 1878. (Benton, 1878, plate V)

Whilst this ilustration actually represents the machine used at Enfield, important features of it are not clearly shown. A second illustration, Fig. 5.27, whilst being for a different rifle, is fundamentally the same and, when enlarged, offers clearer details.

The process is described in *The Engineer*:

> *To effect this, it is inserted into a recess on a long moveable table, A, ...
> To fix the stock in position it is clipped at the five spots made on the long
> side and pressed up against two horizontal lips projecting a little over
> on each side of the top of the recess made to receive it by studs from
> below: this ensures it being in line and level with the trued surface.*[25]

However this raises various questions. Obviously, the 'lips projecting a little over on each side of the top of the recess' would also need to project at each end of the recess since, without that constraint, the stock could assume any angle to the horizontal since the one pair of lips would act as a pivot point.

The second question concerns the spots. The description implies that the spots were simply used as bearing points for 'clips' but, as datum faces, they would have been used to bear against five accurately prepared vertical bearing surfaces and clamped against those, because by this stage the stock had no other flat sides.

Details from the above illustrations are shown below in Figs. 5.26A and 5.27A. Fig. 5.27A, being clearer, has allowed minor editing to highlight the stock and cutters in place but it should be noted that the Enfield machine had only three vertical cutters, not four as shown in the Springfield machine.

The process commenced using one of the cutters, B in Fig. 5.26A, nearest the front. The frame carrying the cutter spindles was moved to the rear so that the

MANUFACTURE OF THE STOCK

Fig. 5.26A. Detail of the Enfield machine showing the cutter spindles, B and the table, A.

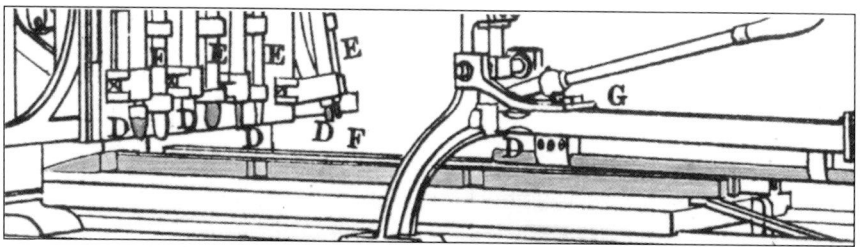

Fig. 5.27A. Detail of the Springfield machine; stock highlighted in light grey the cutters highlighted in dark grey, and the 'pattern' F behind the stock.

guide pin, adjacent to the cutter, could be lowered by a lever to engage with a pattern having a tapered groove, which exactly matched that required from the barrel channel and which was placed immediately behind and parallel to the stock.

By moving the table longitudinally, by means of a rack and pinion, and by moving the frame carrying the cutter, constrained by its guide pin in the pattern channel, both laterally and vertically a tapered groove was cut in the stock. In a similar manner, the second cutter from the front was then used to cut a recess for the heel of the breech pin and then the third cutter, inclined at the appropriate angle, was used to cut the recess for the tang of the breech pin.

The next machining operation was to remove any unevenness left in the barrel channel by the previous tool. The table was moved to the front and:

> *There the stock* [should read barrel channel] *receives its finished taper by means of the cutting-spindle D,* [see Fig. 5.27A] *which is suspended from the cross-bracket, F, and can be indued* [sic] *with a vibratory motion by means of the long handle, which is seen to be in position. The tool which is fixed to this cutting-spindle is merely a curved lip of steel, very similar to the side of a common auger. It is by means of this vibrating motion that the mould or form can be followed as the feed is given, and the side and up and down movements enable the required taper to be given.*[26]

This long cutter, D, is highlighted in grey in Fig. 5.27A.

Again it is a confusing statement and a number of points arise. There is no mention of a guide pin to follow the 'mould or form' which suggests that the cutting spindle with its 'curved lip of steel' was of the same contour as the finished barrel. The use of the term 'vibratory motion' is misleading; it is taken to mean the up and down motion given to the spindle by means of the handle, since any true vibratory motion of the spindle implies it turned eccentrically or moved longitudinally, either of which would have defeated the object of the exercise.

The fifth and final operation was to square the end of the barrel channel against which the breech sat, and to do this the stock was moved to the far end of the machine and according to *The Engineer* the cut was taken:

> *by a similar tool spindle to that just described. This is merely brought down into position by means of handles.*

This part of the machine is hidden in both illustrations and, more importantly, the cutter is not shown. Bearing in mind the nature of the job it has to do, it would have to be able to cut on the sides and exposed end face, as shown in Fig. 5.28.

To enter the barrel channel it would have needed to have been mounted on a pendant-type support able to move only vertically, and both it and the cutter had to be of correct form to match the breech-end of the barrel. The cutter would also have been driven via a pulley mounted on the opposite end of the shaft to the cutter so that, with a belt in place, it did not exceed the width of the channel.

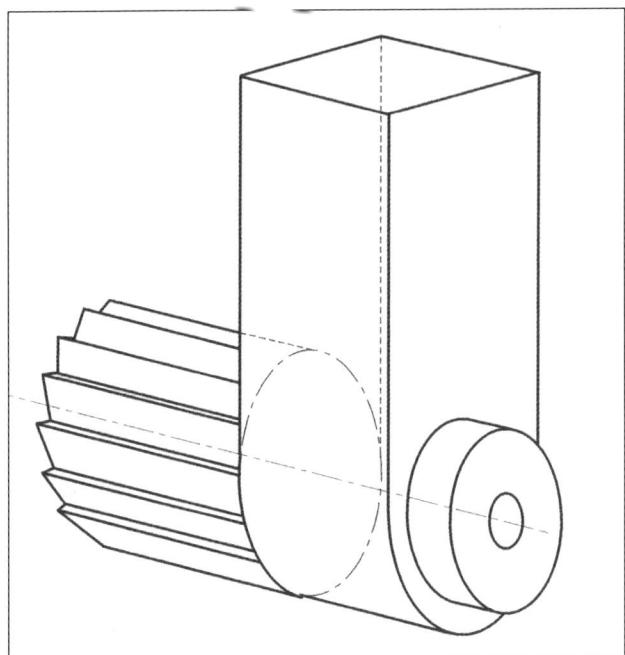

Fig. 5.28. Suggested nature of the rotary cutter and its mounting for squaring the breech face of the barrel channel. (© P. Smithurst)

MANUFACTURE OF THE STOCK

Fig. 5.29. A stock with barrel channel finished - the complex nature of the breech-end is clearly visible. (1885S607.00006 courtesy Birmingham Museums Trust)

Hand-finishing the recess for the breech tang

Not mentioned in *The Engineer* was the need to round the shoulder of the recess for the barrel-tang where it meets the recess for the heel of the breech pin, and to square this recess, both of which were performed by hand.[27, 28] The result of this operation is shown in Fig. 5.29.

Sawing to length at butt and muzzle and bedding the butt plate

After completing the work on the bedding of the barrel, the stock was sawn to exact length at butt and muzzle on another machine. No details are offered in any account but, since the square face of the breech-end of the barrel channel was the only longitudinal datum, the author concludes that the stock would have laid flat, using the spots as horizontal alignment datums, against a model of the barrel accurately positioned so that butt and muzzle would be cut off at their correct distances from the breech face.

This was followed by shaping the butt to receive the butt-plate using a:

> *revolving cutter block which runs horizontally. This block is fitted with steel tools in the form of plane irons with their edges nicked in, so as to give in effect a series of chisels, which leave, of course the spaces between them in the cut unacted upon by simply setting the tools, so that each following one takes away what the former has left, a good surface is obtained, and the cut being across the grain, only half of the weight is put upon each cutter. The curve is given to the butt by a guide pin and mould.*[29]

This again is an inadequate description. It is clear from a surviving butt profile gauge, Fig. 5.30, that the end face of the butt was curved on two perpedicular planes, one convex and the other concave.

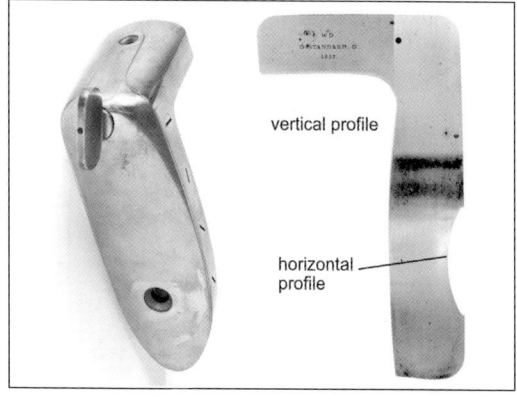

Fig. 5.30. Buttplate fitted to its gauge and the profile gauge for end face of the butt. (part of PR.10142 © Royal Armouries)

The revolving 'cutter block', referred to as 'running horizontally' is taken to mean that its periphery revolves horizontally, i.e., it has a vertical axis, and would need to be of the nature shown in Fig. 5.31, with its cutting edges extending just over the width of the butt and offset on each side to give a 'continuous' cut in two stages.

It is suggested here that the production of the curved surfaces on the end of the butt, based on the description provided, is as shown in Fig. 5.32. To prevent presenting the butt accidentally to the cutter before the guide pin and template were engaged, a safeguard, such as an extended mounting for the guide pin so it could bear against an extension of the template before it engaged with the template proper, would have been needed.

The convex form at the end of the butt would be automatically given by the concave form of the cutter blades and the concave form by the profile of the template.

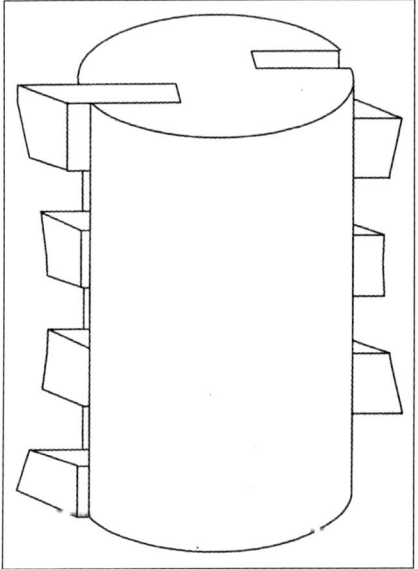

Left: Fig. 5.31. Suggested nature of the cutter block for machining the butt face. (© P. Smithurst)

Below: Fig. 5.32. Arrangement envisaged for machining the curved faces of the butt. (© P. Smithurst)

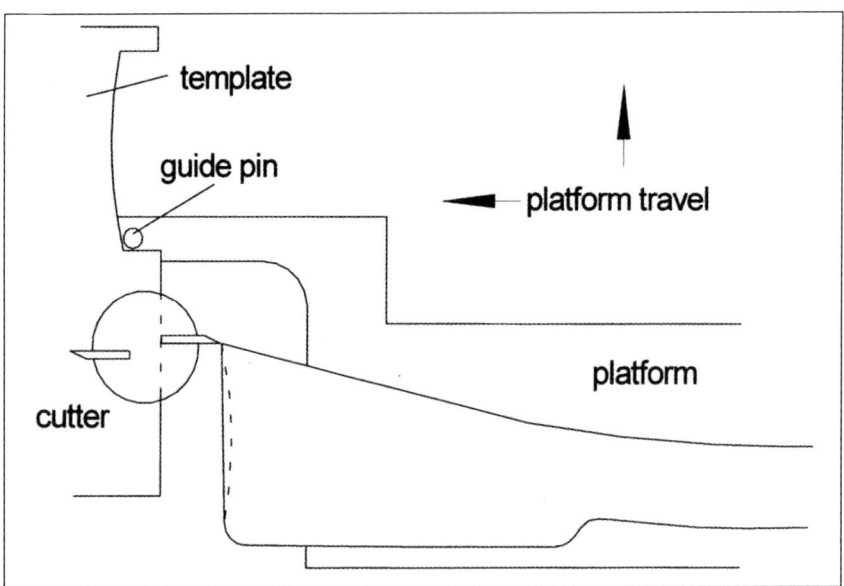

MANUFACTURE OF THE STOCK

Planing the stock

This comprised a number of operations:

1. the two opposing flat faces where the lock was to fit were planed
2. recesses for the cups for the *side-nails* (lock screws) were formed
3. the upper edges of the barrel channel were profiled
4. upper and lower edges of the butt were planed.

The description of these operations in *The Engineer*[30] is once again both inadequate and convoluted and judging by the evidence of the stocks themselves, differs from the itemised list provided by Greenwood in section 5.4.

Planing was carried out on a multi-functional machine. First to be planed was each side of the lock housing, and this was carried out by mounting the inverted stock so that the barrel channel fit on a pattern of the barrel attached to a sliding flat platform. Attached to this platform was a vertical housing into which the two opposing spottings on the butt could be fitted, and the stock then clamped in position. The platform carrying the stock was traversed in turn past cutter blocks fitted with plane blades rotating on vertical axes placed at front and rear, so that each side of the stock at the position of the lock was planed flat, parallel to and equidistant from the centre line.

The stock was then re-mounted on another table sliding independently on two horizontal axes. It was mounted so that it was lying flat, the spots along the right-hand side of the stock seated on raised blocks so that its centre-line was horizontal, and then clamped in position.

By means of a guide pin acting against a horizontal template, the stock was traversed past a cutter which planed a flat surface along the top of the butt. From the evidence of the stocks, Fig. 5.33, this was performed in two stages; one to form

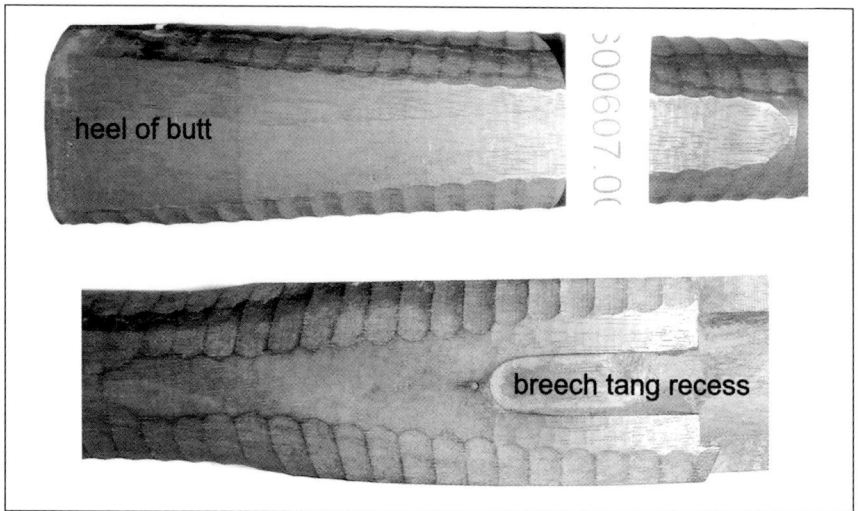

Fig. 5.33. Two parts of the same stock, showing interrupted planing on the top of the butt and at the breech. (1885S607.00009. courtesy Birmingham Museums Trust)

a flat surface at the heel where the tang of the butt-plate would fit, the second to form a flat surface where the breech tang fitted.

Evidence of the stocks also shows, Fig. 5.34, that the underside of the butt was planed in a continuous cut from the toe of the butt to the position of the lower band, covering the position where the trigger guard would be later fitted.

The account in *The Engineer* does refer to 'taking a cut down the back of the butt' which would be correct if the top of the butt was facing the observer.

Following planing the butt, the two recessses on the face opposite the lock were formed to receive the 'side-cups', Fig. 5.35, into which the heads of the lock screws, or 'side nails', fitted. These were simple forms, consisting of a circular disc approximately 0.625 inches (16mm) in diameter and 0.187 inches (4mm) thick having two diametrically opposed lobes and a circular well for the heads of the side-nails.

With the stock laid flat with the lock face downwards, the recesses were simply cut with two appropriately sized cutters, one for the central body and one for the side lobes. A template and guide pin would have undoubtedly been used since the centres of the side-cup recesses had to coincide exactly with the position of the side-nail [lock screw] holes in the lockplate so, again, accurate positioning was essential.

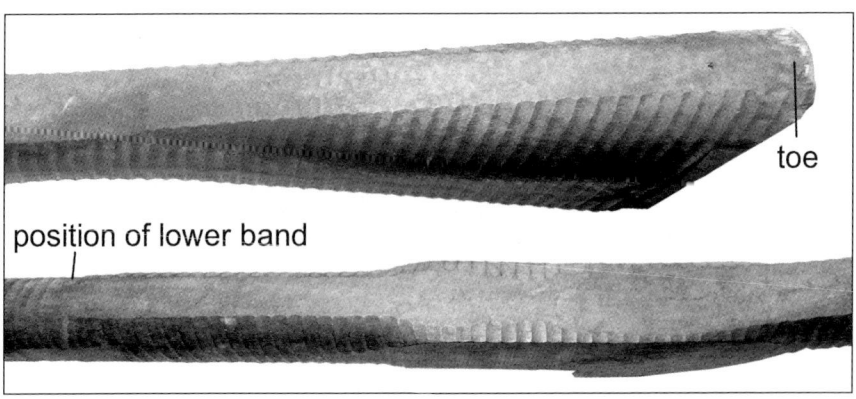

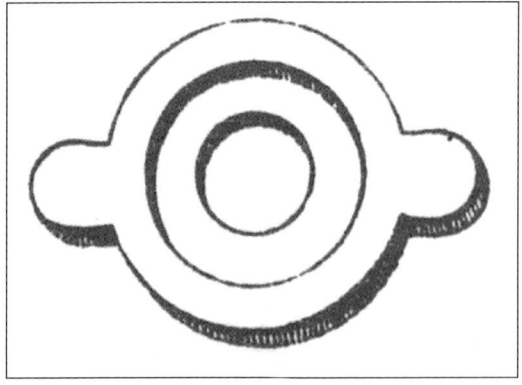

Above: Fig. 5.34. Planed underside from toe of butt to lower band is continuous. (1885S607.00009. courtesy Birmingham Museums Trust)

Left: Fig. 5.35. A side-cup. (Petrie, 1866, plate V).

MANUFACTURE OF THE STOCK

Bedding the butt plate tang, three holes for the screws are bored and the two large holes tapped

These operations were accomplished by a single machine provided by the Ames Co. which is shown in Fig. 5.36.

As it stands this image, reproduced closely to the size as it appeared in *The Engineer*, is difficult to read, the key letters being too small and obscured by the shading of the engraving. Relevant details have therefore been reproduced to larger scale, Fig. 5.37, and new key letters added to assist in understanding the description of its operation. Even then, the description does not make certain features and procedures clear and some extrapolation and deductions have been resorted to, based upon the knowledge of what needed to be accomplished, and these will be highlighted in what follows.

The operation commenced with the fixing of the butt of the stock in position:

> ...the butt-end of the stock being fitted against a plate of iron hung upon a spindle, so as to enable its position to be changed without being released; it is fixed tight in the following manner: The eccentric

Fig. 5.36. The Ames butt-plate tang bedding and drilling machine used at Enfield. (*Engineer*, 1849, p.258)

handle A presses the small plate which rests upon it against the stock and lifts it against the cross-piece, B, and the small portion of a mould the size of the barrel, C, fixed at the end of the side-plate, D; the handle, E, clips the stock sideways in a similar manner. ...One of the bearings of the spindle on which the stock turns is shown by the letter G.[31]

There are various points which the above description omits to mention. At this stage in its manufacture the stock was still in its rough-turned and planed condition. It would thus retain the spots on each side of the butt which would have served as clamping points and positional datums. Also, 'the small plate' which rested upon the lever, A, would have needed to be quite long and narrow, and suitably curved, so that it could bear upon the planed portion on the underside of the stock where the trigger guard would be ultimately fitted.

To clarify what was to be achieved can be done through simple illustration. In Fig. 5.38 is a buttplate from the rifle, showing the tang which had to be let-in and the three screw holes, alongside a drawing, Fig. 5.39, showing these features on the butt.

It will be seen in the diagram, Fig. 5.39, that the tang recess was parallel with the top edge of the butt, its screw hole perpendicular, the upper screw at the heel of the butt angled and the lower buttplate screw horizontal.

To reduce the number of settings, the stock needed to be first mounted with the upper edge of the butt in the horizontal position for cutting the recess for the buttplate tang, drilling its screw hole, followed by drilling and tapping the lower buttplate screw hole.

With the stock fixed in position on the machine:

The tool M is brought down so as to allow the pin behind it to rest in the copy, by following which the recess is cut to the right shape.

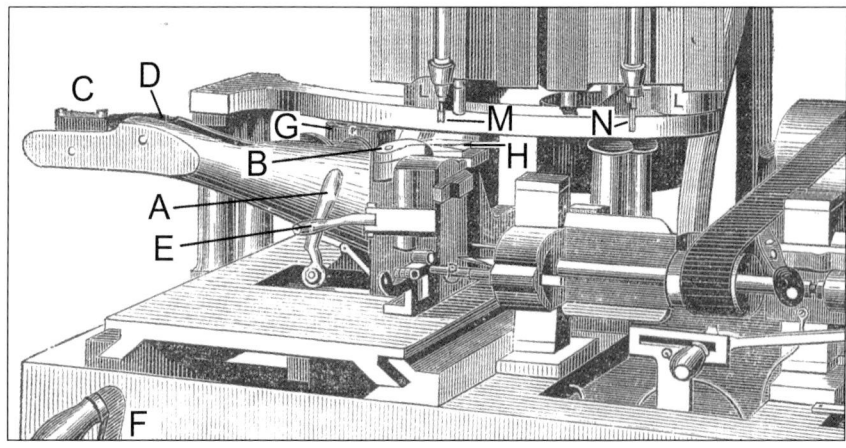

Fig. 5.37. A detail from Fig. 5.36 showing the features referred to in the description.

MANUFACTURE OF THE STOCK

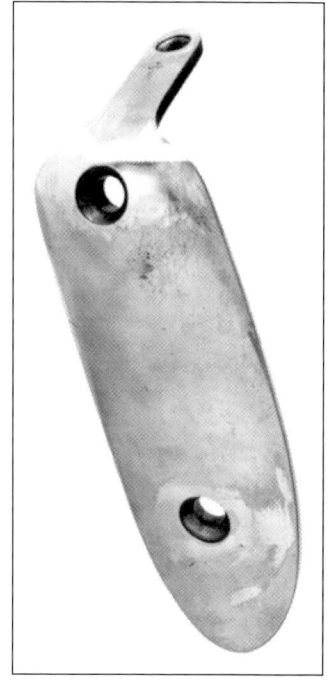

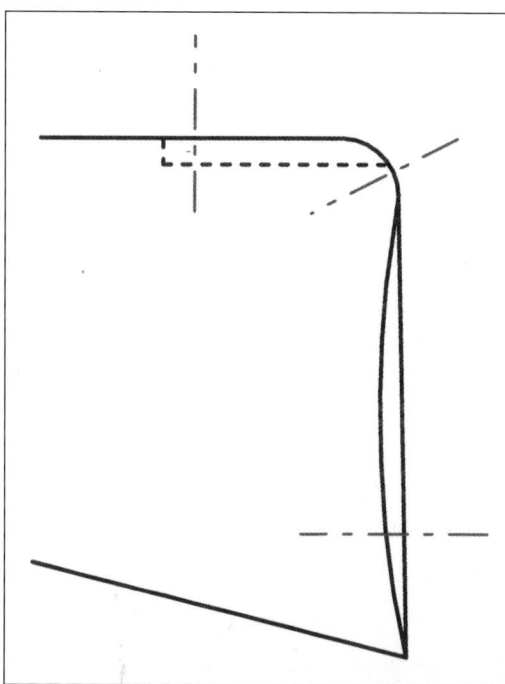

Above left: Fig. 5.38. The buttplate.

Above right: Fig. 5.39. Diagram of the butt showing the alignments of three screw holes and the buttplate tang with the butt horizontal. (© P. Smithurst)

Right: Fig. 5.40. Detail showing the cutter, M, the guide pin and template in grey. B is the fixture against which the stock, shown beneath it, - is pressed.

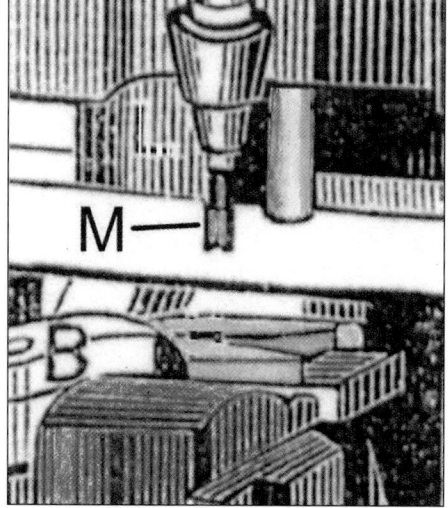

This is shown more clearly in Fig. 5.40 which is a further detail prepared from the original engraving.

The cutting tool, M, was carried at the end of a spindle rotating at high speed. Behind it can be seen the guide-pin, highlighted in grey and beneath this, the template, or *copy*, also highlighted in grey. Obviously, if the template were an exact copy of the recess to be cut, the cutter and guide pin would need to be the same diameter so that the dimensions of the template were accurately transferred to the butt. In this position, the stock was centred beneath the cutter. With the stock fixed, the spindle carrying the cutter was brought down by a lever, at the same time engaging the guide pin with the pattern which also acted as a depth stop. By moving the table carrying the stock fixture on two horizontal axes by means of levers, the guide pin was caused to trace the pattern and cut the corresponding recess in the butt.

The cutter and guide pin were then disengaged and the frame carrying the tool spindles rotated to bring the drill, N, in Fig. 5.37 and its guide pin into the same position, the exact rotation of the frame being determined by a stop. In this instance, the guide pin would have been the same diameter as the semi-circular tip of the pattern so as to ensure that the drill was on the same centre line. By lowering the drill spindle with a lever the tang-screw hole was drilled.

The next step was to drill the hole for the lower of the two large buttplate screws. This utilised a horizontal drilling spindle, again with an associated guide pin which simply entered a correctly positioned hole, despite the apparent misalignment in Fig.5.41.

The table upon which the stock was mounted was moved forward to bring the guide-pin hole nearest the front into line with the guide-pin. This ensured the correct position of the drill bit which was then advanced by means of a lever. After withdrawing the drill, the table with the butt was then moved rearward, the guide pin entered into the farthest hole, allowing the hole in the butt to be tapped.

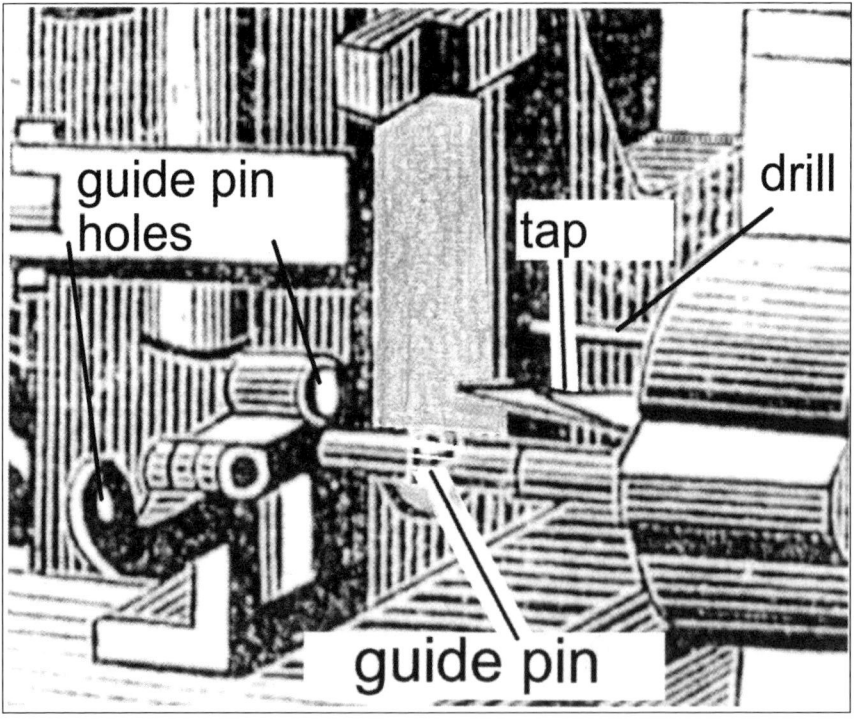

Fig. 5.41. Detail of Fig. 5.37 showing the drill and guide pin with the end of the butt highlighted in grey.

MANUFACTURE OF THE STOCK

This used a self-reversing tapping head which automatically cut the thread to the required depth and then withdrew itself.

The final operation performed by this machine was the drilling and tapping of the oblique hole at the top corner, or heel, of the butt. According to *The Engineer* this operation was *precisely similar to the one just described*. This is patently incorrect. The only way it could be achieved using the same tooling was to first rotate the butt to bring the position of the oblique hole into horizontal alignment with the tooling. This is where the importance of the note:

> *the stock being fitted against a plate of iron hung upon a spindle, so as to enable its position to be changed without being released*

becomes apparent.

Such a feature would need to take into account the accurate geometrical positioning of the centre of rotation. This has been experimented with graphically, Fig. 5.42, and the centre point indicated provides the correct geometry for bringing the centre of the oblique hole into horizontal alignment with the same drill and tap used for the lower, horizontal hole, for this operation.

It will be seen in Fig.5.37 that there is an opening in the sliding table which would make a rotation of this nature possible and, aligned thus, the operations do become 'precisely similar'. The corner under the tang of the butt plate was then finished by rounding by hand.[32]

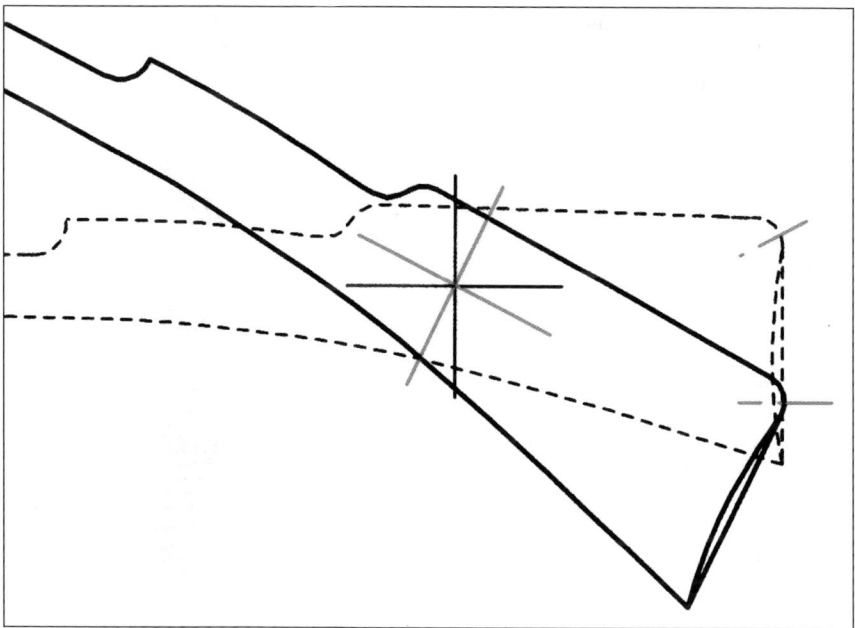

Fig. 5.42. Hypothetical centre of rotation to put the oblique hole into horizontal alignment. (© P. Smithurst)

MAKING THE ENFIELD PATTERN 1853 RIFLE-MUSKET

Bedding the lock

This process was the cutting of the recess in the stock into which the lock fitted using the machine in Fig. 5.43. One of the these machines used at Enfield survives in the Science Museum in London. Another, although from Colt's factory in Hartford, Connecticut, is at the American Precision Museum in Windsor, Vermont. The author, at one time Executive Director at this museum, was able to operate the machine and can attest to its effectiveness.

The illustration Fig. 5.43 does little justice to the machine's complexity. The 'carousel', mounted above the table on which the stock is fixed, carries five powered spindles, each fitted with a cutter specific to the operation it has to perform. This carousel is rotated, bringing each spindle to bear in succession, and at each rotation of the carousel the drive belt is automatically disengaged and re-engaged successively with each spindle. The stock was clamped on its side on the moveable table, A, Fig. 5.44, using the spots on the stock to ensure its correct horizontal position, and by pressing the barrel channel up against a short model of the barrel, it was correctly positioned longitudinally and laterally.

Fig. 5.43. The lock bedding machine used at Enfield. The stock can be seen laid and fixed on its left hand side on a moveable carriage beneath the rotating carousel fitted with a number of cutter spindles. (*Engineer*, 1859, p.294)

MANUFACTURE OF THE STOCK

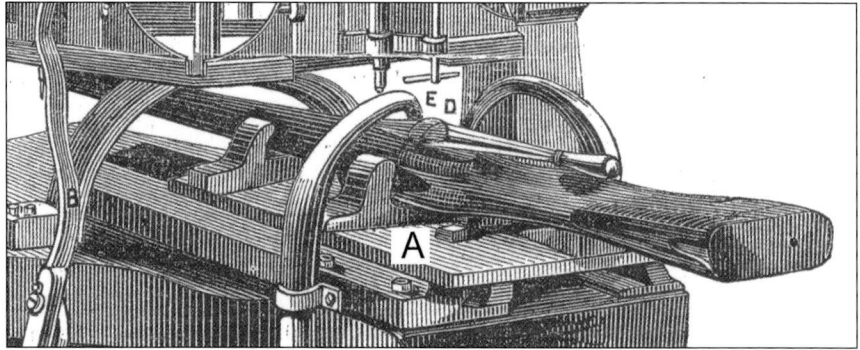

Fig. 5.44. Detail showing stock fixed on the table, A.

Not clearly visible in the illustration in Fig. 5.44 is how the model barrel was arranged, but this detail is provided by a contemporary Greenwood & Batley drawing of aspects of this machine, Fig. 5.45, which has not been previously studied or published.

The key to this machine was the pattern, Fig. 5.46, which was an exact replica of the recess to be cut in the stock.

Examination of an edge-on view of a lock, Fig. 5.47, which although not a Pattern 1853 Enfield lock, is closely comparable in terms of its internal components

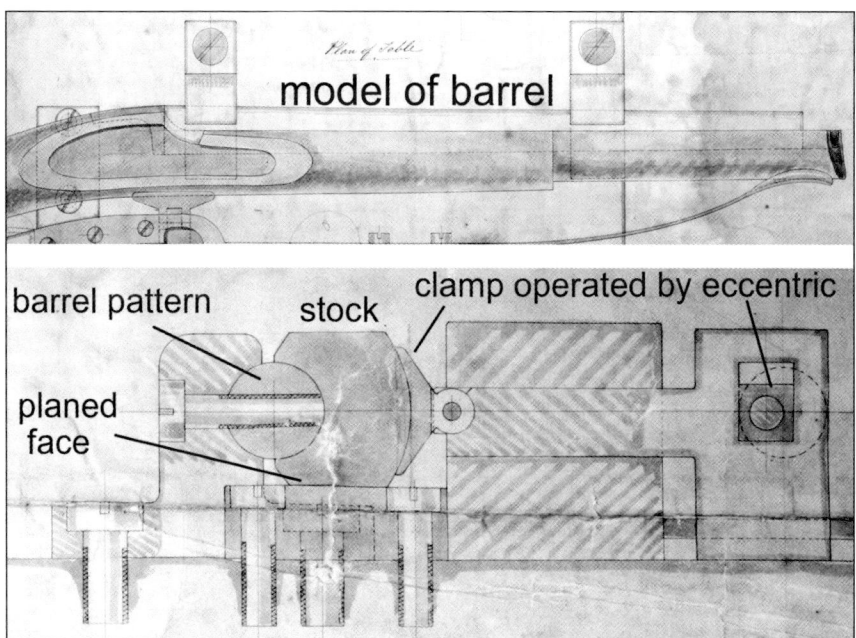

Fig. 5.45. Details highlighting the model of the barrel against which the stock is located. (© West Yorkshire Archives from Greenwood & Batley drawing No. 668)

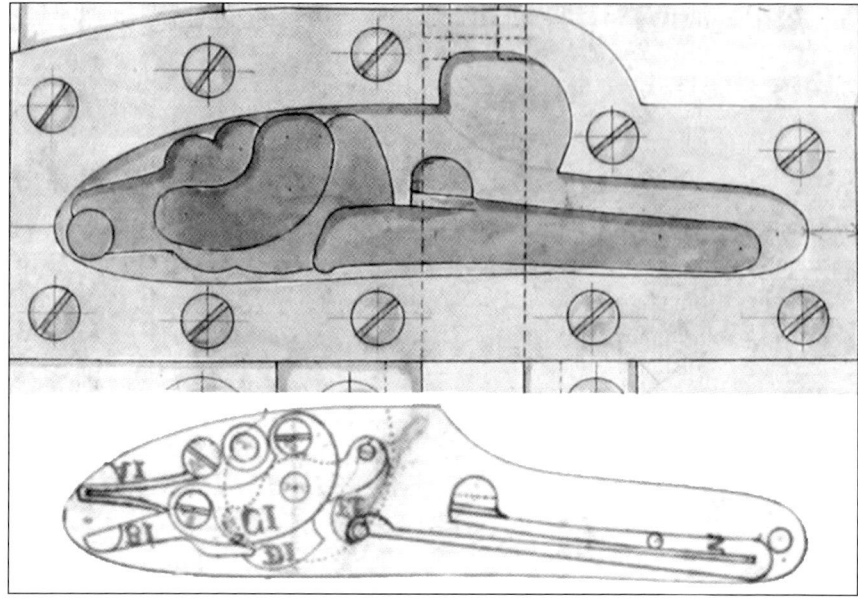

Fig. 5.46. Detail of the lockplate recess pattern (Greenwood & Batley drawing No. 668, © West Yorkshire Archives) compared with a 'reverse' drawing of the lock internal components. (from Royal Small Arms Factory drawing No. 746, © Royal Armouries)

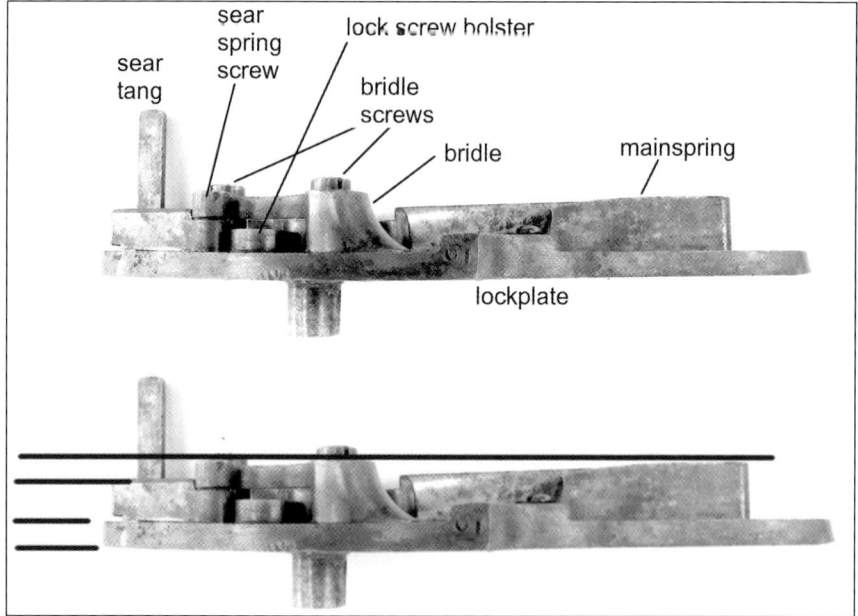

Fig. 5.47. Edgewise view of lock to show varying projections of the components on the inner lockplate face and the three principal depths of the lock housing. (© P. Smithurst)

MANUFACTURE OF THE STOCK

and their arrangement, shows that components project to varying distances from the inner face of the lockplate.

These variations in the projection of components are closely followed in the lock recess of the stock, removing as little material as necessary so as not to further weaken the stock at what is one of its weakest points.

It is suggested that the three different shades of grey within the recess pattern drawing, Fig. 5.46, coincide with the three different depths at various points to accommodate the components, including in certain instances the movement in arcs of varying sweeps and radii of mainspring, tumbler, sear spring and sear. The sear tang is anomalous in having to project deep into the stock to engage with the trigger. It has been noted in Chapter 4 that the lock fits closely against the shoulder of the nipple bolster on the barrel, so it was therefore essential that the stock and pattern be in correct mutual alignment on this machine, Fig. 5.48. This was ensured by the correct positioning of the model barrel against which the stock was clamped.

The description in '*The Engineer*' identifies 5 stages in the lock-bedding process:

1. recess for the lockplate cut
2. recesses for screw heads cut
3. recess for the tang of the sear is cut
4. recess for the mainspring is cut
5. small tool used to finish the corners left by No. 4

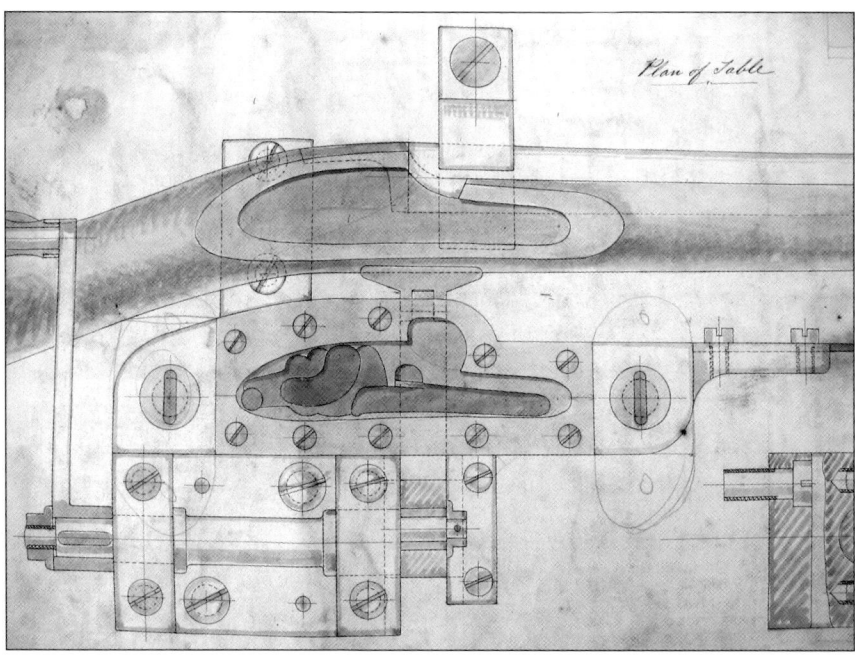

Fig. 5.48. Position of the stock in relation to the pattern on the machine table. (© West Yorkshire Archives Greenwood & Batley Drawing No. 668)

A curious feature of Fig. 5.48 is that, notwithstanding the first lock-bedding operation has already been carried out, i.e., creating the recess for the lockplate, it also shows a stock in which various features appear completed when, in reality, those operations, such as finish-turning, were performed after bedding the lock.

Bedding the lockplate

The first operation, cutting the recess for the lockplate, used the cutter shown in Fig. 5.49.

There are a number of general points worth noting about this and other cutters; it had spiral flutes, like a slot-drill or router bit, and it revolved at 7,000 rpm. A feature of cutters for the 1st, 4th and 5th operations was having their shanks turned eccentrically to the extent of $\frac{1}{64}$th-inch (0.4mm), and this eccentricity was also applied to the tapered sockets in the spindle which were graduated on half of the circumference of their end faces. These features allowed perfect accuracy to be maintained when the cutters were re-fitted after sharpening by minutely adjusting their positions in the sockets.

For the first operation, the cutter and guide-pin were 0.375 inch diameter and the guide-pin was fitted with a cross bar, held in place by a taper key, to act as

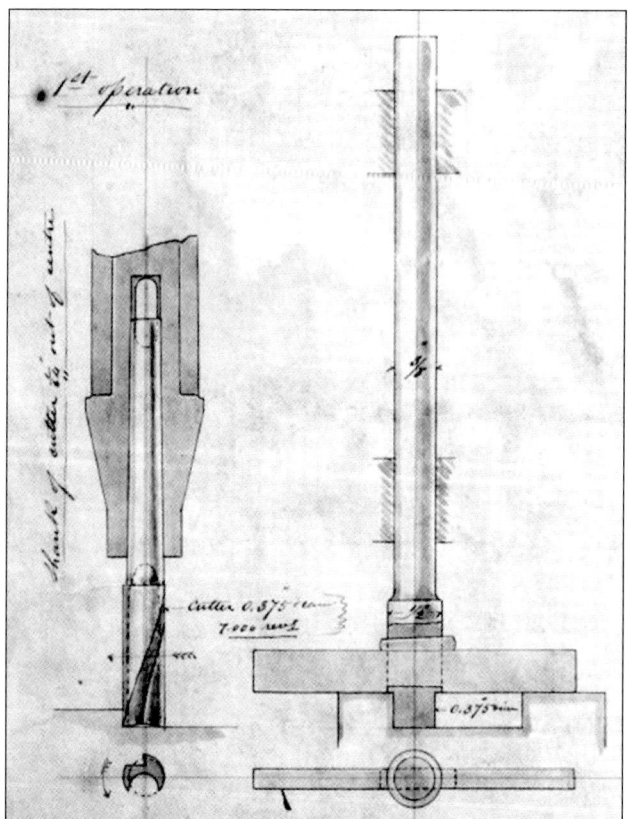

Fig. 5.49. The cutter and guide pin for the first operation. (© West Yorkshire Archives, Greenwood & Batley Drawing No. 668)

a depth stop against the top face of the pattern throughout the full extent of its movement. The first spindle was set in motion and the operator lowered it, entering the guide pin into the pattern and, by means of levers, traversed both table and spindle so that the shallow recess was completely cut out.

Bedding the screw heads

With regard to the next stage, conflicting information is presented by the accounts of '*The Engineer*'[33] and Greenwood,[34] but the evidence of the Greenwood & Batley drawings suggests that it was to recess the screw heads. The carousel was rotated to bring the second spindle into position and the drawing in Fig 5.50 is noted as being for the second operation, using 0.3-inch diameter cutter and a 0.25-inch diameter guide pin.

If this drawing is examined in detail and compared with the lock in Fig. 5.51, it becomes apparent that this second operation cannot have been for the drilling of the hole for the sear tang as stated by Greenwood since the hole would not be deep enough, but was for *cutting holes for the heads of the screws* as noted in '*The Engineer*'.

Bearing in mind that the heads of the screws, by measurement of several examples, were 0.264 inches in diameter, then the 0.3 inches diameter cutter shown

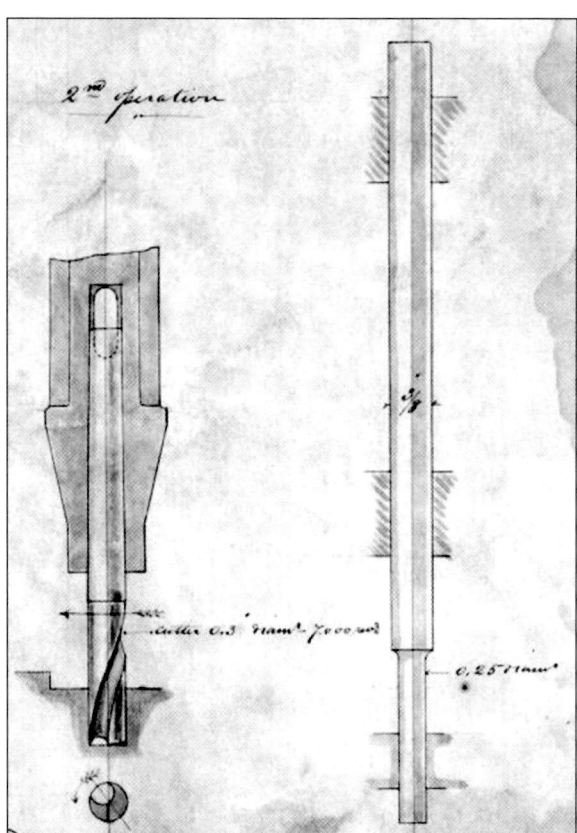

Fig. 5.50. Cutter and guide pin for the second operation. (© West Yorkshire Archives, Greenwood & Batley Drawing No. 668)

MAKING THE ENFIELD PATTERN 1853 RIFLE-MUSKET

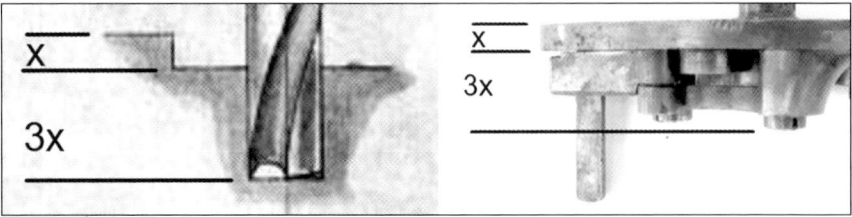

Fig. 5.51. Approximate scaling of the depth of cut shown in Fig. 5.50 and the lock components in profile.

provided clearance. These recesses are not apparent in the drawing of the lock-bedding pattern but are evident in the actual lock recess of the stock in Fig. 5.52.

A question which arises is, that in the absence of holes for the guide pin in the pattern, how was it aligned with the centres of the screws? The absence of these holes in the pattern may simply be an error in the drawing but, considering the meticulous care taken with the remainder of it, this seems very unlikely.

Examination of Fig. 5.50 shows that in addition to the two mounting points for the guide pin, the pin itself passes through a third item at its lower end, Fig. 5.53.

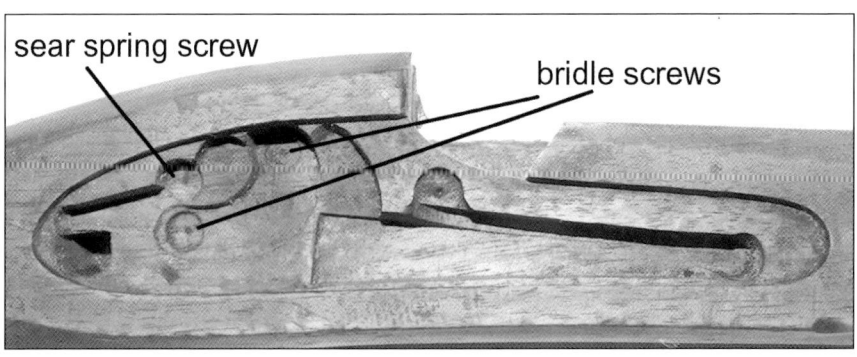

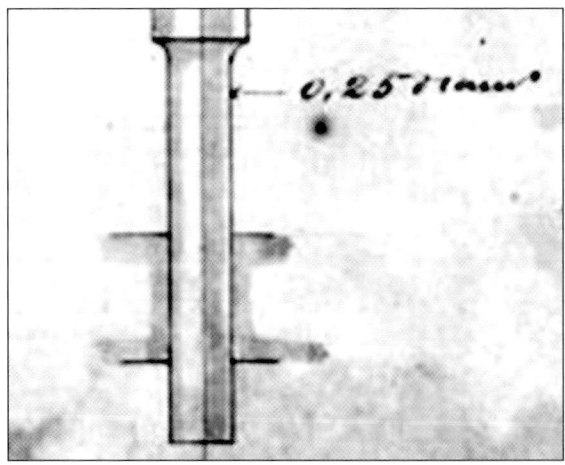

Above: Fig. 5.52. The recess in the stock showing cavities for the screw heads. (© P. Smithurst)

Left: Fig. 5.53. Detail of Fig. 5.50 showing guide pin passing through an un-named feature. (© West Yorkshire Archives, Greenwood & Batley Drawing No. 668)

MANUFACTURE OF THE STOCK

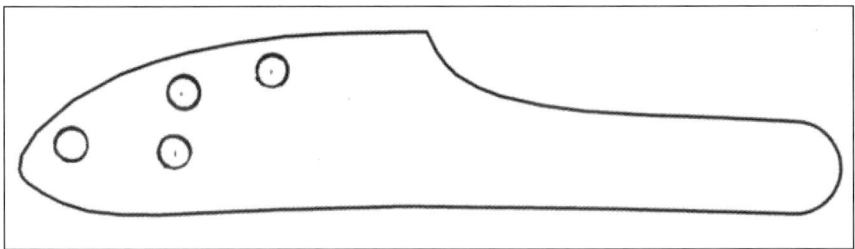

Fig. 5.54. The guide plate or 'jig' postulated for use in conjunction with the pattern for drilling the holes in the 2nd and 3rd operations. (© P. Smithurst)

In the absence of any other evidence, it is suggested here that this may be a drilling jig, an exact copy of the lockplate in outline, and which could be fitted in the lockplate recess in the pattern, and having holes exactly located for the drilling of the holes for the bridle and sear spring screw heads.

However, such an item is not mentioned in any of the accounts, nor does it feature in the contract with Ames for the lock-bedding machine and, as far as is known, no such device has been encountered. At first sight a device of this nature seems an unecessary complication but, bearing in mind that the only recess created in the stock at this stage was the shallow one for the lockplate, when the operator was trying to locate the guide pin against a shoulder in the pattern, the cutter would be following its movements and ran the risk of being pulled off course and cutting where it should not. A guide plate for the guide pin, as is suggested, would have eliminated that risk.

Drilling the hole for the sear tang

The third operation was the drilling of the deep hole for the sear tang. It was a similar arrangement to that for the second operation but using a longer drill. Again, comparison of approximate scaling of the drawing with the lock profile, Fig. 5.55, confirms that it was used for this purpose.

This used a guide pin of 0.25-inches diameter as in the second operation but a longer drill of 0.415-inches diameter. It could, therefore, have used the same guide

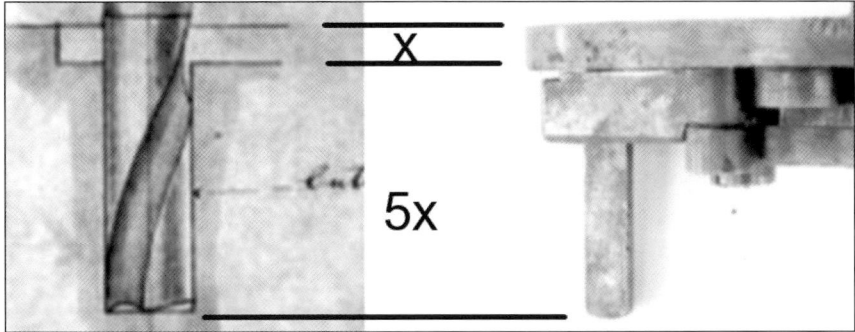

Fig. 5.55. Approximate scaling of the depth of cut and the sear tang. (part © West Yorkshire Archives, Greenwood & Batley Drawing No. 668)

plate postulated for the second operation in which a fourth hole, coinciding with the centre of the sear tang hole, had been provided, as shown in Fig. 5.54.

Recessing for the mainspring

A 0.3-inch diameter cutter and guide pin were used and their function becomes apparent by applying a similar scaling of the lock components alongside the cutter and guide pin, Fig. 5.56.

Greenwood[35] notes that this fourth operation *cut out the principal recesses below the lockplate and partially cuts out the curved recess for the cone seat* [author's underlining].

It can be seen that in bedding the barrel and the breech pin, there remained two portions of the timber left by the 'oblique cut' of the slabbing process.

A portion on the lockplate side would be removed during the lock-bedding operation. The only way this operation *partially cuts out the curved recess for the cone seat* would have been during bedding the lockplate by moving the guide pin into the 'bulbous' extension from the lockplate recess on the pattern. This is shown in detail below alongside the cut-out in the stock for the 'cone seat' or 'nipple bolster', Fig. 5.58.

Therefore, if the previous interpretation of the colouring of the drawing is correct, the removal of timber for the opening for the 'cone seat' would only be to the same depth as the lockplate recess, and therefore only partial, leaving a small portion of the wall of the barrel channel, Fig. 5.59.

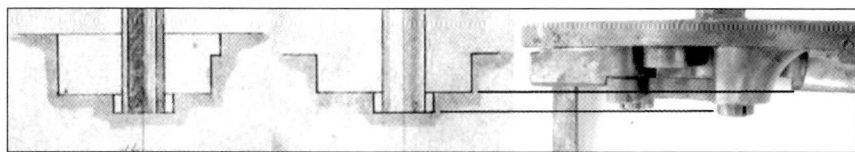

Fig. 5.56. Scaling of the depths of cut to receive the bridle, mainspring and sear spring. (part © West Yorkshire Archives, Greenwood & Batley Drawing No. 668)

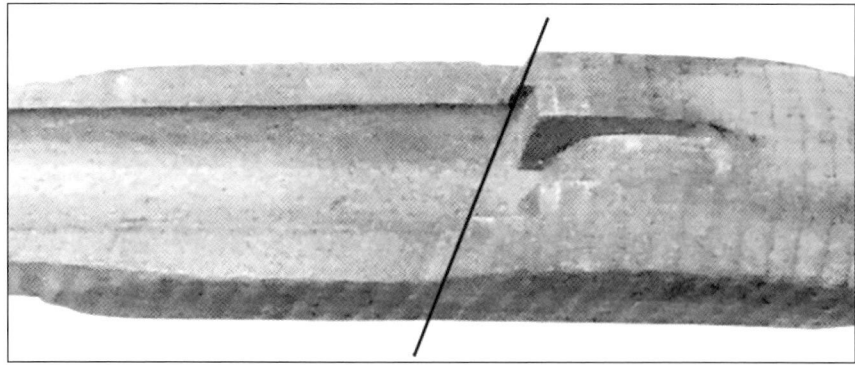

Fig. 5.57. Portions of the 'oblique cut' left after barrel-bedding. (607.00006 courtesy Birmingham Museums Trust)

MANUFACTURE OF THE STOCK

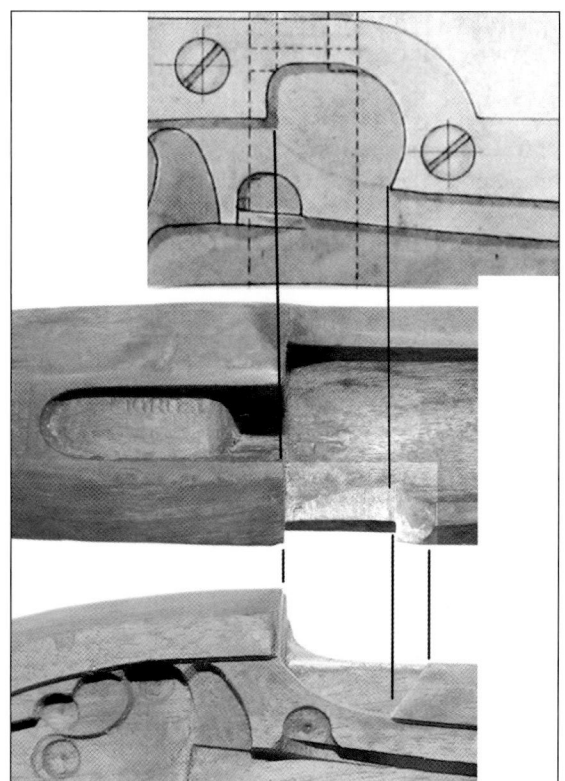

Right: Fig. 5.58. Outward extension of the lockplate recess in the pattern lock housing, top, shown in conjunction with the cut-out for the 'cone seat' or 'nipple bolster'. (part © West Yorkshire Archives, Greenwood & Batley Drawing No. 668)

Below: Fig. 5.59. Timber left behind after machining the opening for the 'cone seat'. (© P. Smithurst)

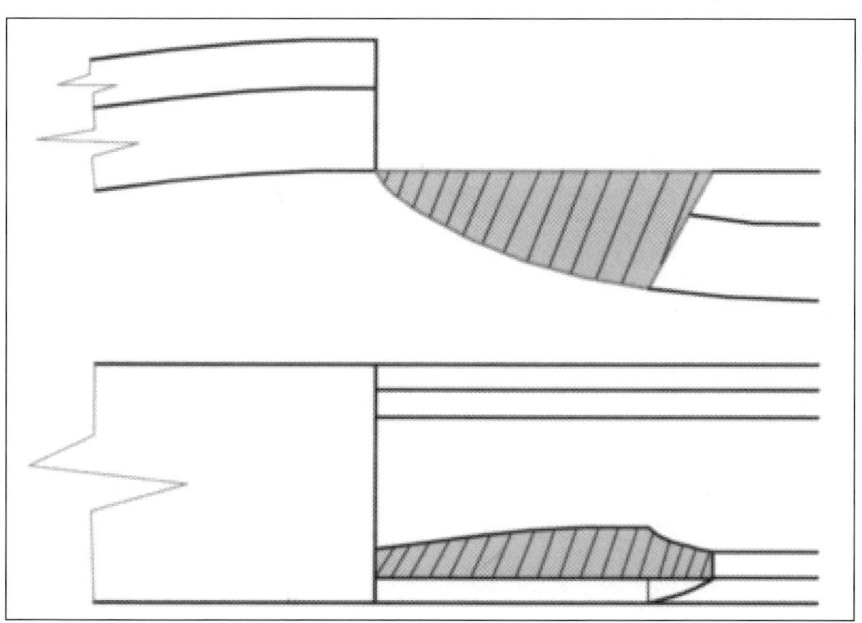

This is felt to be an oversight in what was a carefully planned operation and it is argued that this could have been achieved if the corresponding portion of the lock-bedding pattern was deeper than has been indicated or deduced, thereby allowing the bulk of this timber to be removed and leaving only a small fillet, Fig. 5.60.

This would better satisfy the comment by Greenwood that:

> *The end of the curved recess for the cone seat (nipple bolster) where it joints against the lockplate is squared by hand.*[36]

At some unspecified stage, the remains of the timber left by the oblique cut in the 'slabbing' process was removed to leave a quadrant fillet between the left-hand wall of the barrel channel and the vertical breech face, Fig. 5.61.

The fifth operation was very much a repeat of the fourth but using a 0.2-inch diameter cutter and guide pin to remove the last vestiges of timber left in corners of the lock recess by the larger cutter used in the previous operation.

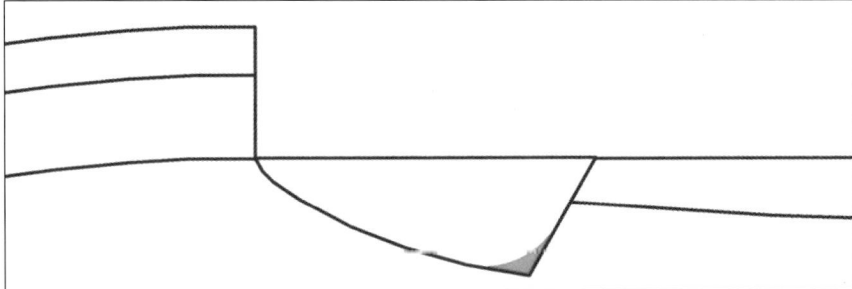

Fig. 5.60. Fillet left after cutting out the opening for the 'cone seat'. (© P. Smithurst)

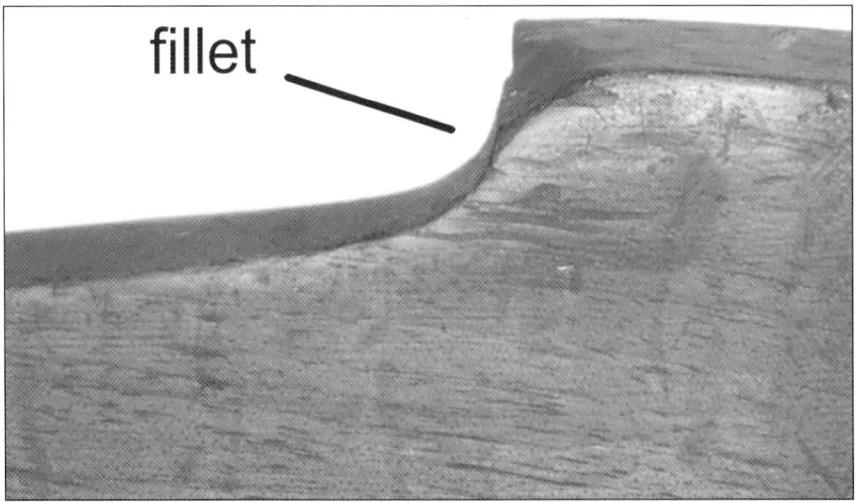

Fig. 5.61. Fillet on the right-hand wall of the barrel channel. (© P. Smithurst)

MANUFACTURE OF THE STOCK

Bedding the trigger guard, drilling screw holes, bedding trigger plate and the stop for the ramrod

All operations were carried out on one machine. In Fig. 5.63 the other features created at this stage can also be seen and will be dealt with in order.

As can be seen from the illustrations, the mating surface between trigger guard and stock follows a curve which approximates to the arc of a circle.

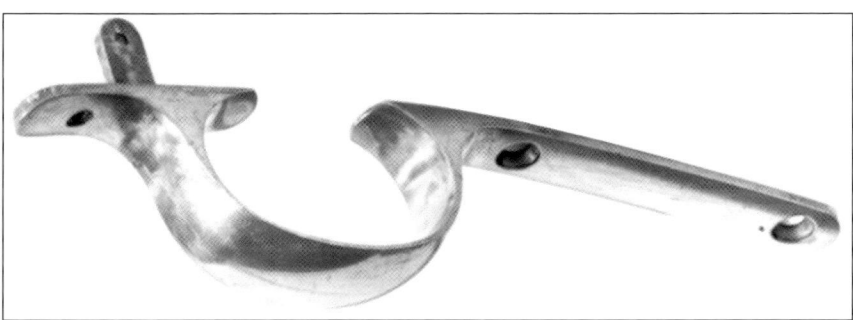

Fig. 5.62. The Enfield Pattern 1853 trigger guard. (© P. Smithurst)

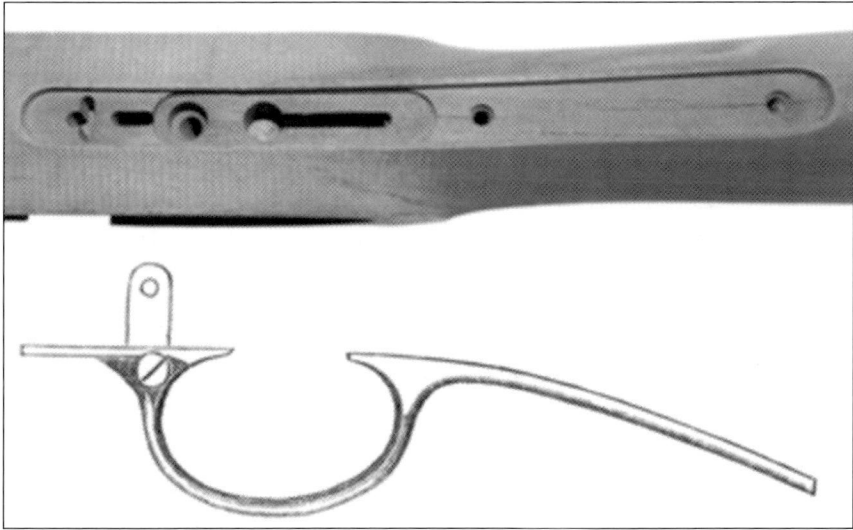

Fig. 5.63. Features of the Enfield Pattern 1853 trigger guard and its bed on the stock.

Fig. 5.64. The Enfield Pattern 1853 stock showing the extremities of the trigger guard. (author's collection)

MAKING THE ENFIELD PATTERN 1853 RIFLE-MUSKET

This gives some indication of the primary requirements of a machine to perform this operation and the machine is illustrated in *The Engineer*, but gives an overall impression rather than any real detail. A clearer illustration of this machine is provided by Benton, a detail of the main functional elements of which are shown in Fig. 5.65.

In the caption to Benton's illustration, the item H is described as *an oscillating fixture*. It is difficult to understand how it can 'oscillate' since it appears fixed at each end, but it would act as a guide if its centre of curvature was concentric with the arc of the stock, G, on which the trigger guard is bedded. Thus, the mounting on which the stock is fixed is constrained to follow that guide when traversed longitudinally by the rack and pinion beneath it. Inferrence suggests that since the recess for the trigger plate is the same as that for the trigger guard, it was cut in this same operation. The item identified as E in the illustration is a cutter and F is the guide pin which followed a pattern which is not visible.

After 'bedding' the guard, the tool carousel was rotated to bring the next tool and guide-pin into position for drilling the holes for the trigger-plate 'bosses', Fig. 5.66.

These bosses were projections on the trigger plate, Fig. 5.67, serving two important functions.

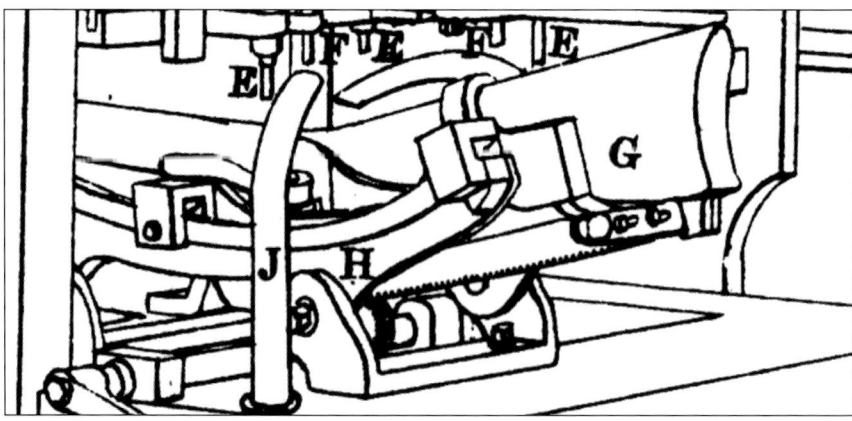

Fig. 5.65. The trigger guard bedding machine illustrated by Benton. (1878, Pl. XI)

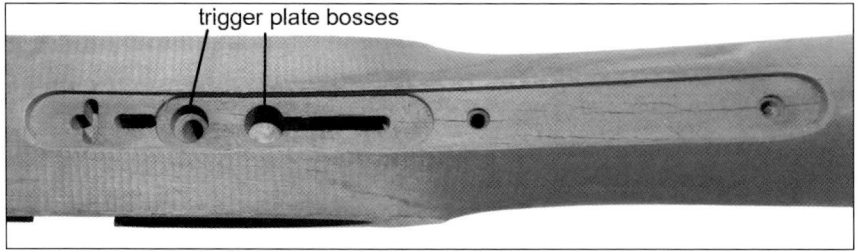

Fig. 5.66. Holes for the trigger plate bosses. (1885S607.00022 courtesy Birmingham Museums Trust)

MANUFACTURE OF THE STOCK

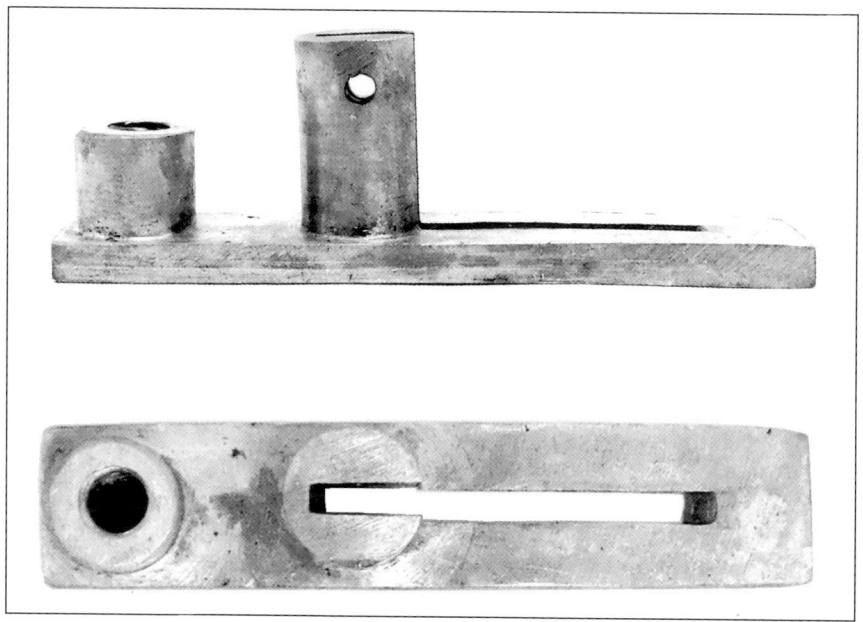

Fig. 5.67. The Enfield Pattern 1853 trigger plate. (author's collection)

The shorter one at the front had a threaded socket into which the breech tang screw entered, firmly anchoring both components. It therefore had to be concentric with that screw hole. The taller one at the rear had a threaded cross hole into which the trigger screw fitted, acting as a pivot for the trigger, and was cut into by the slot for the trigger.

The third operation was to drill the three holes for the trigger guard screws, using a third cutter and guide pin.

Lastly, the fourth operation was to form the slot for the trigger, trigger guard tenon and the ramrod stop, a transverse slot just behind the front screw hole. Although the machine in Fig. 5.65 only shows three sets of cutters and guide pins, the machine at Enfield had four (*Engineer*, 1859, p.294).

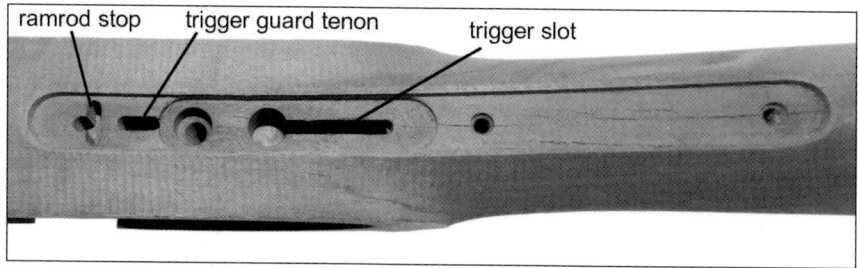

Fig. 5.68. Further details of the trigger guard bedding operations. (1885S607.00022 courtesy Birmingham Museums Trust)

MAKING THE ENFIELD PATTERN 1853 RIFLE-MUSKET

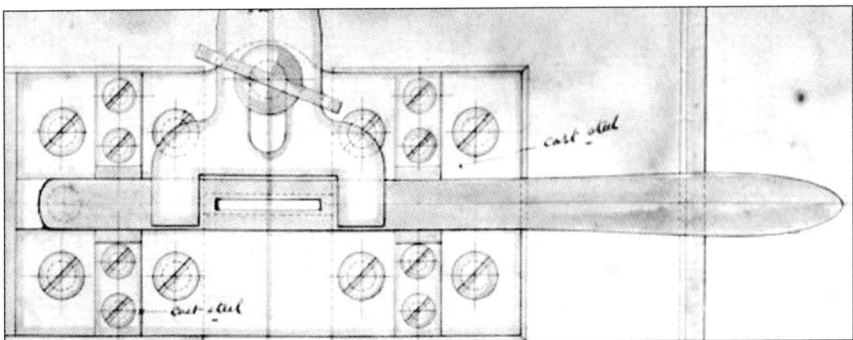

Fig. 5.69. Detail showing a pattern/guide for cutting the trigger recess in a stock. (© West Yorkshire Archives, Greenwood & Batley drawing No. 731)

A surviving Greenwood & Batley drawing, Fig. 5.69, shows the pattern for performing the slotting of the stock for the trigger but is for a different weapon since it has a swelling at the end of the tang, and does not incorporate a guide for forming the slot for the ramrod stop, nor the longitudinal slot for the tenon on the front of the Enfield trigger guard.

Stock cut for the bands and the nose cap let-on.

The purpose of the first process was to machine the rough-turned stock to create the squared seatings against which the barrel bands were fixed and this is shown in Fig. 5.70.

The only source of information regarding the machine for this purpose and process is not aided in its description by the lack of illustrations. However, it is stated that the next operation, turning between the bands, *the machine is similar, in many respects, to the one just described*[87] and for this the details provided by Greenwood[38] can be drawn upon.

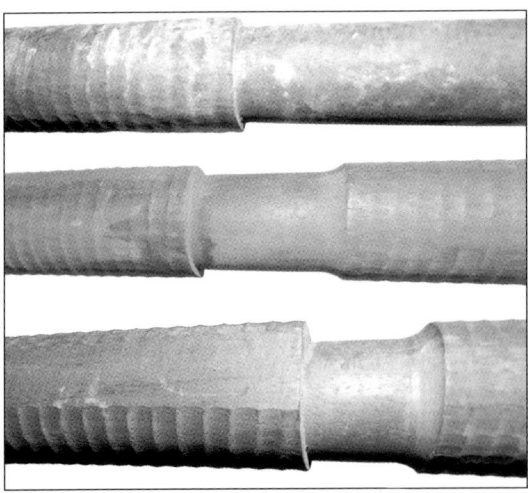

Fig. 5.70. Rough-turned stock cut for the bands with their square shoulders to the left and in descending order: top band, also recessed for nosecap (not shown); middle band; bottom band. (1885S607.0015 courtesy Birmingham Museums Trust)

MANUFACTURE OF THE STOCK

Stock turned between the bands

The only difference between the machines would seem to be that, apart from different widths and numbers of cutters, the stock needed to be clamped in different places so as not to interfere with the cuts to be made. Within those constraints, the machines and processes were the same and 'turning between the bands', which removed the remaining surplus timber to create the result shown in part in Fig. 5.71, will be described.

The machine used is shown in Fig. 5.72 and after its functions have been described, it will be understood that it also provided the basic requirements of the previous operation for *cutting under the bands*.

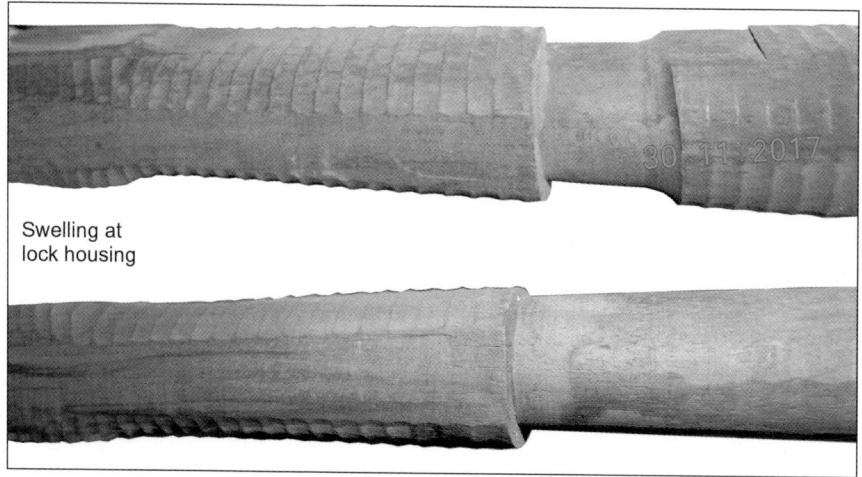

Fig. 5.71. Turning between bands showing removal of the timber left after cutting for the bands. (this only partially shows the turning between the bottom and middle bands). (top:1885S607.0015 and bottom; 1885S607.0016. courtesy Birmingham Museums Trust)

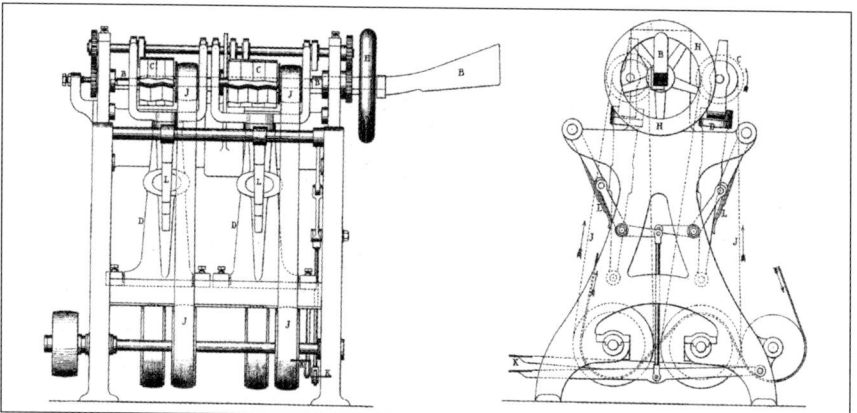

Fig. 5.72. Greenwood & Batley machine for turning between the bands. (Greenwood, 1862, Plate 82)

MAKING THE ENFIELD PATTERN 1853 RIFLE-MUSKET

It is another copying machine but, because the two portions of the fore-end of the stock are uniform in cross-section along their length, the pattern takes the form of a cam, rather than a complete model of the fore-end. Also, instead of a cutter which traversed the workpiece, two pairs of cutters were used, each pair combined extending the full length of the parts to be machined.

Unlike the previous copy-turning machine described in *The Engineer*, in this machine the stock was inserted through hollow bearings which only carried the model barrel to which the stock was fixed; the cams were mounted on a separate shaft.

An examination of a cross-section through the top portion of the machine, Fig. 5.73, shows the arrangement of the principal components.

A plan view, Fig. 5.74, shows the nature of the cutter blocks and their relationship with the stock.

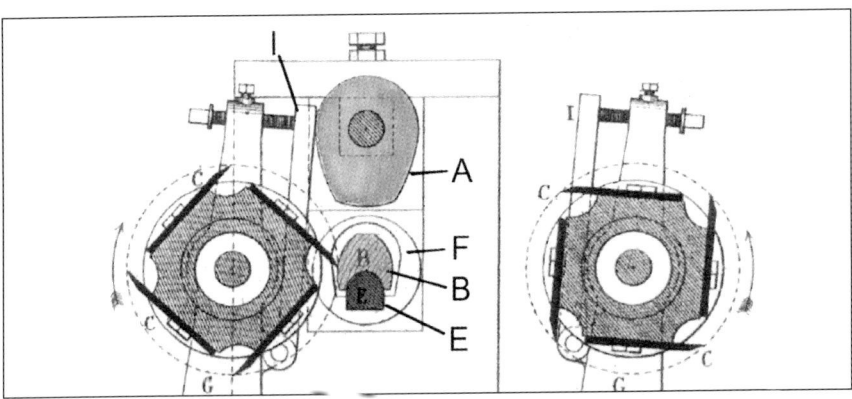

Fig. 5.73. Transverse cross-section through the working elements. The stock, B, mounted on a model of the barrel, E, carried in hollow bearings, F. Above it is a shaft carrying the cams, A and the tracers or followers, I, fixed to a rocking frame, G, carrying the rotating blocks into which the cutters, C, are fixed. (adapted from Greenwood, 1862, Plate 91)

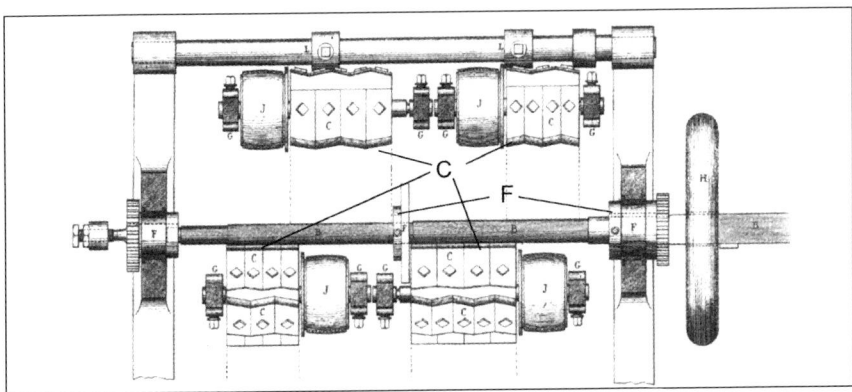

Fig. 5.74. Plan view showing the hollow bearings, F, carrying the model barrel on which the stock, shown in grey, is fitted, and the four cutter blocks, C. (adapted from Greenwood, 1862, Plate 92)

MANUFACTURE OF THE STOCK

The rotations of the shafts carrying the cams and the stock were synchronised by two identical gears in mesh with an 'idler' gear, so that when rotated by hand, both rotated in the same direction and at the same speed.

Two pairs of cutter blocks, C, were carried on 'rocking' frames on each side of the stock. In addition, each 'rocking' frame was fitted with a tracer or cam follower, I, in Fig. 5.73. Its position on the rotating cam determined the depth of cut taken on the stock and was finely adjustable to allow for wear and readjustment after re-sharpening the blades. The cutter blocks were of sufficient length so that the matching pairs on opposite sides cut the full length of the part of the stock they worked on, with a small overlap in the middle which is evident in Fig. 5.74. It should be noted that where the stocks had been cut for seating the bands in the previous operation, these areas then acted as bearing and clamping points, F, for turning between the bands.

One pair of rotating cutters was brought into contact with the stock by the operator pressing a foot pedal and the stock, mounted on its model barrel, was rotated by hand. This was followed by engaging the pair of cutters on the opposite 'rocking frame', and in this way the whole of the stock between the lower and top bands was machined.

The cutter blocks were fitted with a number of plane blades whose cutting edges were parallel to a vertical plane through the centre line of the stock but were inclined in the horizontal plane. Thus, when rotating and presented up to the stock, the point of contact with the cutting edge moved along the stock providing a 'paring' action along the grain, rather than a 'tearing' action across the grain, leading to a smooth finish. The final step was simply the removal of the sharp corner at the end of the stock to allow the 'nosecap', Fig. 5.87, to sit firmly on the wood when finally fitted.

Butt end and fore-end of stock between lock and bottom band finish-turned in copying lathe

Both Greenwood[39] and Benton[40] note these as being two separate operations whereas '*The Engineer*'[41] makes no distinction and illustrates only one machine.

In the processes described by Greenwood and Benton, the stock would have to be transferred to another machine to complete the work. It is also clear from the illustration of the machine at Enfield that it could not accommodate the stock to allow all the machining to be accomplished in a single stage. That raises the question of why the machine described by Benton and noted earlier for 'rough turning' this portion of the stock was not adapted also to 'fine turning', Fig. 5.76.

It could clearly accomplish this in a single stage and by changing the cutter and arranging a fine feed to the leadscrew, 'fine turning' would have resulted.

One of these Enfield machines was presented to Springfield Armory and is on loan to the American Precision Museum in Windsor, Vermont. The author, when Executive Director at that museum, was able to operate it and can attest to its functionality. The photograph below shows the arrangement of the principal operational features.

Fig. 5.75. Copying lathe used at Enfield for finish-turning the butt. (*Engineer*, 1859, p.348)

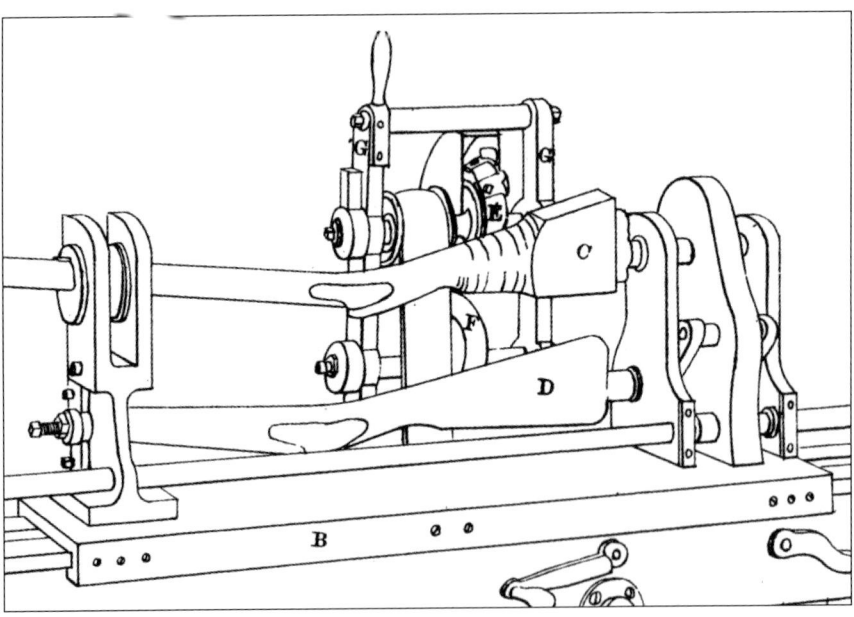

Fig. 5.76. Benton's machine for 'rough turning' the stock from butt to bottom band. (Benton, 1878, Pl. XV)

MANUFACTURE OF THE STOCK

Fig. 5.77. The Enfield butt finish-turning machine. (Courtesy Springfield Armory / American Precision Museum, Windsor, VT.)

The cutter wheel is visible and is fitted with conical cups sharpened on their rims and held in place by set-screws, allowing them to be rotated as occasion demanded to allow for wear or slight damage. Also visible is the hollow bearing through which the stock is passed and clamped in place. The frame carrying the follower and cutter wheels is moved along the bed by a leadscrew as the pattern and stock are rotating. Having finish-turned the butt, the portion of the fore-end between the finish-turned butt up to the position of the first band was turned in a similar machine. By extending the bed of this machine, it could have accommodated the whole of the stock to be worked upon, from butt to bottom band, in a single operation.

One point noted in '*The Engineer*'[42] was that to maintain the sharp corners between the planed surfaces of the lock seat and the flat face on the oposite side, where they meet the curves of the stock, the pattern was enlarged at this point so the follower wheel could move over a rounded corner while the cutter wheel was not in contact with the stock. Continued rotation of the stock and pattern brought the cutter wheel back into contact with the stock at the appropriate point.

It is clear from surviving specimens, Fig. 5.78, that this process, by the very nature of the cutters, left a very slightly ribbed surface which would need further hand-finishing with sandpaper to create a perfectly smooth stock.

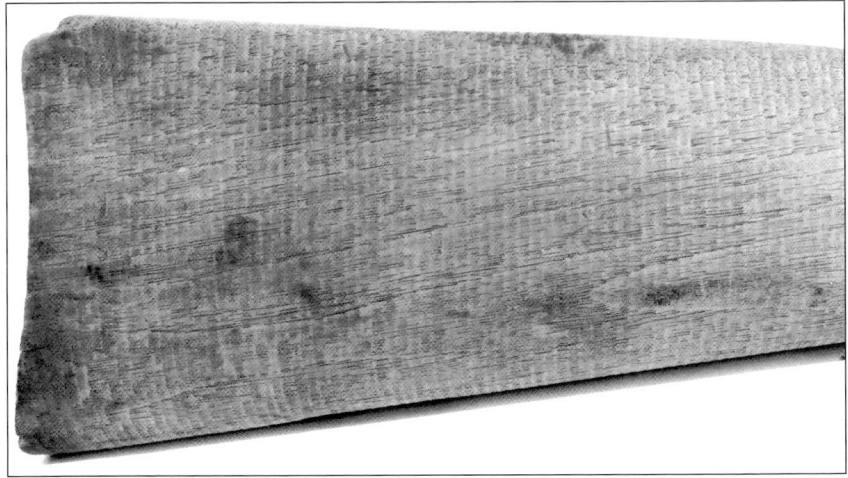

Fig. 5.78. A 'finish-turned' butt of similar period showing the fine furrows left by the cutter. (author's collection)

Stock grooved for ramrod

This was a straightforward operation in comparison with others and involved cutting a groove from the end of the fore-end down to the first, or bottom, band. It was carried out on a machine similar to a vertical milling machine or router with a high-speed spindle to which the cutters were fixed. The stock was inverted and mounted vertically on a model of the barrel to allow accurate positioning. This model barrel was fixed to a table which could slide and move the stock longitudinally. The first groove to be cut used a square-ended cutter. This groove was then enlarged slightly at the bottom using a ball-ended cutter of sufficient size so that the ramrod was a loose fit and would remain so if the wood swelled or shrank due to changes in humidity. The effects of these operations are shown below, Fig. 5.79.

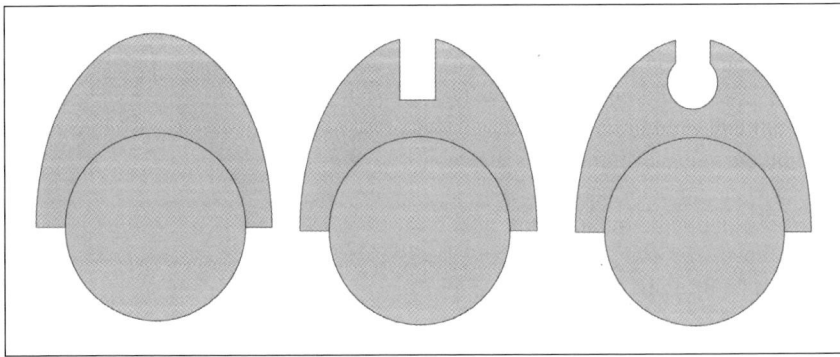

Fig. 5.79. Sequence in cutting the ramrod groove with the inverted stock mounted on a model of the barrel. (© P. Smithurst)

MANUFACTURE OF THE STOCK

Recessed for the ramrod spring and transverse hole for fixing pin bored

The first ramrods for this rifle had a 'swell' close to the head and the head was a hollow mushroom shape. This swell caused the rod to be 'sprung' as it was forced past the lip of the nose-cap to hold it in place. However, experience in the Crimean War showed that with cold, greasy or sweaty hands, this form of rod was difficult to grasp firmly enough to remove it. This, combined with the fact that it needed a 'jag'- a cylindrical block with serrations to enable it to grip a piece of cloth – attached to the opposite end for cleaning the bore, led to its being replaced at some time between 1857 and 1860. The new version had a plain rod without a 'swell' and was equipped with a 'jag-head' which also provided a good surface to grip when removing the rod.

The plain shaft of the new ramrod meant it could not be simply held in place by friction and a spring was devised to bear upon the lower portion of the rod when in place in the stock. It is commonly refered to as 'Burton's spoon', suggesting that it was devised by James Burton, Chief Engineer at Enfield, although a similar device was used earlier on French, and Russian muskets.

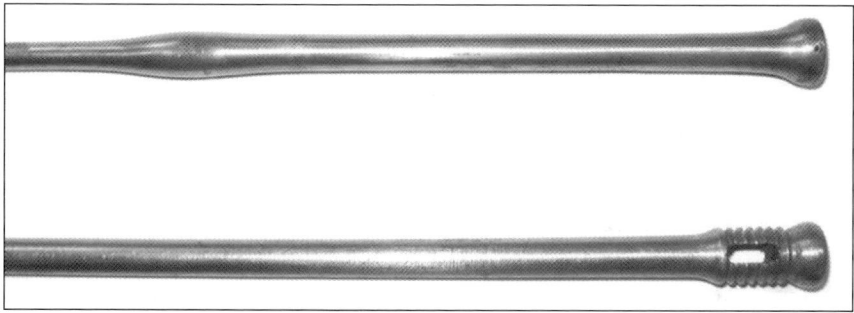

Fig. 5.80. Top: First pattern ramrod with a swell on the shaft and a 'mushroom' head. Bottom; second pattern with a plain shaft and a jag-head. (author's collection)

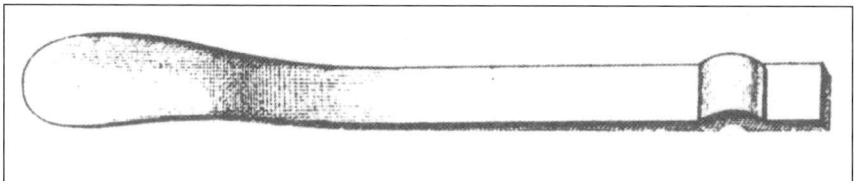

Fig. 5.81. 'Burton's spoon' ramrod retaining spring. (Petrie, 1866, plate V)

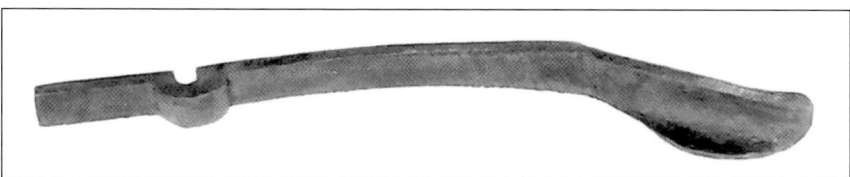

Fig. 5.82. Ramrod retaining spring from French M. 1822 musket. (author's collection)

MAKING THE ENFIELD PATTERN 1853 RIFLE-MUSKET

Fig. 5.83. Recess for rod spring. (1885S00607.00020 courtesy Birmingham Museums Trust)

Having to accommodate a spring meant two additional operations in the manufacture of the stock. The first was to use a profile milling machine with a spindle and a guide pin to follow a pattern, similar, for instance, to that used to cut the recess for the buttplate tang, and circular cutter marks can be seen at the bottom of the recess in Fig. 5.83.

The second operation was to drill the cross hole for the spring-retaining pin. Because of the small diameter of the hole and the long slender nature of the drill to encompass the full length of the hole, it was found advisable to drill from both sides. This used a very ingenious machine with two exactly opposed drilling spindles on vertical slides.[43] The stock was mounted exactly horizontally on its side and in correct alignment.

The lower spindle on its slide was raised by a lever. Attached to this slide was a spring having a boss, serving as a latch, at its extremity. As this slide was raised and the lower hole drilled, this boss engaged with a corresponding boss on the upper slide and latched on to it. Thus, as the bottom slide and spindle was lowered, the upper slide with its spindle was brought downwards to drill the second half of the hole. At the appropriate point, this latching was automatically disengaged and the upper slide was returned to its starting position by a spring. Unfortunately, no illustration of this machine has been found.

Hole for ramrod bored

This hole was a continuation of the channel already prepared. The only detail of this operation which has been discovered is a Greenwood & Batley drawing of the machine but which is in very poor water-stained and grimy condition. Even with digital enhancement it has proved impossible to obtain a good image of the details of the machine. The only clear feature in the drawing is that the stock was inverted and was mounted on a model of the barrel for accurate positioning. The hole was drilled as far as the slot for the ramrod stop shown in Fig. 5.68.

MANUFACTURE OF THE STOCK

Holes for lockplate fixing screws, for breech tang screw, for nosecap screw and pin hole for tenon of trigger guard are drilled

All of these holes had to be drilled in their exact positions to coincide with holes in the components they engaged with. It seems strange that holes for the lock screws, or 'side nails', were not drilled at the same time as the recesses for the side cups were made since both would have required the same centres.

Regarding the other holes, the two unpublished official drawings of the Pattern 1853 rifle referred to at different points, R.S.A.F. No. 458 and 746, are interesting. Careful scrutiny of R.S.A.F. drawing No. 458, Fig.5.84, clearly shows the hole for the trigger guard tenon pin, but reveals a fact so far overlooked: the boss on the trigger plate for the breech tang screw is misaligned and may account for the drawing never being completed.

Another fact that has been overlooked by anyone previously studying the second drawing, R.S.A.F. No. 746, Fig. 5.85, is that the trigger plate is incorrect and the boss to receive the tang screw is shown in completely the wrong position.

The drilling and counterboring at the tip of the barrel channel for the nose cap screw, which might be regarded as the last of the machining operations on the stock, also required care in setting-up. Not only did the hole have to align with the hole in the nose cap, the wood at this point was thin so the counterboring to receive the head of the screw had to be accurately controlled.

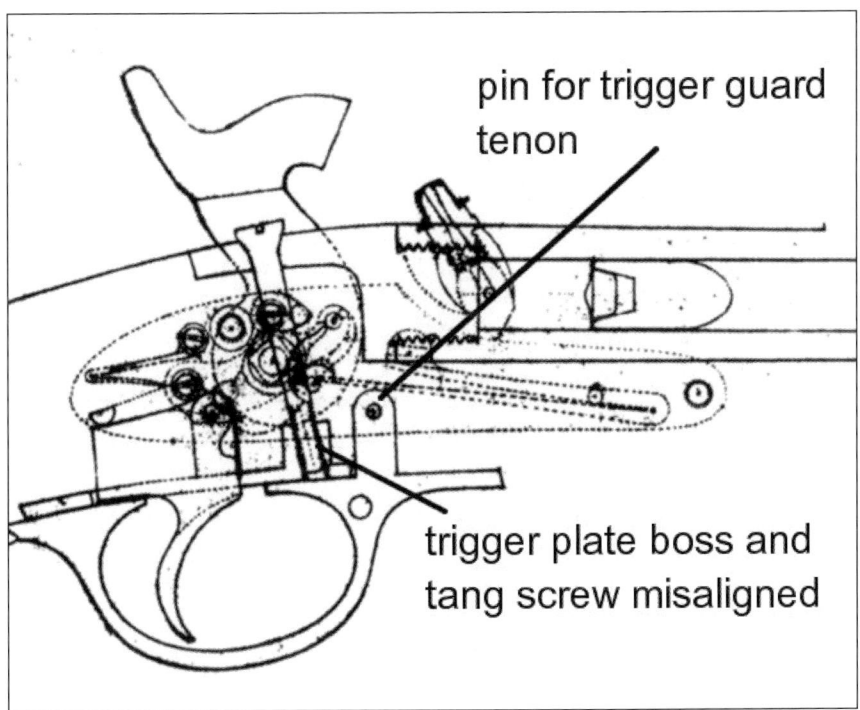

Fig. 5.84. Detail, Royal Small Arms Factory Drawing No 458. (© Royal Armouries)

MAKING THE ENFIELD PATTERN 1853 RIFLE-MUSKET

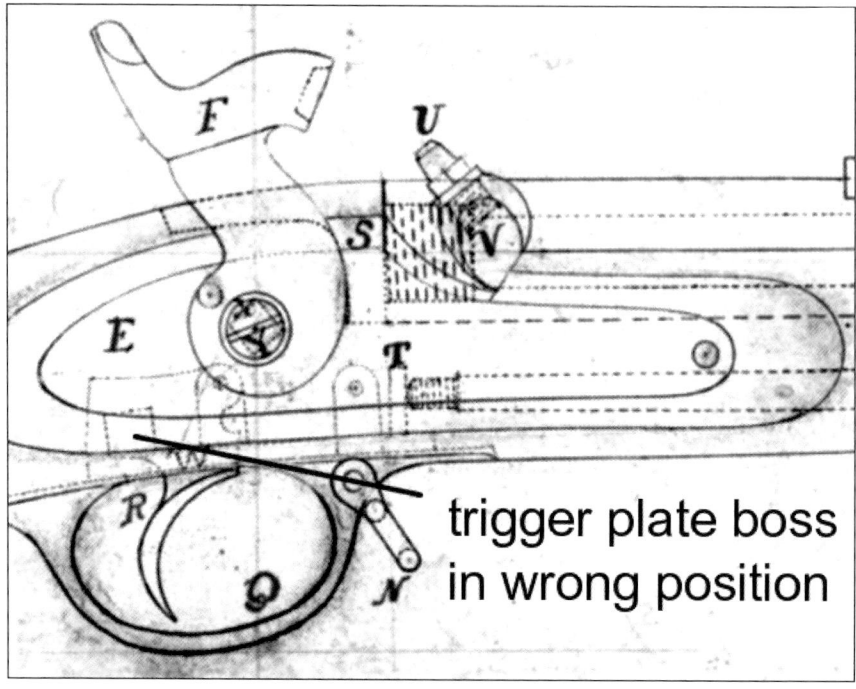

Fig. 5.85. Detail, Royal Small Arms Factory Drawing No 746. (© Royal Armouries)

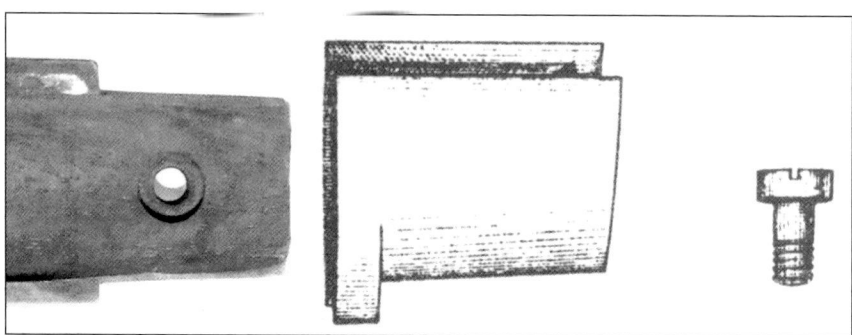

Fig. 5.86. Left: Plan view of the tip of the fore end showing the counterboring for the head of the nose cap screw in the barrel channel, (1885S00607.00022 courtesy Birmingham Museums Trust) and Right, the nosecap and screw. (Petrie, 1866, Plate V)

    All that remained to be done was to lightly sandpaper then oil the stock before putting into store ready for assembly with the other components as a finished weapon. In contrast, Benton[44] records that before *boring for the tip screw* [nosecap screw] *the entire surface of the stock* was finished and smoothed *by the use of spokeshaves, scrapers and sandpaper* which either indicates a lesser quality of finish resulting from the use of the softer American walnut, or less attention given to keeping the cutters sharp.

MANUFACTURE OF THE STOCK

Gauging

The intricacies of the stock are reflected in both the large number of machines and individual operations they performed in its manufacture and also of the many gauges supplied to check it. Of these, only three items within the Royal Armouries' collections relate to the stock.

The first is the butt-profile gauge already mentioned and shown in Fig. 5.30. The second is a 'working model' lock issued to the 'Stocking Department', Fig. 5.87.

The 'working model' is clearly marked on its outer face, Fig. 5.88.

It is correctly assembled and able to be 'cocked', as befits a working model, to enable it to fully test all aspects of the lock-bedding operation.

What is additionally clear, and is to be expected considering the nature of its purpose, is that this lock has no 'assembly marks', showing it to be one of the interchangeable series produced at Enfield from 1857 onwards.

The third item was used by the London Armoury Company and is inscribed "L.A.C. / For Barrel Gauges / Standard". Taken literally, this would appear to be

Right: Fig. 5.87. 'Working Model' of lock issued to the Stocking Department at Enfield. (part of PR.10142. © Royal Armouries)

Below: Fig. 5.88. Marking on the 'Working Model'. (part of PR.10142. © Royal Armouries)

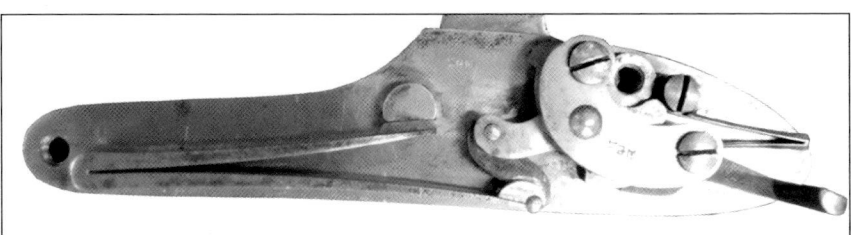

Fig. 5.89. Interior of the 'Working Model'. (part of PR.10142. © Royal Armouries)

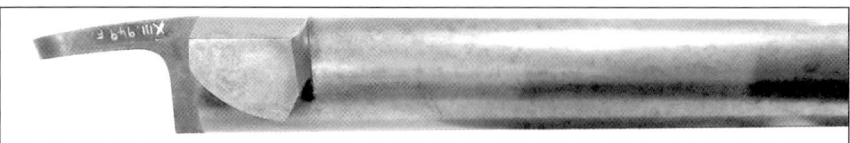

Fig. 5.90. Model of breech-end of a barrel marked 'For Barrel Gauges'. (XIII.949F © Royal Armouries)

Fig. 5.91. Breech end of stock ready to receive the barrel with breech tang finished and seating for underside of nipple bolster correctly shaped. (607.00022 courtesy Birmingham Museums Trust)

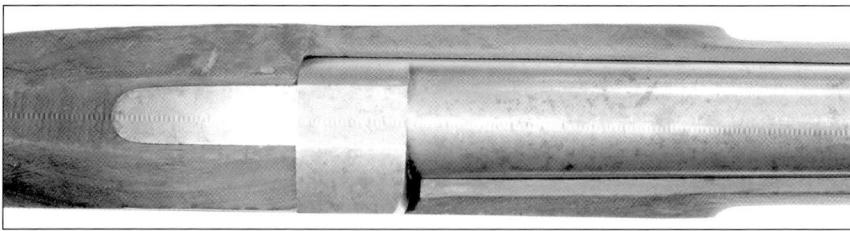

Fig. 5.92. Breech end of stock fitted with breeching gauge, (XIII.949F © Royal Armouries; stock author's collection)

used for testing gauges for the barrel, but it will be argued later, Chapter 7, that such a use is incompatible with the known barrel gauges.

It would, however, serve for testing the ability of the stock to accept the finished barrel by checking the correct bedding of this complex portion of the barrel, Figs. 5.90–5.92.

5.5 Conclusion

This account of stock manufacture has utilised the study of, and drawn upon, a variety of artefacts and documents not previously studied or published. In combination with the few existing published accounts, a fuller picture of this topic has been presented. The machines developed for stock manufacture exhibit high degrees of ingenuity to duplicate many of the hand-skills and enabled an unusual degree of dimensional accuracy in creating a finely engineered wooden product.

Chapter 6

Manufacture of the Barrel
Part 1 – creating the tube

6.1 Introduction

At the end of this stage of its manufacture, the barrel of any rifle or musket needs to satisfy one criterion: it has to be able to withstand the pressure generated by the explosion of propellant within it. This fell within the province of those involved in making the iron tube from which the finished barrel would be created. Traditionally, the 'raw' tube had been made by hand and details of the methods employed can be found in book 1. Those hand methods were slow and highly labour intensive but remained in use in many places until the second half of the 19th century.

In Britain, however, the process of making the raw tubes by rolling had been pioneered and the technology used was principally based on Osborn's patents of 1813[1] and 1817[2] and Heywood's patent of 1814.[3] Osborn's patent of 1813 related to a rolling mill for producing tapered round tubes and his 1817 patent to a rolling mill for welding a seamed tube. Heywood's patent was for a rolling mill to create those seamed tubes.

6.2 Specifications for the Enfield barrel

The Report of the Select Committee on Small Arms included a brief specification and an excerpt is given below:

> *Specification for barrels in the filed state for Rifle Musket Pattern of 1853, .577 bore.*
>
> *The barrels to be rolled of the best charcoal iron, and the lumps forged of the same, without steely or hard parts to obstruct the rifling; to be free from greys or flaws from end to end; no wires to be admitted, and if any be afterwards found, the whole expense incurred in setting up to be charged against the barrel-masters;*
>
> *7th January 1854.*
> *J. A. Pellatt, for the Inspector of Small Arms.*[4]

> **Barrels.**
>
> The Barrels to be rolled from wrought iron moulds of the best quality, and to contain no injurious greys or flaws either inside or outside. The bore of the

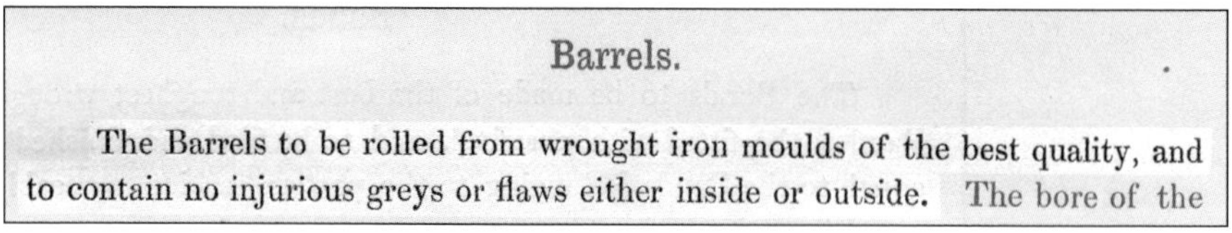

Fig. 6.1. Excerpt from a contract specification for the manufacture of Pattern 1853 Artillery Carbine barrels. (see Appendix 2, item 5. Author's collection)

Recently discovered documents show that at the onset of manufacturing, the wording changed in the specifications issued to contractors for the production of the 1853 Pattern Artillery Carbine and the 1856 Short Rifle, which would not have differed fundamentally from those applied to the long 'three-band' rifle barrel.

The 'moulds' referred to are not what we might envisage but simply refer to something that has been roughly formed, such as a plate of iron transformed into a U-shape prior to converting into a tube. The 1856 Pattern Short Rifle contract (appendix 2, item 6) for instance simply added that the *breech pin and front-sight* were also to be made of the best wrought iron, suggesting that prior to 1856, contractors for the rifle and the carbine may have been tempted to use an inferior, and therefore cheaper, quality of iron for breech-pin and front-sight. It is inconceivable that manufacture at the Enfield factory, once it began production, would have been any less stringent.

In studying the manufacturing operations applied to the barrel, various accounts have been drawn upon. All the accounts misinterpret or omit details and even some stages in the process. Drawing on information contained in these various accounts, the stages in making the 'raw' tube are deduced to comprise:

'Skelp' – passed through rolls to produce a 'mould'.
'Mould' – passed through rolls to produce a tube with open seam.
'Welding' – seamed tube rolled to weld the seam.
'Tapering' – Welded tube passed through rolls to produce full length tapered tube.
'Lumping' – iron lump welded on and shaped for the nipple bolster.

6.3 Manufacture of the barrel tube

Creating the 'mould'

The 'skelp' – a flat plate of iron – was the starting point of this process and the size of this varies with the source of information:

- *sixteen inches long and four inches wide and approximately ten pounds*[5, 6]
- *twelve inches long and five and a half inches wide and a half inch thick*[7]
- *twelve inches long by four inches wide and three-quarters of an inch thick*[8]
- *thirteen inches long, five and a quarter inches wide at one end and five inches at the other with the long edges bevelled to provide a better joint when welded, and a half inch thick*[9]

- *thirteen inches long, five and a quarter inches wide at one end and five inches at the other with the long edges bevelled to provide a better joint when welded, and nine-sixteenths inches thick*[10]
- *A rather pointless reference simply suggests that it was twelve inches long*[11]

While sizes obviously varied what mattered most was that there was enough metal so that the finished tube was of sufficient dimensions for conversion into a finished barrel. There was agreement that the iron be of the very best quality and this was supplied by Marshall & Co. of Wednesbury.[12]

The first step was to convert this skelp into the channel-shaped 'mould', and this was partly formed by hand prior to rolling.

One account[13] states the skelp was *moderately heated and subjected to the action of a pair of rollers which bend it into a shape called a 'mould'*. This mould was produced by a set of rolls shown in Fig. 6.3, using the three grooves, or 'passes',

Fig. 6.2. The roughly pre-formed *skelp* ready for rolling to form the 'mould'. (2002.D.0048 courtesy Birmingham Museums Trust)

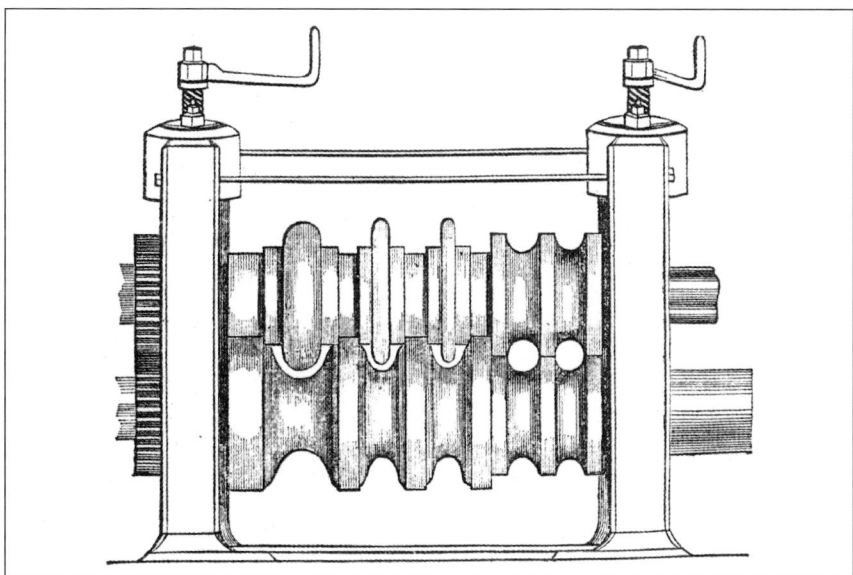

Fig. 6.3. Rolls for producing the 'mould' and converting it into a seamed tube. The shapes obtained can be seen from the 'daylight' between the rolls. (Jervis, 1854, p. 3)

MAKING THE ENFIELD PATTERN 1853 RIFLE-MUSKET

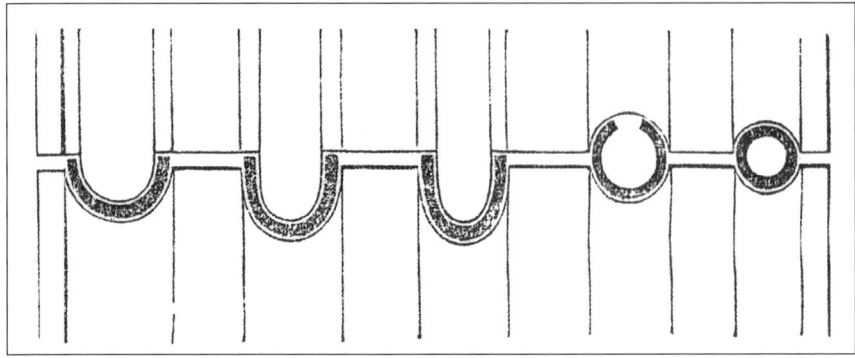

Fig. 6.4. Creating the mould. (I.L.N. 1855, p. 410)

on the left. 'Passes' was a common term used by those in the trade for the mating grooves in the rolls through which the metal was passed; it could also refer to the number of times the metal went through the rolls.

This creation of the mould by the rolling mill in Fig. 6.3 is more clearly shown in Fig. 6.4, in which the original published image, shown upside down, has been inverted to show the correct orientation.

In any rolling mill the rolls contra-rotate so that their mating surfaces are both travelling in the same direction. The metal, after passage through the rolls, was manually returned to the receiving side of the rolls, often via a reheating furnace, for passage through successive grooves. The 'mould', having been initially formed by passage through the first opening, or 'pass', in the rolls was re-heated in a furnace before being fed through the next two 'passes' which lengthened it and reduced its width.

Creating the seamed tube

Further re-heating prepared it for being converted into a tube by the last two passes and which, again because they decrease in diameter, lengthened it even further. The collection in Birmingham Museum does not have specimens of all the early stages but does have an example showing a seamed tube which is the result of the first stage of 'turning the edges over' to bring them together, Fig. 6.5.

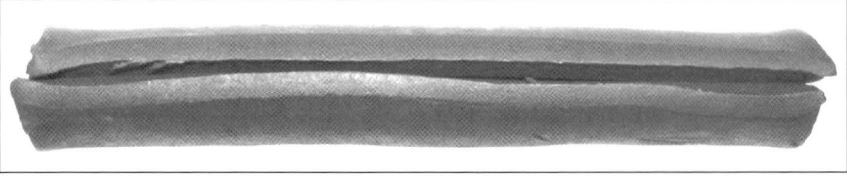

Fig. 6.5. The first step in converting the 'mould' into a tube. (2002.D.484 courtesy Birmingham Museums Trust)

Welding the seam

Having obtained the seamed tube, it required welding, drawing out to full length and tapering, with a small allowance for trimming at the muzzle and breech ends to bring it to finished length. These actions were performed using a second set of rolls, Fig. 6.6.

The seamed tube was heated to welding temperature, at which the surface became a fluid film. It was withdrawn from the furnace on a mandrel with a circular guard to prevent the white-hot metal slipping down the mandrel onto the worker's hand, and inserted into the far left-hand 'pass' of the rolls.

Any rolling mill has to have the periphery of the rolls flanking the grooves in contact to maintain the correct geometry and sizes of the passes, but often during operation, the reaction to the pressure exerted on the metal causes the top roll to lift slightly, especially if the bearings are worn, leaving a small gap. Metal is forced into this gap, creating a rib or 'fin' on the rolled metal, and these are visible in Fig. 6.7.

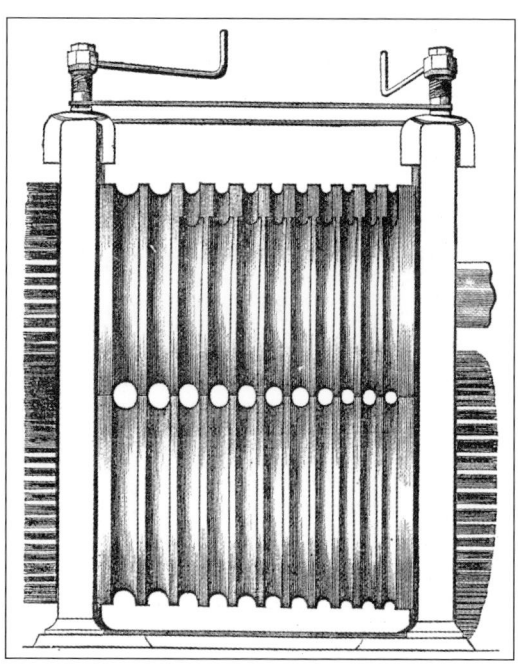

Right: Fig. 6.6. The second set of rolls incorporated two welding 'passes' on the left and nine tapering 'passes' on the right. (Jervis, 1854, p. 4)

Below: Fig. 6.7. The seamed tube after passing through the first 'pass' of the welding rolls (top) and the second 'pass' (bottom). The 'fin' created by the small gap between the contact faces of the rolls is clearly visible. (*top* 2002.D.322; *bottom* 2002.D.293 courtesy Birmingham Museums Trust)

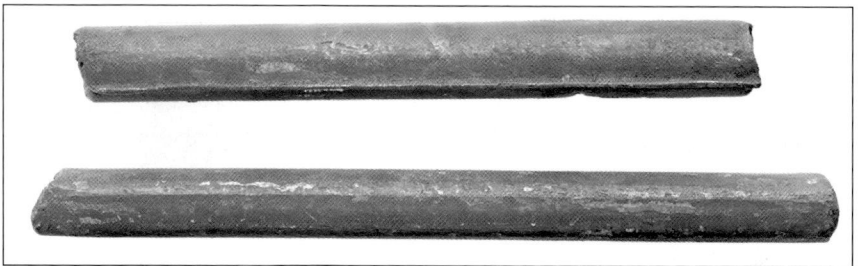

MAKING THE ENFIELD PATTERN 1853 RIFLE-MUSKET

In the next passage through the rolls the barrel tube would be rotated through 90 degrees so that this 'fin' was forced back into the body of the metal, a system repeated until, in the last pass of the barrel through the rolls, this fin was eliminated.

Barrel rolling is shown in Fig. 6.8, but it implies that the mandrel was simply held by hand as the barrel was drawn off it and passed through the rolls. This would not only create the risk of having both the tube and the mandrel drawn into the rolls but negated an important function of the mandrel, making it an almost impossible undertaking if held by hand. The barrel with the mandrel inside it had to be placed square to the groove so the barrel was drawn off the mandrel to pass through the rolls.

An illustration in an American report, Fig. 6.9, although of later date, exhibits close similarity to the barrel rolling mill shown in Osborn's 1817 patent since, as Fitch notes,[14] English barrel rolling technology had been introduced into America in 1860.

Left: Fig. 6.8. Barrel rolling at Birmingham, showing a stage in the production of an Enfield Rifle barrel – not shown is a means for regulating the position of the mandrel. (I.L.N. 1855, p. 410)

Below: Fig. 6.9. Barrel rolling mill with cross bars shown in place to ensure correct positioning of the mandrel with its white-hot tube. (Fitch, 1882, p. 10)

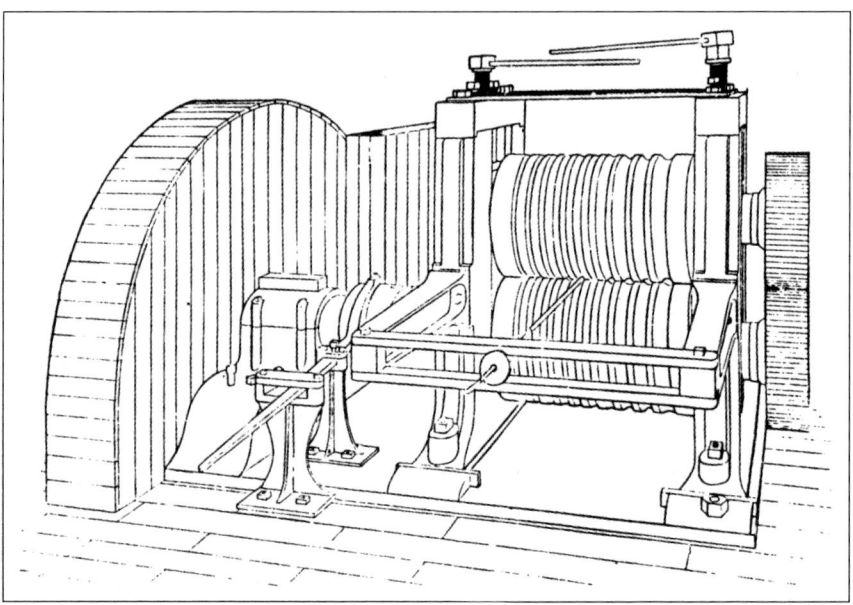

MANUFACTURE OF THE BARREL

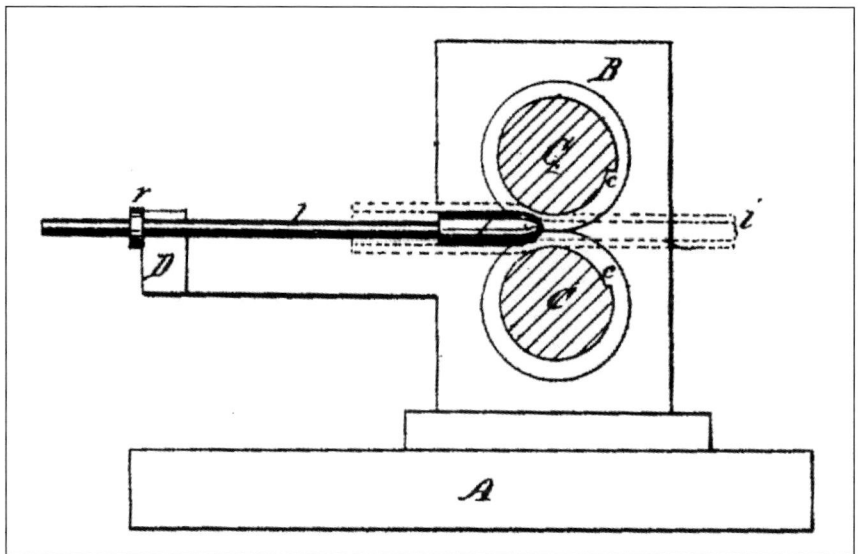

Fig. 6.10. Illustration from Burton's patent of 1860.

In this illustration, guard rails can be seen in front of the rolls, against which the disc on the mandrel rested, and are carefully adjusted so that as Fitch comments:

> *It is important to have the rods extend just to the center of the rolls, for if they go too far the hardened bearing of the mandrel will be torn off, and if not far enough, the barrel will be crushed together... The proper length of the mandrel is determined by a stop, with washers upon the rod, bringing up against a bar in the frame.*[15]

This requirement for the correct positioning of the mandrel is also made clear in Burton's patent of 1860, Fig. 6.10, which shows a section of the barrel with the mandrel in place.

The 'hardened bearing' referred to by Fitch above is taken to mean the enlarged 'head' of the mandrel as illustrated by Burton. Based on the U.K. patent record, a mandrel with a head was first used in the making of iron gas pipes in 1824[16] and became standard practice. Burton's drawing is almost an exact copy of that in Russell's patent. Being only short, the head eliminated the risk of having the barrel tube cool sufficient to become a 'shrink-fit' on a full length mandrel that filled the bore.

Tapering the tube

After passage through the first two cylindrical-section welding passes of the rolls, the barrels then had to be tapered. Closer examination of Fig. 6.6 reveals that the nine passes on the right in the top roll have steps in them because the grooves are tapered and the starting and finishing diameters are different. [Fig. 6.11]

However, it has to be pointed out that there is an important error in Fig. 6.6; the bottom roll should be a mirror image of the top roll so that as the rolls rotate, the steps pass each other at the same point. The rolls should therefore appear as shown in Fig. 6.12:

This same error was repeated in another account[17] which borrowed extensively from Jervis. The rotation of the rolls had to be synchronised through gearing so

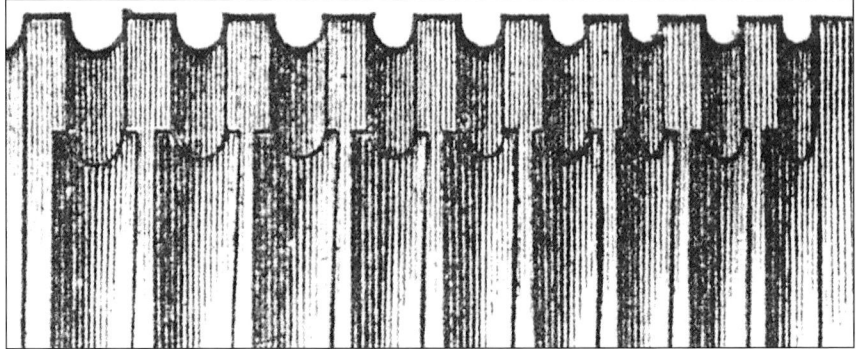

Fig. 6.11. Detail from Fig. 6.6 shows the steps in the tapering rolls where the muzzle and breech diameters meet. (Jervis, 1854, p. 4)

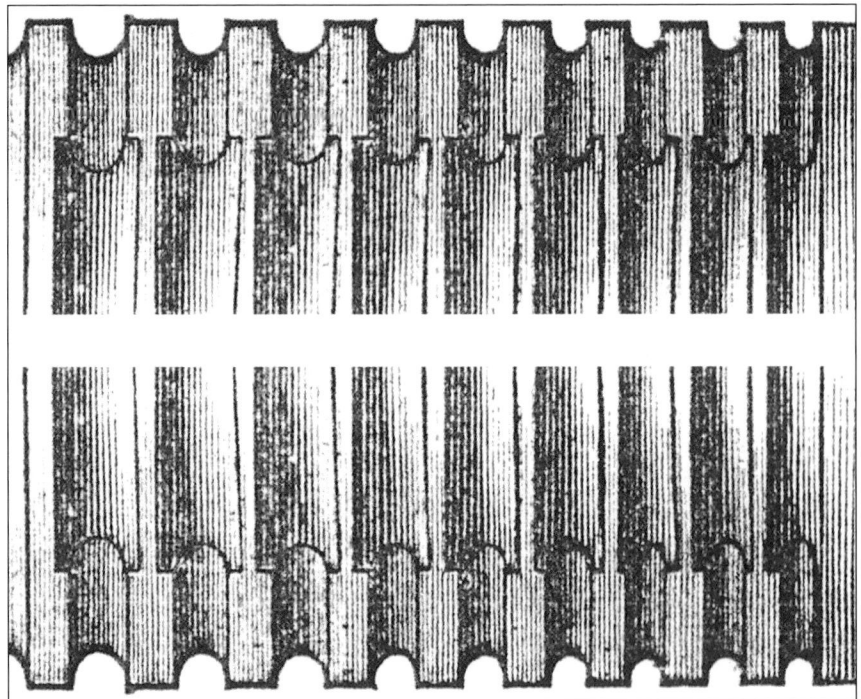

Fig. 6.12. Photographic reconstruction showing how the top and bottom rolls should be mirror images if tapered tubes are to be produced. (© P. Smithurst)

MANUFACTURE OF THE BARREL

that at each revolution, the steps coincided exactly, otherwise the taper would be deformed. The pre-heated barrel tube had to be inserted when the wide portion of the groove appeared, and the use of tapering rolls required very good judgement on the part of the workmen:

> *This operation... is a very difficult one: the grooves of the rollers not being concentric, the workman has to watch his opportunity very nicely and thrust in the barrel at the exact moment the proper part of the roller comes round.*[18]

This is reflected and amplified in another comment:

> *The breech end of the barrel was first formed, the roller watching his opportunity, and inserting the bar of iron when the broad section of the groove presented itself.*[19]

An important point to note is that the tube being tapered had to be such that when it was rolled to full length, its length did not exceed the length of the grooves.

Taper rolling would have been greatly aided by the use of what are often referred to as 'segmental rolls' in which a part of the circumference is removed to leave a gap between the start and finish of the grooves. The functioning of segmental tapering rolls is more clearly explained in the diagrams. Fig. 6.13 shows the segmental nature of the rolls, the tube entering the deeper groove at A and exiting the shallow portion at B. They would also have accommodated an over-length tube.

A side view, showing the passage of the tube through the rolls is given in Fig. 6.14.

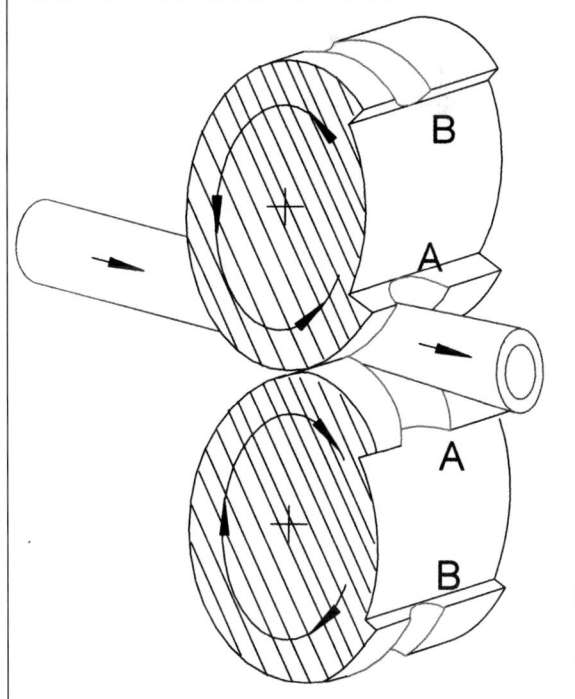

Fig. 6.13. Perspective sketch of segmental tapering rolls.
(© P. Smithurst)

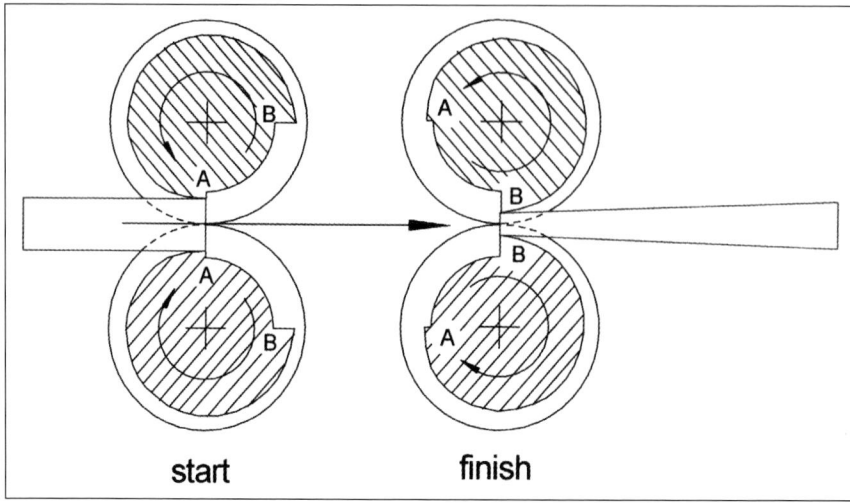

Fig. 6.14. Sketch showing transformation of a cylindrical to a tapered tube on its passage through the rolls from A to B. (© P. Smithurst)

Fig. 6.15. The tube after the first pass through the tapering rolls. (2002.D.294 courtesy Birmingham Museums Trust)

Fig. 6.16. The final stage of taper rolling, creating a longer than needed barrel tube. (Courtesy Birmingham Museums Trust; number not visible)

Nowhere in the descriptions of taper-rolling is there mention of a mandrel being used but it would have been essential to maintain bore size while the outside was being progressively reduced in diameter.

The result of the first pass through the rolls is shown in Fig. 6.15 and the final product, a full-length tapered tube, in Fig. 6.16.

Straightening

After rolling, the practice at both Birmingham[20] and Enfield[21] was to insert an iron bar into the hot barrel and roughly straighten it by striking it on a large iron plate, at the same time dislodging any 'fire-scale' – flakes of iron oxides – coating the surface, Fig. 6.17.

MANUFACTURE OF THE BARREL

Fig. 6.17. Iron plate used for straightening the barrels at various stages. (I.L.N. 1855, p. 410)

Fig. 6.18. The tube after straightening. (Courtesy Birmingham Museums Trust; number not visible)

A further operation at Enfield used a machine referred to as a 'squeezer'[22] in which the lower anvil was in the form of a die the full length of the barrel and having a semi-circular channel matching the larger diameter of the barrel and a similar upper die attached to an eccentric which caused the two dies to open and close. The workman moved the cold barrel backwards and forwards in the lower swage, turning it as necessary so that the dies would remove any bends, producing a tube as straight as could possibly be achieved by this means. Fig. 6.18.

'Lumping'

Having prepared the barrel tube, the next stage was the welding-on at the breech end of a small lump of iron, sometimes referred to as a 'bolster', of the same quality iron from which the barrel was made in the process known as 'lumping'. The lump of iron and the breech end of the barrel were heated to welding heat and the lump of iron, held in tongs, was attached by being 'dabbed-on' by hand. A good weld was vital to the service and safety of the rifle, so, whilst still at

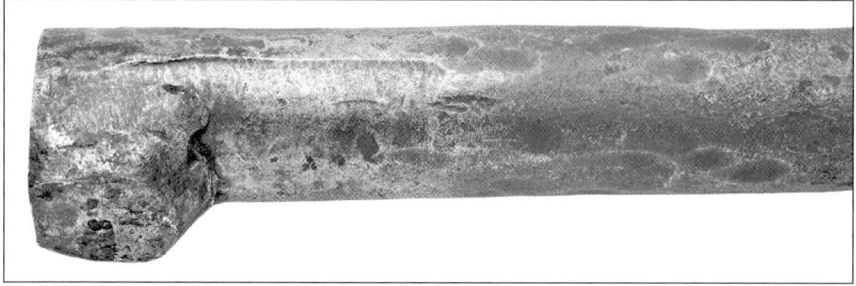

Fig. 6.19. A 'lumped' barrel, the marks from the die visible as a slight ridge projecting beyond the 'lump'. (2002.D.307.4, courtesy Birmingham Museums Trust)

welding temperature, the breech-end was placed between two dies and given blows from a trip-hammer to consolidate the weld and, at the same time, roughly shape the lump, Fig. 6.19.

6.4 Conclusion

The development of barrel rolling technology was a great improvement over the previous 'manual' methods. While it still required hand labour, this was drastically reduced. The nature of the skills required also changed and high levels of different skills were required, especially in the use of tapering rolls. Nowhere has any record of the number of barrel tubes produced daily or weekly been found but it has to be presumed that it was considerably faster than the traditional hand methods. In consideration of the welding of the seam for instance, a single pass through the rolls could be accomplished in a minute or less and be more certain of producing a sound and uniform weld. Performing this work by hand would have required several stages in the welding stage alone with several re-heats in between and likewise in drawing out to length and tapering.

Chapter 7

Manufacture of the Barrel
Part 2 – finishing the barrel

7.1 Introduction

By the early 19th century various mechanical methods had been developed for finishing the tube inside and outside. It is important in this context, therefore, to quote from a memorandum sent by James Burton to Colonel Dixon, Inspector of Small Arms, on 10th September 1855 covering this very topic:

> *I beg leave respectfully to submit for your information the accompanying brief statement, exhibiting the numbers, description and probable cost of the machines yet required for this establishment, in order to complete the system of machinery for the manufacture of small arms. No facilities having as yet been provided for the finishing of the barrels.* (see Appendix 3)

In the later 18th century through to the 1840s various attempts were made at the Enfield Lock barrel mill to replace grinding with lathe turning and a detailed study of this initiative has been made.[1] It only met with limited success and grinding held sway, partly because of deep-rooted tradition in which skilled craftsmen could carry out the work with a sufficient accuracy and speed born of long experience. Such was the importance of grinding that it persisted both at Enfield and even more so in the private trade for many years.

7.2 Equipping the Enfield Factory

Dixon authorised Burton to purchase the necessary machines and on November 14th, 1855, an order was placed with Robbins & Lawrence (see Appendix 2). This was at a crucial time in the fortunes of Robbins & Lawrence and in December 1855, Captain Jervis, who had been sent to America to oversee a contract between Fox, Henderson & Co. and Robbins & Lawrence for Enfield Pattern 1853 rifles, reported that Robbins & Lawrence had become insolvent and almost a year later, in November 1856, Robbins & Lawrence had stopped work (Appendix 8).

MAKING THE ENFIELD PATTERN 1853 RIFLE-MUSKET

It has already been recorded in Chapter 3 that any doubts as to whether Robbins & Lawrence were able to supply any machines contained in their extensive contracts before their demise are dispelled by the comment in the report of the American 'Military Commission to Europe' in 1855-56:

> ...the the names of "Ames", of Chicopee, Massachusetts, and "Robbins & Lawrence", of Windsor, Vermont, &c, are accordingly to be read on most of the machines at Enfield.

No records of receipts have been located and the nature of the machines actually supplied can only be gained through inference from oblique references. Nevertheless, a study of Burton's order (Appendix 2) sheds some light on the machining operations he envisaged for finishing the barrel:

> *8 Rough boring machines for the first cut after the barrel is welded, of 3 spindles each, and each to be furnished with one set of tools called nut augurs.* [nut augurs (sic) were basically spiral-fluted reamers, not the augers used in woodworking]
>
> *4 Second boring or reaming machines of 1 spindle each, and each to be furnished with one set of reamers.*
>
> *6 Finish boring machines of 2 spindles each, each to be furnished with one set of reamers.*
>
> *1 Lathe for turning the barrel whole length, with one set of tools.*
>
> *2 Double hand lathes, each to contain two head and tailstocks, with all the tools &c for squaring & cutting the barrel to exact length.*
>
> *12 Milling machines for milling the barrel (breech &c).*
>
> *17 sets of fixtures for the above 12 milling machines, and 21 sets of mills or saws for the various cuts upon the barrel &c.*
>
> *4 Drill presses for drilling & counterboring the tang & cone seat; each to be furnished with drills and fixtures for holding the barrels.*
>
> *2 Machines for milling the end of the breech screw to diameter & length, with tools, cutters & fixtures for holding the tang.*
>
> *2 Machines for counterboring the breech of the barrel for breech screw; each with tools and fixtures for holding the barrel.*
>
> *2 Machines for cutting the thread upon the breech screw, each to be furnished with one set of Dies and taps.*
>
> *2 Machines for tapping and countersinking the cone seat, with bits and taps complete.*

MANUFACTURE OF THE BARREL

1 Machine for polishing the interior of the barrel; to be furnished with one lot of rods, connecting rod, balance wheel and all the overhead works complete.

4 Machines for polishing the exterior of the barrel of 5 spindles each, with all the overhead works complete.

2 Machines for clamp milling the muzzle of the barrel for bayonet socket, with dies and reamers complete.

1 Machine for grinding the exterior of the barrel, complete.

Notably absent from this list are rifling machines which will be discussed later.

7.3 Operations

First rough-boring

In consideration of the whole process of manufacturing any barrel, creating an accurate bore at this stage would facilitate later operations. Indeed, the observation in relation to activities at Springfield Armory is applicable:

> *It is a general principle however that the inside work is kept always in advance of the outside, as it is the custom with all machinists and turners to adopt the rule that is so indispensable and excellent in morals, namely, to make all right first within, and then to attend to the exterior.*[2]

The barrel first had to be annealed to eliminate any work-hardening that had developed during rolling, lumping, and straightening before work could commence. In describing manufacture in Birmingham, one account[3] records that *rough boring* was carried out:

> *by steel 'bitts', properly hardened and tempered, tapering in the first instance from about half to three-eighths of an inch, the cutting part being about eighteen inches long. These revolve at the rate of 500 times per minute, and three or four of them are used...*

This 'bitt' approximates to the boring bit in use at Enfield in 1854, Fig. 7.1.

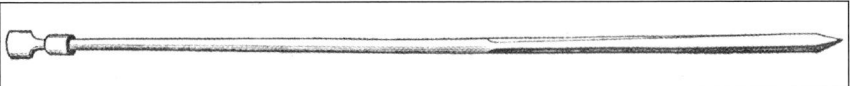

Fig. 7.1. The boring bit used at Enfield prior to 1857. (Jervis, 1854, p. 5)

175

MAKING THE ENFIELD PATTERN 1853 RIFLE-MUSKET

This type of tool had become universal but later, at Enfield, it is clear that new machines had been introduced and rough boring had become a two-stage process, the second stage removing less metal.[4] Whether these machines were from Robbins & Lawrence is not known but there is an intriguing order from Colonel Dixon at Enfield in the Greenwood & Batley order book of 1859 (Appendix 4) for:

> *2 machines for first boring barrels with self-acting feed motion and <u>same as special machine at Enfield</u>*
> *1 machine for second boring barrels with self-acting feed motion and <u>same as special machine at Enfield</u>* [author's underlining]

This suggests that some different machines, possibly from Robbins & Lawrence, were already in use at Enfield. That the machine at Enfield accorded with the items on Burton's list is supported by the observation that:

> *The tools themselves are shaped like a screw, tapering slightly away from the cutting edge, so as to ensure the borings being left behind in the barrel as they creep along their course.*[5]

This is at least suggestive of a Robbins & Lawrence origin.

These 'bits', the 'nut augers' of Burton, had spiral cutting heads six inches long welded to a shaft sufficiently long to pass through the barrel. In America this process was referred to as 'nut-boring' and Benton also comments that at Springfield Armory the barrels were bored with two 'twist augers'.[6] Burton specified the 'rough boring' machines *to be furnished with one set of tools called nut augurs.* Unfortunately, none of these accounts illustrates one of these bits or augers and the only illustration located is that in the US Patent granted to Pettibone in 1814,[7] Fig. 7.2.

In 'rough boring', the nature of the crude bores in which rough boring commenced meant that barrels could only be set approximately to the centre-line of the machine but, as will be seen in some images to follow, there was sufficient metal to provide a 'machining allowance' to accommodate slight misalignment. *The Engineer*[8] and *Mechanic's Magazine*[9] report that at Enfield the barrels:

> *...are fixed in sets of four in an iron trough half-filled with water. The four spindles or rods, to which the boring tools are fixed, are inserted into the rough barrels, and as they revolve the trough is drawn, by a self-acting motion, slowly back, taking the barrels with it; by this means the boring tools make a very slow cut.*

Fig. 7.2. The boring tool illustrated in Pettibone's patent.

MANUFACTURE OF THE BARREL

This differs from the machine specified by Burton, which had three spindles, but it was noted in *Mechanic's Magazine* that this four-spindle machine was *of American extraction* and is again suggestive that part of the order with Robbins & Lawrence may have been fulfilled, though in modified form. Two of these augers, or bits, of different diameters were used in the first rough boring stage. No illustration of the machine is provided but they were undoubtedly closely similar to those in use at Springfield Armory which are described and illustrated by Benton in 1878,[10] Fig. 7.3, and which Fitch notes[11] as being similar to that designed by F. Howe at Robbins & Lawrence in 1850. The major difference was that the Springfield machines were fed by hand, not by 'self-acting motion'. Another difference in American practice was that Fitch also states the trough contained 'soda-water' – not the popular drink, but a solution of sodium carbonate or sodium hydroxide which had the advantage that, like many alkaline solutions, it would not corrode ferrous metals. A similar practice was followed in the Sheffield tool and cutlery trades where, after grinding, blades were usually dipped in lime-water and left to dry, leaving a thin protective white film of lime on the blade.

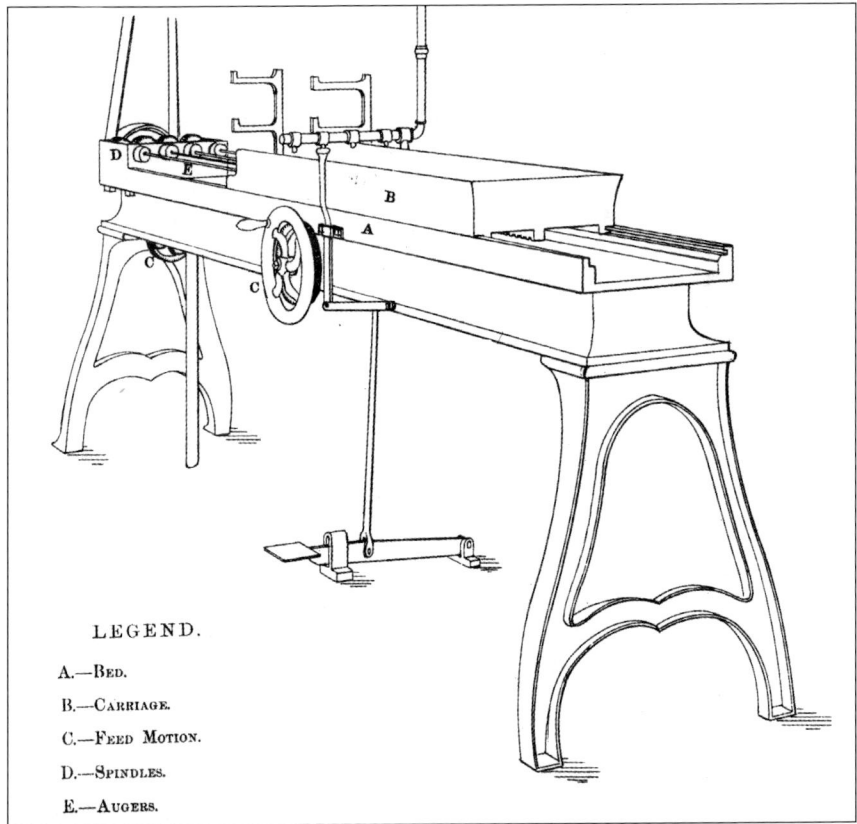

Fig. 7.3. A 4-spindle barrel 'rough' or 'nut boring' machine used at Springfield Armory in 1878. (Benton, 1878, plate. XVII)

At Springfield as recorded by Fitch, and at Enfield also, judging from the description above, in the first rough-boring the bits were pulled through the barrel which, being under tension, would help to keep them straight. The use of a four-spindle machine is recorded elsewhere,[12] and also notes that after rough boring, the barrel was again 'set and straightened' with hammers by hand.

Second rough-boring

Burton's memorandum and subsequent order to Robbins & Lawrence (Appendix 2) makes it clear that there was to be a second stage in the rough-boring process:

> *Second boring or reaming machines of 1 spindle each, and each to be furnished with one set of reamers*

and this view is supported by the 1859 order from Dixon to Greenwood & Batley:

> *1 machine for second boring barrels with self-acting feed motion and same as special machine at Enfield* (Appendix 4, item 826)

This second stage is not noted in contemporary commentaries and the use of the term 'reamers' shows this to be an altogether different process and similar to those used in 'fine boring' which followed later.

Barrel cut to length

At Springfield this was the point the *rough ends left by the circular saws* on the barrel were *squared up in a milling machine.*[13] However, Burton's order to Robbins & Lawrence on 14th November 1855 called for:

> *Double hand lathes, each to contain two head and tailstocks, with all the tools &c for squaring & cutting the barrel to exact length.* (Appendix 3)

This would have fulfilled the same requirements as a milling machine. Performing this operation at this point would have certainly been logical in view of the comment, *turning the exterior is the point now to be attended to*[14] when describing the process at Enfield and which was also the next stage at

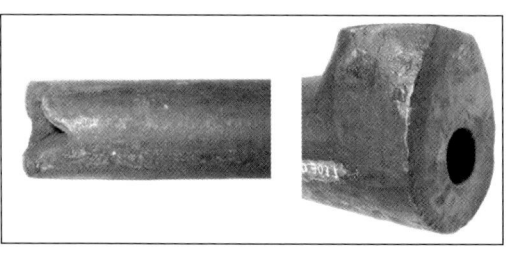

Fig. 7.4. Muzzle and breech of a barrel after the 'hot processing'. (2002.D.703.1 courtesy Birmingham Museums Trust)

Springfield.[15] The need for cutting to length and 'squaring' muzzle and breech is clearly shown by muzzle and breech after the hot processing and in which state they were rough-bored.

Rough turned outside

In fixing the barrels in a lathe for the purpose of turning the outside, they are centred by means of a cone inserted in the end.[16]

The 'cones', more commonly referred to as lathe 'centres', were of hardened steel and having the muzzle and breech square with the bore were prerequisites for setting up the barrel between 'centres' so that the bore was as coaxial as possible with the centre line of the lathe. This, in turn, ensured the external surface was coaxial with the bore and the criterion of a uniform wall thickness would also be met.

Unfortunately, Burton's order only states the purpose of the machine - *1 Lathe for turning the barrel whole length, with one set of tools* (Appendix 3) - and offers no enlightenment of its operational characteristics. However, from two descriptions of this operation[17, 18] it is clear that this was a taper-turning operation. With the barrel set up in the lathe, the slide rest carrying the turning tool followed a flat guide set at an angle to the lathe axis to create the correct external taper and a similar method was also used to allow the 'steady', which supported the barrel opposite the turning tool to avoid any flexure, to match the changing barrel diameter as turning progressed.

No illustrations of this lathe have been found but two machines used elsewhere can be mentioned. In 1818 in America, Sylvester Nash[19] patented a self-acting lathe for taper-turning musket barrels. However, it differed in a number of ways from the lathe described in use at Enfield and what are considered its flaws have been discussed in a companion volume:

1. It was the mandrel on which the barrel was mounted, not the barrel itself, which was supported on centres at each end.
2. The steadies were stationary and only brought into use when the saddle carrying the turning tool had passed them.
3. The steadies were only applied to the slender part of the barrel, not over its full length.
4. The turning tool was lifted by the guide bar to create a taper and would therefore not have had a consistent centre-height, leading to difficulty in cutting metal at some points in its travel along the barrel.

In contrast the lathe used at the Tula weapon factory in 1826, although a close contemporary of Nash's lathe, was more sophisticated and corresponds more closely with that described as being used at Enfield.

In the lathe at Tula the carriage was constructed with sliding front and rear cross-slides, each guided by equally-angled guide-bars which acted so as to

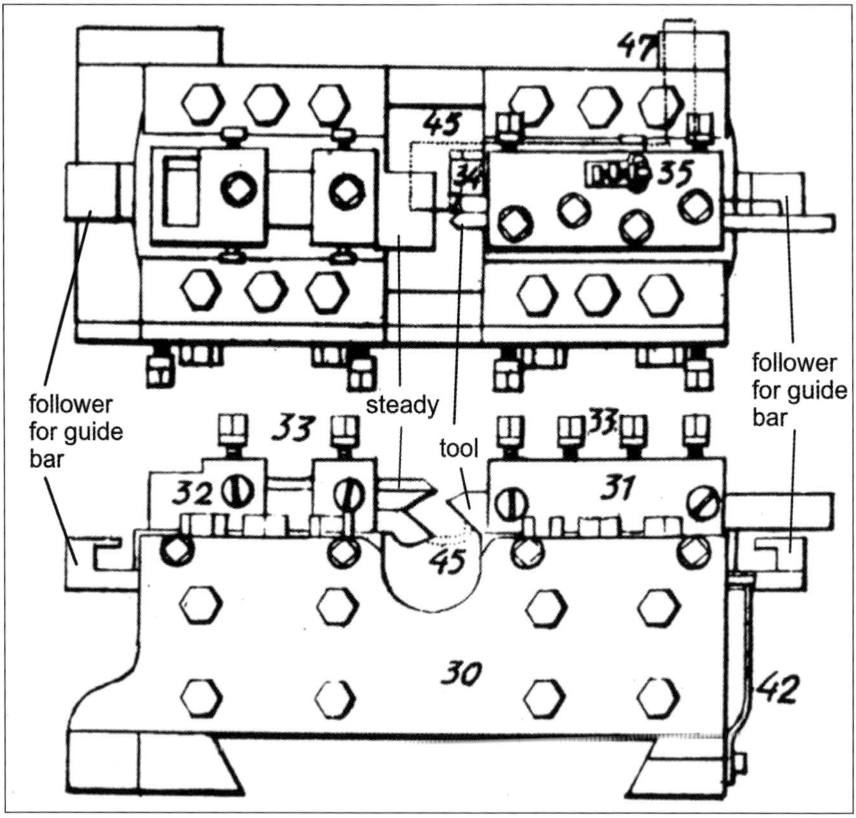

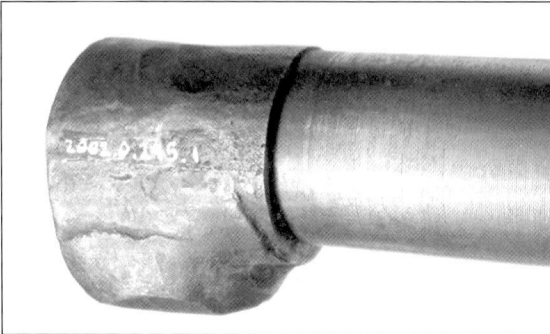

Above: Fig. 7.5. Annotated elevation and plan views of the carriage of the barrel turning lathe at Tula. (from Gamel, 1826, plate VII)

Left: Fig. 7.6. Barrel turned up to the lump. (2002.D.295.1, courtesy Birmingham Museums Trust)

enable the turning tool to impart a taper and the steady to follow that taper as the carriage traversed the barrel, Fig. 7.5. However, as with Nash's lathe, the barrel was mounted on a mandrel which was carried on the centres. Unlike the Tula flintlock barrel, which had no projections, the Enfield barrel could only be turned up to the 'lump'.

This illustration also shows the excess metal in the raw barrel as noted earlier.

MANUFACTURE OF THE BARREL

Grinding

After turning barrels were again cold-hammered to straighten them if necessary and were then ground against large and wide sandstone grinding wheels to remove tool marks. In Birmingham, barrels were traditionally held loosely at each end and allowed to rotate in contact with the grindstone. A *lumped-barrel*, as opposed to a plain flintlock barrel, would have required extra care but the barrels of the 1851 and 1853 rifles[70,21] were 'hand-ground' in Birmingham.

At Enfield the barrels were also ground but were fitted to a mandrel and rotated by means of a small crank handle. This method was used at Springfield Armory in the United States in 1852, Fig. 7.8, and is reminiscent of 18th century French practice.[22]

Fig. 7.7. Grinding barrels for the 1851 'Minié' rifle at Birmingham. (I.L.N., 1851, p. 85)

Fig. 7.8. Barrel grinding at Springfield Armory. (Abbott, 1852, p. 8)

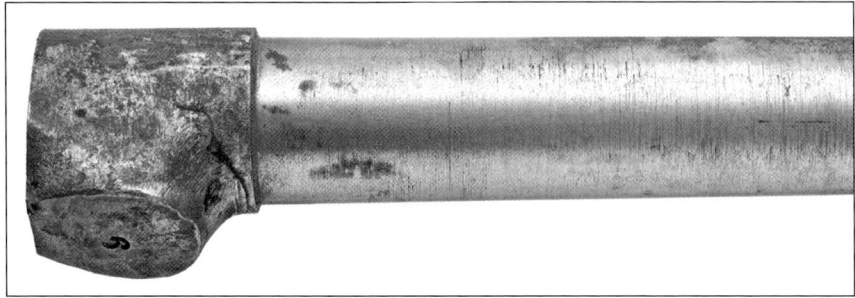

Fig. 7.9. Barrel after grinding. On the actual item it is possible to clearly see the 'swirled' marks left by moving the barrel side to side across the face of the rotating grindstone. (2002.D.299.2, courtesy Birmingham Museums Trust)

Grinding might seem detrimental to the accuracy achieved by turning but it was claimed that grinders could approach lathe-turned accuracy.[23] As noted at the commencement of this chapter there were abortive attempts to replace grinding by lathe-turning at Enfield during the period 1780–1840.

After grinding, the bore was viewed again and, if necessary, the barrel straightened by a judicious blow or two from a hammer.[24]

Fine boring

As this term implies, it was a more delicate process than 'rough boring' and gradually brought the bore close to its finished size and condition. It was carried out in two stages, before and after finishing the outside of the barrel. The earliest description of this process applied to the pattern 1853 barrel in Birmingham contents itself with a simple commentary:

> ...the barrel comes into the hands of the fine borer, who works with a different kind of 'bitt', not taper, and having only one cutting edge, about fourteen inches long.[25]

However, a description of the operations at Enfield at a later date provides more details:

> A square tool or rimer, some 15 in. long, is passed through them, the barrels, still being fixed and the tool revolving. The spaces left by the square fitting into the round hole are filled up with strips of wood, and the cut is regulated by the interposition of a thickness of paper.[26]

This was superficially similar to the second rough-boring operation but incorporated a simple technique for enlarging the bore by very small increments. After the first passage of the square tool through the bore it was 'spilled-up', meaning that it was packed with a 'spill', or sliver, of oak so as to cause the boring bar to have its cutting

MANUFACTURE OF THE BARREL

edge pressed against the walls of the bore. This process acquired the traditional term 'spill-boring'. The spill was held in place on the boring head by a ring and, after being well greased, was entered into the bore, the ring being gradually pushed off and the barrel bored once more.[27]

The action of a 'spilled' boring tool is shown in Fig. 7.11. It only cuts on one corner, the other two being covered by the spill, while the fourth has been rounded and polished to burnish the bore.

Very fine adjustment of bore size was accomplished by the insertion of slips of paper between the spill and the tool and it was reported that this process allowed the bore diameter to be regulated to one-thousandth of an inch.[28, 29] As to the nature of the machine used, Burton again specifies its purpose in the order to Robbins & Lawrence, but not its functional characteristics:

6 Finish boring machines of 2 spindles each, each to be furnished with one set of reamers. (Appendix 3)

In 1861 it was recorded that *one barrel only is operated upon at a time*[30] and is indicative that the two-spindle machines for fine boring ordered from Robbins & Lawrence may not have been delivered. The reamers referred to can be taken to mean the square spill-boring tools since that is how they basically functioned – as reamers – and revolved at 70 rpm. It has been stated that the carriage to which the

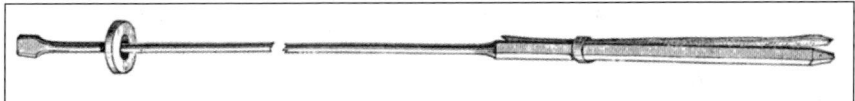

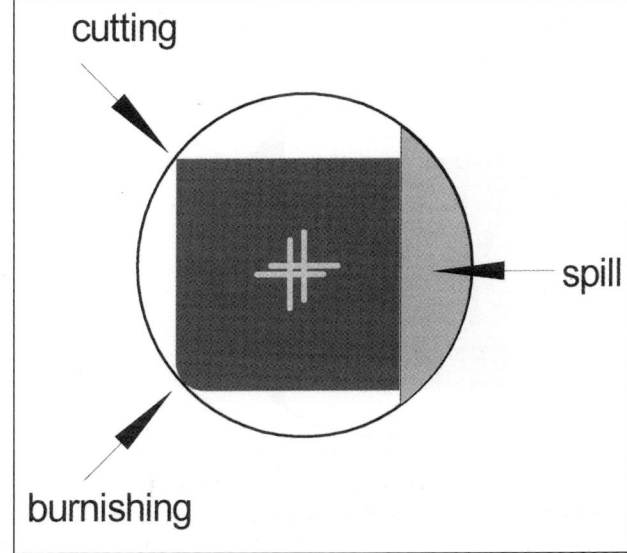

Above: Fig. 7.10. A boring bar 'spilled-up' with the spill retaining ring. (Greener, 1884, p. 273)

Right: Fig. 7.11. The action of a 'spilled' boring tool. (© P. Smithurst)

Fig. 7.12. Fine-boring machine used in the private trade. (Greener, 1884, p. 272)

barrel was fitted was drawn along by a 56 lbs. (25.4 kg) weight.[31] This may seem crude but had the advantage over a 'self-acting', i.e., mechanical, feed in that the rate of feed was self-regulating and allowed the speed of passage of the boring tool to adjust itself if it encountered 'hard-spots' in the bore and not be forced through with possibly detrimental effects. No illustration of one of these machines used at Enfield has been found but a cruder version used in the private gun trade is illustrated by Greener, Fig. 7.12.

Whilst lacking the sophisticated cast iron construction of the Enfield machines, it operated on the same principles and the chains to which the weights are attached are clearly visible.

Breeching

This includes a number of processes on both barrel and breech-pin:

- Counterboring the breech and tapping to receive the breech-pin.
- Machining the cylindrical body of the breech-pin to size and threading.

Various points become obvious from an examination of a finished barrel:

1. The vertical centre line of the tang and heel of the breech pin is parallel with the side face of the bolster.
2. The abutment faces of tang and heel are seated firmly on the end face of the barrel.

These features are apparent in Figs. 7.13 and 7.14 and become important reference points for the fitting of front and rear sights.

MANUFACTURE OF THE BARREL

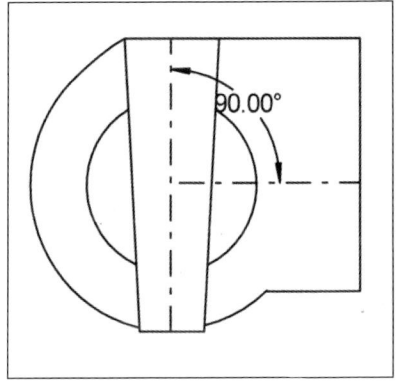 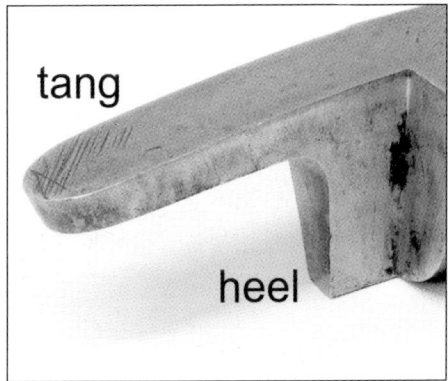

Above left: Fig. 7.13. End elevation of barrel with breech pin fitted. (© P. Smithurst)

Above right: Fig. 7.14. Joint between breech pin heel and tang and rear face of the barrel. (© P. Smithurst)

Two more features which are not visible but of equal importance were:

3. The threaded portion of the breech pin had to be seated on the shoulder formed in the counter-bore.
4. The threads in barrel and on breech pin had to be a very close fit.

Neglect of either of factors 3 and 4 would create crevices for the accumulation of powder residues which would be, in one case difficult and in the other, impossible

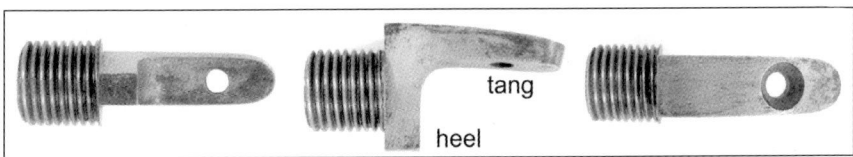

Fig. 7.15. The Enfield Pattern 1853 breech pin [plug] in its finished machined state. (author's collection)

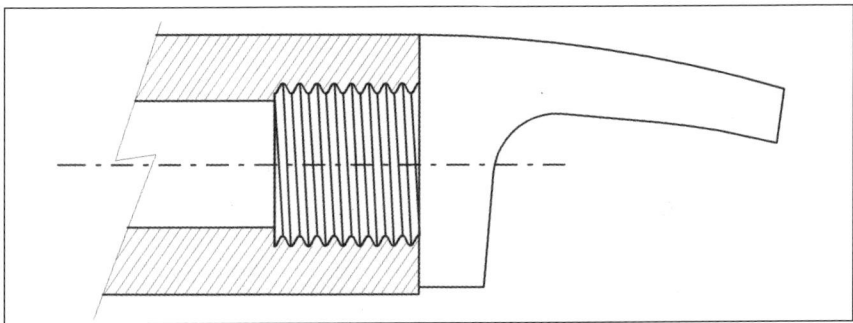

Fig. 7.16. The 'fit' requirements of the Enfield Pattern '53 breech pin/plug. (© P. Smithurst)

to remove by standard cleaning procedures, leading to corrosion at these most highly-stressed parts of the barrel. Item 4 is embodied in the specification by Pellat quoted at the beginning of this chapter – *the breech pin to be cylindrical, clean screwed, and not to shake on the second thread.*

To these criteria can be added two more:

5. The face of the breech had to be at 90° to the axis of the bore and counter-bore.
6. The faces of the tang and heel of the breech pin had to be coplanar and at 90° to the axis of the threaded body.

Individually, these six criteria might appear simple but collectively they presented serious manufacturing challenges which have not been addressed in any account to date.

The problem lay in the very nature of screw threads. It is a simple fact that in fitting a bolt into a nut, one or other has to be rotated until the threads engage. The same is true of fitting the breech-pin into the breech. The difference lies in the fact that when the breech pin is screwed in tightly, the end face of the breech pin had to meet the shoulder at the end of the counterbore and the tang and heel of the breech pin must not only be tightly up against the breech face of the barrel, but they must also be in correct alignment with any chosen datum face of the barrel. This is like saying that when the nut is screwed as far as it will go on the bolt, a chosen face of the hexagonal nut must coincide exactly with a chosen face of the hexagonal bolt head.

Any screw thread has a 'starting point'. A screw of 20 threads per inch which is exactly 0.5 inches long will, obviously, have 10 threads. Entering it into a hole with a matching 20 threads per inch, it will rotate exactly 10 times and enter exactly 0.5-inches. However, reverting to the nut and bolt analogy, unless the threads on both items start in exactly the correct place, it is impossible to obtain alignment of any chosen datum faces. In the example shown, Fig. 7.17, the possible starting points of threads in nuts and bolts are shown as a grey line and their angular variations are infinite.

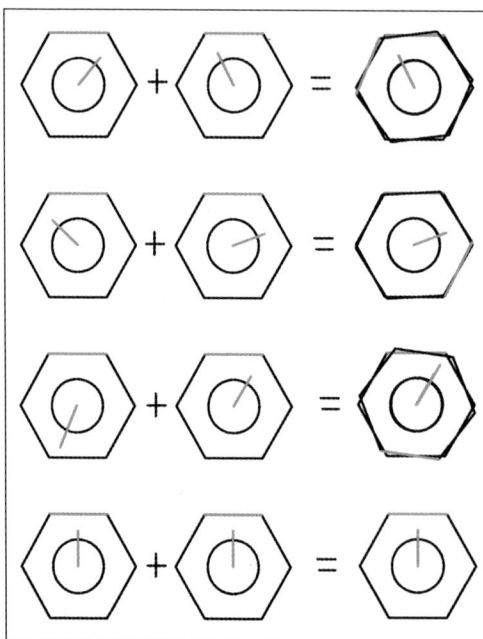

Fig. 7.17. Effect of thread starting points on positional accuracy of two components. (© P. Smithurst)

MANUFACTURE OF THE BARREL

Some may appear to give alignment of the faces but this as a misconception. In only one case, the last in the illustration, in which the thread starting points coincide, does true alignment result. The chance of correct alignment, unless very special care was taken whether the bolt and nut are threaded by hand or machine, would be infinitesimal. It is that special care which is the crux to solving this problem.

This alignment was the requirement of breeching and could only result from very accurate machining. It was not something that could be achieved by hand-threading, despite the statement in at least one account,[32] and was certainly recognised by Burton in his memorandum and subsequent order to Robbins & Lawrence:

2 Machines for milling the end of the breech screw to diameter & length, with tools, cutters & fixtures for holding the tang.

2 Machines for counterboring the breech of the barrel for breech screw; each with tools and fixtures for holding the barrel.

2 Machines for cutting the thread upon the breech screw, each to be furnished with one set of Dies and taps. (Appendix 3)

For the reasons referred to earlier, it is unlikely that these machines were delivered from Robbins & Lawrence, leaving the question of how this was carried out.

Machining the breech pin

That question is partly answered by an entry in the Greenwood & Batley order book on 5th September 1857 for:

1 m/c for threading breech screw (Appendix 4)

'Breech screw' is taken to mean the 'breech-pin'. Considering the nature of the work and the workpiece, this order would undoubtedly have included a feature referred to by Burton in his order to Robbins & Lawrence, namely 'fixtures for holding the tang'. Without any further details of the Enfield machines used for making the breech pin, but considering the nature of the required criteria, the only comparisons that can be made are with those used at Tula (Gamel, 1826). These, with additional refinements, provide a logical basis for postulating methods employed at Enfield.

At Tula, the breech-pin was forged by hand. Given its unusual shape, whilst some preliminary shaping might have been done by hand at Enfield, it would almost certainly have been finished using dies in a drop-stamp. This would have the advantage of creating an item of reasonably uniform size for subsequent work. Also, any forging needs to be sufficiently over-size to allow for subsequent machining to bring it to correct size and form and this 'over-size' became important in later stages. Dedicated machines were used at Tula for turning and threading the breech pin and this would certainly have been the practice at Enfield where repeatable accuracy was being sought. Burton had envisaged milling machines for bringing the 'body' of the breech pin to correct length and diameter for subsequent threading. This implies the use of a hollow end-mill to form the cylindrical body and face the heel and tang faces, followed by a second operation to face the end of the breech pin body to correct length. In both cases, the workpiece would have to be accurately positioned.

At Tula, the breech pin had small conical centre holes made, coinciding with the axis of the cylindrical portion, Fig. 7.18.

These centre holes were for mounting the breech-pin forging on centres in the lathe, one forming part of a fixture to accommodate the tang, but would have served equally well for accurate location in a milling machine. Burton's reference to 'fixtures for holding the tang' is reflected in the Tula machine, Fig. 7.19, taking the form of a special fixture on the nose of the lathe mandrel, 2, accommodating the tang of the breech pin, 13, which was used to rotate it while the centre holes created on the pin were used to locate it and support the outer end.

Such an arrangement allowed the machining of the cylindrical 'body' to correct diameter and length and also the faces of the heel and tang, ensuring a coplanar and 90° interface between the two. As noted earlier, the same outcome would have been possible using a dedicated milling machine at Enfield.

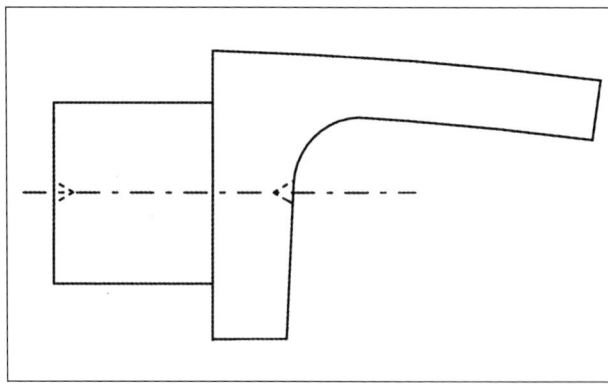

Left: Fig. 7.18. Drawing showing centre holes in the Tula Breech-pin. (© P. Smithurst)

Below: Fig. 7.19. The fixture for the breech pin in the Tula lathe. (Gamel, 1826, Fig. XII)

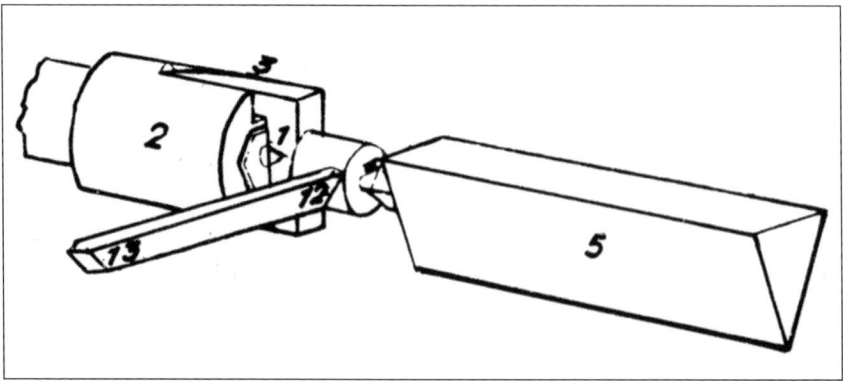

Threading the breech pin

The threading of the breech pin at Tula was carried out using a similar machine but in which the tailstock took the form of a set of adjustable dies.

The breech pin was securely held in a similar fixture on the end of the lathe mandrel, which was capable of sliding in its bearings, Fig. 7.20. The mandrel was set to revolve slowly and the breech pin fed into the dies, 4, held in a special fixture, 20, for the thread to be cut. Once the threading had begun, the sliding mandrel allowed the rotating breech pin to be drawn into the die. This machine was fitted with an automatic disengage and reverse mechanism actuated when the threading had proceeded to the required extent.

A similar machine would have fulfilled Burton's requirement for:

> *Machines for cutting the thread upon the breech screw, each to be furnished with one set of Dies and taps.*

However, there is no indication that the Tula machine allowed every breech pin to enter the dies at exactly the same point, so the thread could have started anywhere on the periphery, a factor of little consequence in a flintlock barrel without a 'lump', but unacceptable in the case of the Enfield breech pin.

It would have been possible to overcome this deficiency by the application of 'indexing', that is, a repeatable and accurate radial alignment of dies and breech pin at the commencement of threading. The fixture holding the breech pin would have required a means of ensuring that every breech pin was always presented to the die at the same position. Then, setting the lathe in motion whilst pressing the pin against the dies, the threads would be cut, starting at the same point on the breech pin body, 'square' with the axis and of the same length to match the recess they had to fit.

Machining and threading the breech counter-bore

Burton's concept of performing this operation is embodied in his order for:

> *2 Machines for counterboring the breech of the barrel for breech screw; each with tools and fixtures for holding the barrel.*

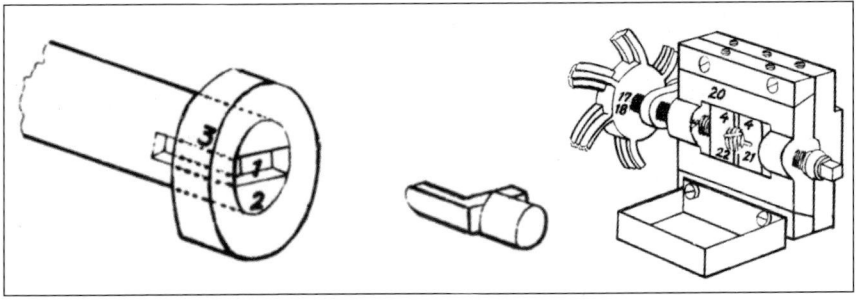

Fig. 7.20. Details of machine for threading the breech pin at Tula. (Gamel, 1826, Fig. XIII)

MAKING THE ENFIELD PATTERN 1853 RIFLE-MUSKET

Any machines would need to ensure the counterbore was machined to the same depth and be threaded to match the starting point of the thread on the breech pin. Apart from what the machines were required to accomplish, further details are not recorded. However, it is again possible to draw comparisons. Machines for performing these operations were used at Tula[33] and later at Liège.[34] Both utilised similar principles, each of which would have been applicable to the Enfield barrel. However, while the barrels made at Liège, being 'percussion' and having a nipple bolster, had closer similarities with the Enfield barrel, the nature of the breech pin did not require any particular orientation since it did not have a tang. On the other hand, the barrel in the Tula illustrations is clearly octagonal at the breech and would be expected to require special care in threading the breech and the breech pin to achieve correct orientation with one of the octagonal faces, but there is no reference to that requirement in the text.

The principal features of the machine used at Liège are shown in Fig. 7.21.

In this drawing, F is a flange integral with the lathe mandrel which has a bore of sufficient size to receive the barrel. M is a plate attached to this flange by bolts, L and has a hole shaped to match the cross-section of the breech-end of the barrel. The barrel was inserted and passed down the hollow mandrel until the breech face was approximately flush with the face of M and was then clamped in place by the bolt, K, bearing on the 'lump' or bolster. The bolts, L, screwed into the flange, F, passed through oversized holes in M, allowing M to 'float' on the flange, F, and enabling the barrel to be aligned on the pilot, P, which was exactly coaxial with the mandrel centre-line. The bolts, L, were then tightened and the centred barrel was ready for subsequent operations. The bar with the pilot was then replaced by the counterbore having an equal square shank and a pilot extension to guide it, Fig. 7.22.

Had all the tools been carefully made with square shanks of the same size, they would have been interchangeable on the machine.

Counterboring was followed by threading using a tap with a pilot extension, to maintain alignment with the bore and, a square body, Fig. 7.23.

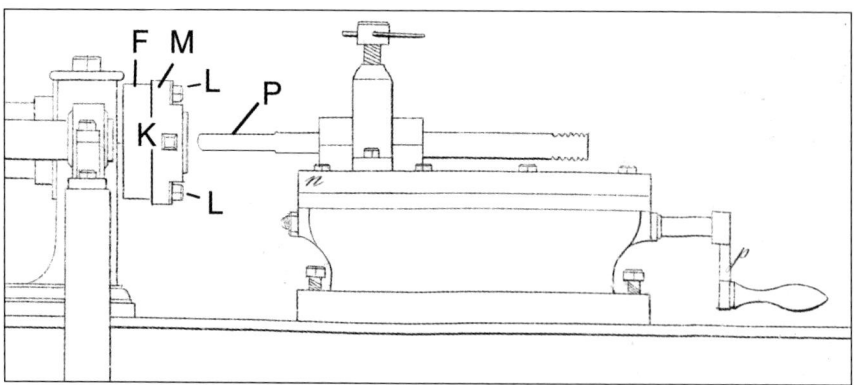

Fig. 7.21. Annotated detail of the machine at Liège for counter-boring and threading the breech. (from *Memoria*, 1850, plate 12)

MANUFACTURE OF THE BARREL

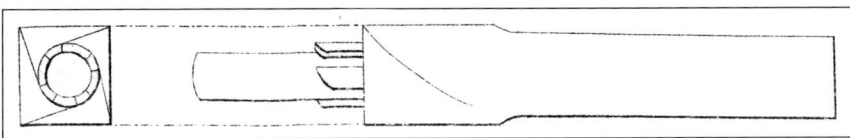

Fig. 7.22. Counterbore used at Liège with a square head and shank. (*Memoria*, 1850, plate 13, Fig. 3a)

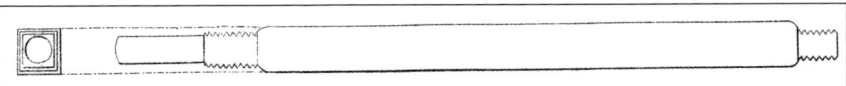

Fig. 7.23. The first tap used at Liège with its pilot. (*Memoria*, 1850, plate 13, Fig. 4a)

It was then returned to the machine and the breech faced with a face-milling cutter with a pilot. To face the breech *after* counterboring and threading would have created counterbores of different depths. The Spanish musket had a totally different breech pin from the Enfield rifle and such variations were of little consequence but would not have been acceptable for the Enfield rifle.

At Tula, a more sophisticated machine was used. The barrel, previously having had the breech faced and counterbored to correct depth in another machine, was mounted in a special fixture, centred by a pilot, and fixed by clamping screws.

This machine also had a sliding spindle and threading was done under power but intermittently, each partial rotation of the spindle screwing the barrel onto the tap, the spindle sliding forward at each stage. The threading machines for the breech at Tula and at Liège both used taps with a square shank which could be manufactured in such a way that the threads started at a specific point in relation

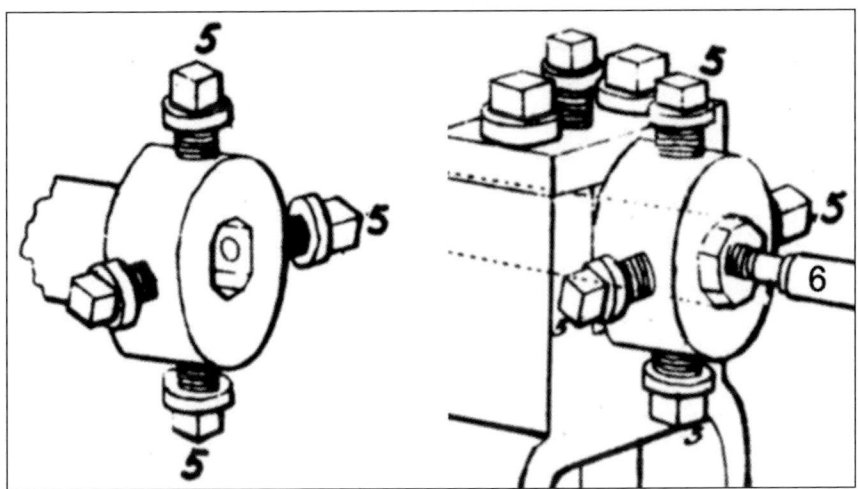

Fig. 7.24. Threading the breech at Tula - the barrel centred and secured in a special fixture on the lathe mandrel nose. (from Gamel, 1826, Fig. X)

to a datum. Both also worked on a barrel with an identifiable datum feature. Both could therefore be provided with an indexing facility to ensure that the threads formed in the breech started at the same point in each barrel. Combining this with a modified Tula machine for cutting exactly-positioned threads on the breech pin would have performed the required tasks at Enfield.

To have enabled this to have worked at Enfield, some preliminary machining of the 'lump' or 'nipple bolster' would have been necessary to create the required datum face/s for successful operation. The contention that this actually took place between the second 'rough' boring and turning operations and before breeching commenced is supported by the images below. The first, Fig. 7.25, shows barrels after the first and second 'rough' borings to have an un-machined 'lump'

The next image, Fig. 7.26, shows two barrels after counterboring and threading. Note in the left-hand image the burr left by milling the top face of the lump, indicating this took place after machining the barrels to length and fine boring; likewise, the burr raised on the end face after threading shows threading took place after the end face was machined.

To hold the Enfield barrel, a device similar to that used at Liège is envisaged and in which there is a shaped recess to accommodate the 'lump', using the top and side machined faces of the lump to act as a datum faces which would be of importance when it came to 'starting' the threads.

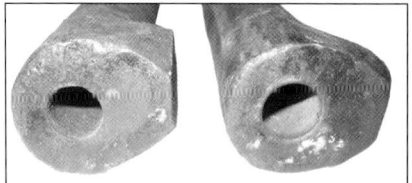 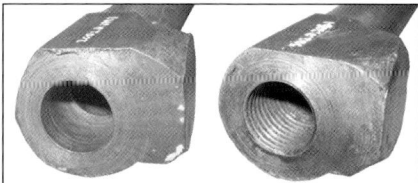

Above left: Fig. 7.25. First and second 'rough' boring of the Enfield barrel with the 'lump' still in its forged state. (Courtesy Birmingham Museums Trust)

Above right: Fig. 7.26. Enfield barrels after counterboring, left, and threading, right. (Courtesy Birmingham Museums Trust)

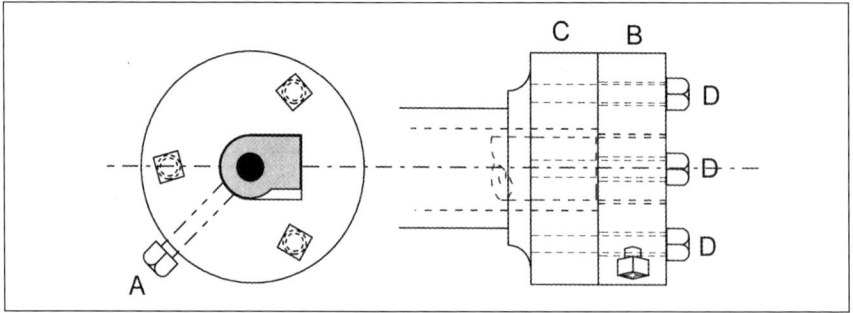

Fig. 7.27. Suggested fixture for a lathe spindle to accept the breech of the Enfield barrel (shown in grey) (© P. Smithurst)

MANUFACTURE OF THE BARREL

The top and side faces of the 'lump' would be machined square with each other and at a set distance from the centreline of the bore. That would allow the barrel to be clamped in place using the bolt, A, in the front part of the fixture, B. This part of the fixture would 'float' on the mounting plate, C, by virtue of its securing bolts D, passing through over-sized holes, to allow the bore to be aligned with a guide pin in the tailstock, the guide pin entered into the bore and the bolts, D, tightened.

The barrel could then be counterbored followed by tapping. By establishing a suitable datum on the barrel mounting that coincided with a suitable datum on the tap, threads would always have closely matching starting points with those of the breech pin. Following these procedures, the breech pin would, when tightened, approach correct alignment.

In any machining operation there are always sources of small errors due to tool or machine wear. This could have been allowed for in the breech-pin forging being oversize so that any small machining error, which placed the breech-pin heel and tang slightly askew, as shown in Fig. 7.28, could be corrected.

The anticipation of error is reflected by some tools ordered from the Ames Company: *two filing jigs for breech.*[35]

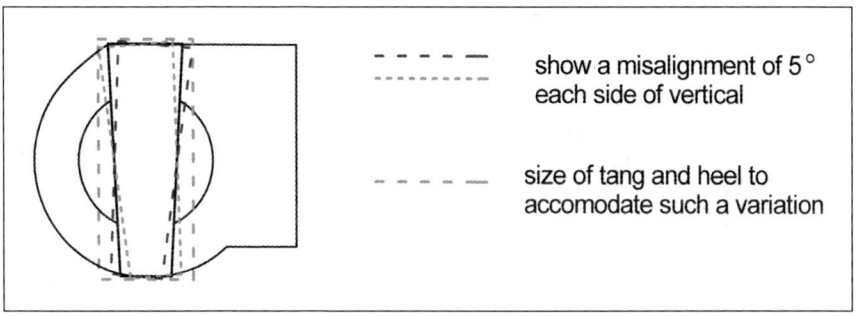

Fig. 7.28. Any misalignment from machining errors could be allowed for in an over-sized breech pin forging. (© P. Smithurst)

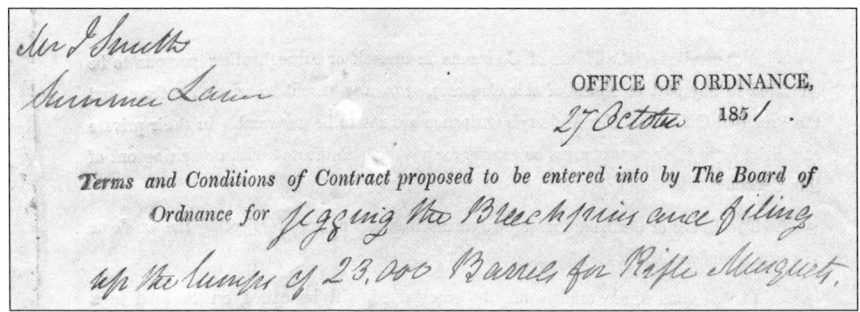

Fig. 7.29. Excerpt from contract to Joseph Smith, 1851, for *Jegging breech pins and filing up the lump* of the Pattern 1851 rifle barrel. (Appendix 2, item 1, author's collection)

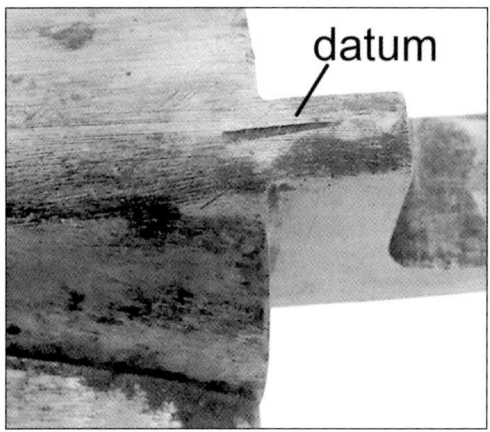

Fig. 7.30. Datum line on breech pin and barrel. (author's collection)

It is another instance showing that 'mechanisation' did not totally eliminate hand skills and reflects an earlier private contract.

Indicative of the challenges faced in 'breeching' is that, even after the rifle became 'interchangeable', that description never applied to the breech pin. Once fitted to a barrel, they became an entity in their own right and, in view of the need to separate them on occasions during manufacture, it was standard practice to inscribe a datum line across the heel and the barrel so they could be re-fitted accurately.

Anyone who has tried to fit a replacement breech pin to any barrel of this nature will understand the difficulties involved.

Gauging the 'breeching'

Various gauges were supplied by the Ames Company for the breeching process:

> *gauge for testing the counterboring of the breech*
> *gauge for testing the tapping of the breech*
> *gauge, profile, for testing underside of tang*
> *gauge, profile, for top of tang*
> *gauge plate for testing diameter of breech tenon*
> *gauge nut for testing the thread of breech screw and tap*
> *gauge, receiving, for breech*[36]

The only gauge of this group which appears to survive may be the *Gauge, receiving, for breech*, for checking the result of the work on the finished breech pin and is dated 1857. However, the surviving gauge could only be used after the nipple bolster had been finished, suggesting it may be different, and will be discussed later.

The front sight and muzzle

The front sight may seem a small and insignificant block of iron, but it performed two vital functions. Its primary purpose was to enable the rifle to be accurately aimed at its target and to achieve that it needed to be accurately positioned so that its vertical centre line lay on the vertical plane through the centre line of the bore and was at the correct distance from where the rear sight would be later fitted. Any discrepancy in either of those criteria would have resulted in serious inaccuracy and negated the whole point of long-range accuracy endowed by the rifling. Its second

MANUFACTURE OF THE BARREL

Fig. 7.31. The Enfield front sight. (XII.979 © Royal Armouries)

important function was to secure the bayonet, for which purpose it had to be at the correct distance from the muzzle and its base had to be a close fit in the slots of the bayonet socket.

Despite the obvious care needed in fitting and forming the front sight, of all the accounts covering the manufacture of this rifle it is only mentioned *en passant* in one, which notes its being fitted after grinding:

> *The barrel is now "sighted". Being laid horizontally on a metal plate, a line is drawn along it, coinciding with the axis of the barrel. The muzzle bead, or sight, is now brazed on by means of a composition of borax, brass clippings, and water.*[37]

This description of the process, lacking in so much detail, is a nonsense. Laying the barrel *horizontally on a metal plate* and scribing a line along it would have produced a line at an angle to the centre line of the bore because the barrel was tapered. Also, the front sight could only be fitted after the top and side faces of the lump had been machined and the breech pin fitted since these provided the necessary reference features for the sight to be accurately placed.

Whereas it is possible to conceive methods by which a line parallel to the centre line of the bore could be accurately scribed on the top of the barrel, working to a longitudinally scribed line would not achieve all the requirements and would introduce sources of error. A fixture which would enable this to be done could equally act as a fixture in a milling machine for the sight bed to be milled parallel to the horizontal plane through the centre line of the bore and the correct distance from breech and/or muzzle.

This would seem to be the approach adopted at Springfield[38] and is reflected in the specimen barrels in Birmingham museum on which no scribed lines are visible.

In his description of the process, Benton[39] simply states that the barrel is placed in a milling machine and a seat for the front sight milled out.

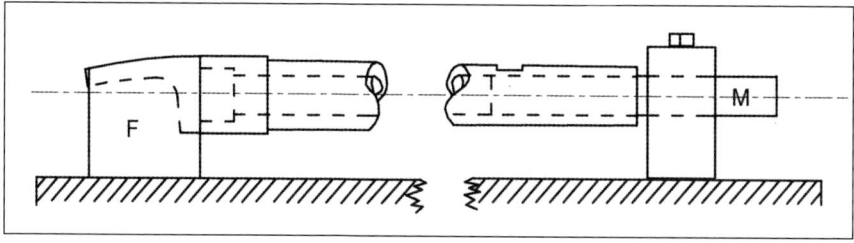

Fig. 7.32. Possible fixture for marking or milling the front sight bed. (© P. Smithurst)

Benton also notes that the ends of the sight bed were undercut, indicating the use of a 'dovetail' end mill. It would have been a useful aid, insofar as these undercut edges were 'turned-up' with a chisel, the foresight block placed in position, and then the edges driven down again, clamping the block in place ready for brazing. Bearing in mind there were no gas torches at this time, brazing had to be done in the forge hearth or a furnace and the sight block needed to be secured in place otherwise any movement would easily dislodge it.

The absence of any undercut on the Birmingham barrel suggests the sight block was held in place by other means, probably the use of a thin iron 'binding wire' which is still a common practice. Such a method slowed down the process and would not have been in keeping with Enfield's ethos of 'system' and 'efficiency' so it is suggested that a system similar to Benton's was probably used. Once the sight block had been secured, a mixture of borax and brass clippings, made into a thick paste with water, was applied to the joints and the muzzle either thrust into the hot coals of the forge hearth or into a furnace. It was common practice also to plug the muzzle of 'fine-bored' barrels with clay to prevent heat damage to the inner surface. In passing it is worth mentioning that borax has the advantage of reacting with and dissolving iron oxides, thereby cleaning the surface and helping the molten brass 'flow' into the joint in addition to preventing oxidation in the heating process. Incidentally, this facility for dissolving metal salts resulted in the 'borax bead test' in qualitative inorganic chemical analysis where a metallic compound was fused with a small quantity of borax to produce a 'glass bead' whose colour is indicative of the metal present.

In his list of machines, Burton included:

2 Machines for clamp milling the muzzle of the barrel for bayonet socket, with dies and reamers complete.

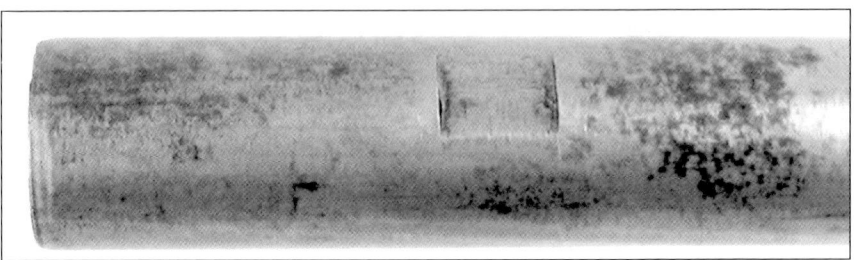

Fig. 7.33. The Enfield front sight bed milled. (2002.D.299.3. courtesy Birmingham Museums Trust)

Fig. 7.34. The front sight block brazed on. (2002.D.299.4. courtesy Birmingham Museums Trust)

MANUFACTURE OF THE BARREL

It was necessary for the muzzle to be cylindrical and of correct diameter to fit the bayonet socket, but clamp milling could only be easily applied before the front sight block was attached. There is no evidence to indicate that these machines were ever supplied, or that clamp milling was applied to the muzzle. While the Birmingham specimens do not necessarily reflect practice at Enfield, they do show that bringing the muzzle to correct size and form in the absence of 'clamp-milling' was achieved by turning which left a raised belt contiguous with the sight block.

Milling of the 'blade' on the foresight block also commenced at this point with a transverse cut to shape the front of the blade, followed by a second operation to shape the flanks and to reduce its width.

The raised belt produced by the turning operation was removed by filing.

Further filing then finished the foresight block to its correct dimensions, an important factor since it had to be able to enter the slot in the bayonet socket.

Fig. 7.35. The blade of the foresight milled, and muzzle turned. Note the step at the termination of the cylindrical portion of the muzzle and the raised 'belt' left under the foresight. (2002.D.296.5 courtesy Birmingham Museums Trust)

Fig. 7.36. The 'belt' under the foresight left after turning in the process of being removed by filing. (2002.D.296.4 courtesy Birmingham Museums Trust)

Fig. 7.37. The finished foresight and muzzle. (2002.D.296.2 courtesy Birmingham Museums Trust)

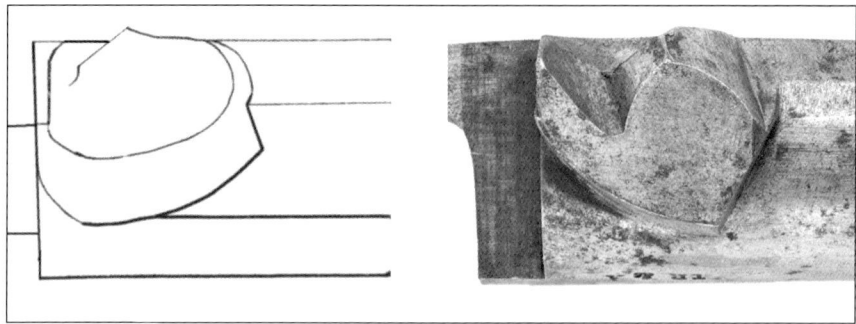

Fig. 7.42. Comparison of the bolsters / nipple seats of the Spanish musket, left, and the Enfield rifle, right. (adapted from *Memoria*, 1850, Plate 1, fig. 1 and; 2002.D.295.5, courtesy Birmingham Museums Trust)

No reason can be conceived why similar machines with appropriate cutters could not have been used to machine the bolster / nipple seat on the Enfield rifle. If the two nipple seats are compared, Fig. 7.42, it will be seen that they have similar features which support such a view.

Percussioning

'Percussioning' was the drilling and threading of the hole into which the nipple was to be screwed, and drilling an angled vent at the bottom of this hole to connect with the bore.[42]

Various tools for these operations were supplied by Ames as noted earlier:

> *drilling tool for vent*
> *drilling tool for cone-seat*
> *tapping tool for cone-seat and taps, reamers.*

These could have utilised the same machine which milled the base for the flange of the nipple. Tapping the hole for the nipple followed drilling the vent to avoid thread damage.

That the breech pin was fitted by the time percussioning was carried out is supported by the face of one shown in Fig. 7.44 upon which a 'scallop' had been left by the tip of the drill used to drill the vent.

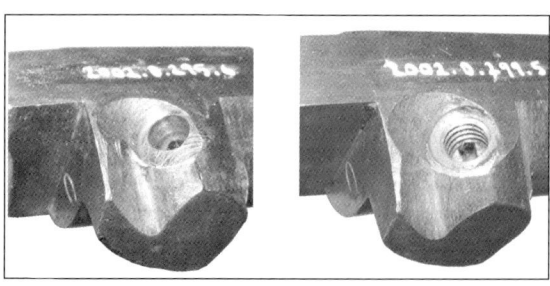

Fig. 7.43. *Left:* The holes for the nipple and vent drilled. *Right:* hole for nipple tapped. (2002.D.295.6 and 2002.D.299.5 courtesy Birmingham Museums Trust)

MANUFACTURE OF THE BARREL

Above left: Fig. 7.44. Face of breech pin showing mark from vent drill. (author's collection)

Above right: Fig. 7.46. The junction of the lockplate with the underside of the bolster highlighted. (© Royal Armouries, XII.1918, Sealed Pattern, dated 1859)

Below: Fig. 7.45. The finished nipple seat / bolster. (2002.D.299.6, courtesy Birmingham Museums Trust)

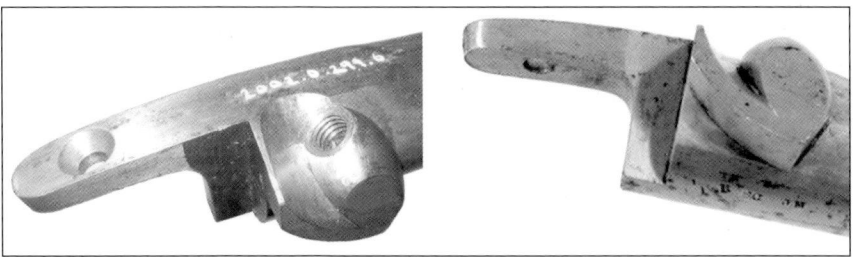

When all the machining of the bolster had been carried out, the nipple seat bolster was carefully finished by filing.

It should also be noted that the nipple had to be placed so that it was centred on the nose of the hammer, and this would be determined by the underside of the bolster being an accurate and close fit with the front concave edge on the lockplate, Fig. 7.46. Great care was therefore needed in the finishing of this part of the bolster.

Polishing

The outside of the barrel was polished by machine. Although there are discrepancies in the various accounts as to which stage in the operations this took place it would have been sensible to delay it as long as possible to avoid any blemishes arising from other work carried out. Five barrels, using the threaded breeches, were screwed onto an upright frame in the machine. In use, the barrels were caused to rotate and reciprocate vertically. Spring loaded clamps probably faced with oak, which could follow the taper, were 'loaded with emery and oil'. They were 'loaded with emery and oil' and applied to the barrels. These bear some resemblance to the *draw-polishing* machines described by Fitch[43] except in those machines the barrels were clamped on the exterior of the breech.

Rifling

Before rifling commenced, the barrel was fine-bored again to bring it as close as possible to .577 inch diameter and for which gauges were used; this will be discussed later. The bore was then polished.

At some stage during the processes performed up to this point, the muzzle of the barrel was 'crowned' - rounding-off the exposed front edge of the barrel walls. No reference to this has been found in any of the contemporary accounts but it is verified by a later letter written by James Burton in 1860 after his return to America which is included in its entirety in Chapter 8. As Burton explains, crowning protected the rifling grooves and lands from damage and, by providing what is effectively a funnel-mouth, made inserting the bullet easier.

The rifling consisted of three equal grooves, making one half turn in the length of the barrel, thirty-nine inches. In the *Illustrated London News* account[44] it is stated that the rifling grooves were of uniform depth and Jervis[45] states they were .262 inches wide and .014 inches deep. These comments apply to non-interchangeable rifles and as will be discussed shortly, criteria changed after 1857. Barrels made by private contractors were often rifled by hand using a simple machine as shown in Fig. 7.47 being used on what are stated to be 'Minié' barrels but further details in the article indicate they are clearly for the Enfield Pattern 1853 rifle.

Fig. 7.47. A 'hand' rifling machine in use in Birmingham. (*Illustrated London News*, 1855, p. 410)

MANUFACTURE OF THE BARREL

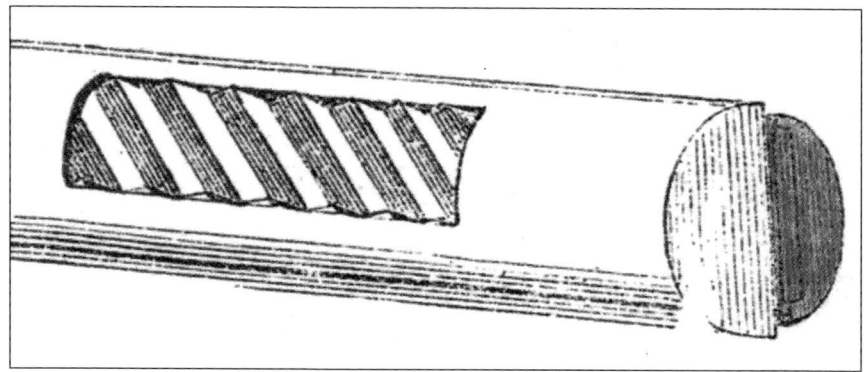

Fig. 7.48. Rifling cutter head for 'float-rifling' with 'spale' fitted. (*Illustrated London News*, 1855, p. 410)

The barrel was clamped in two upright supports positioned along the centre-line of the machine. The rifling tool was carried on a long rod attached to a spindle set in a sliding frame at the opposite end of the machine. This spindle carried a pinion which engaged with a transverse rack which rested on a horizontal inclined plane set so that as the sliding frame moved forward, this rack caused the rifling rod to rotate one-half turn in the length of the barrel. The cutter, which took the form of a short narrow section of a 'float' or file, was set in the 'rifling head', Fig. 7.48. In America, the process was referred to as 'float-rifling' or 'floating-out' or 'file-tool rifling'.[46]

After one groove had been cut, the rifling rod was withdrawn from the barrel and the spindle into which it was fitted rotated by one-third of a revolution and the process thus repeated until all three grooves had been cut. To assist cutting to the correct depth, a sliver of wood, referred to as a 'spale', but variously known as a 'spill' when used in the boring process, could be fitted at the back of the cutter head, as shown in the illustration.

However, when production at Enfield began, the process was more sophisticated and cut 'progressive rifling', that is, grooves of decreasing depth from breech to muzzle, and, depending upon the source, these were – *about a quarter inch in width, and a depth of .02in. at the breech and .005in. at the muzzle*[47] or *.235in. in width, .015in. in depth at the breech, and .005in. at the muzzle.*[48] Clarification on the rifling specifications will be provided later in this chapter.

It was reported[49] that rifling at Enfield was carried out using a Belgian machine but it is more likely to be the machine patented by Manceaux of Paris in 1852, Fig. 7.49, a view confirmed by the comment – *It appears that machinery for rifling muskets has been little used in England; but a French machine has been lately applied to this purpose.*[50]

Manceaux's machine was the only one at this time capable of cutting progressive rifling.

In Manceaux's machine, a rotary motion is given to the rod carrying the rifling cutter as it moves through the barrel by means of a pinion on the rifling rod being acted upon by a rack engaging with the inclined 'sine bar' below the machine.

MAKING THE ENFIELD PATTERN 1853 RIFLE-MUSKET

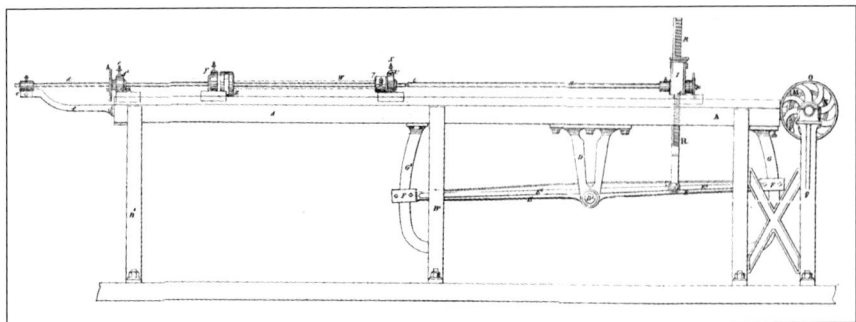

Fig. 7.49. Manceaux's rifling machine as shown in the patent. (Manceaux, 1852)

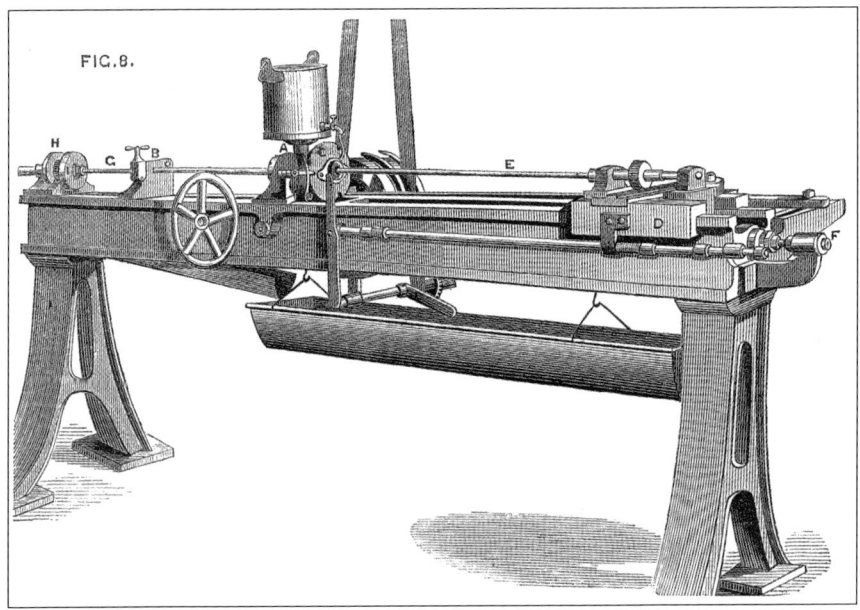

Fig. 7.50. The rifling machine used at Enfield. (*Engineer*, 1859, p. 348)

The Enfield machines, Fig. 7.50, differed in their construction but embodied the same principles.

A plan view of the machine shows its major difference from Manceaux's machine in having the inclined 'sine bar' positioned horizontally at the side of the machine.

The rifling cutter was held at the end of a rod which rotated as it was drawn through the barrel, and at the same time the position of the cutter was adjusted to progressively make a deeper cut as it moved from muzzle to breech. These motions were achieved in the following way:

The muzzle of the barrel was fixed in the clamp, B, and its breech fitted into a housing, A, on the spindle of a headstock which also carried a 'division plate', C.

MANUFACTURE OF THE BARREL

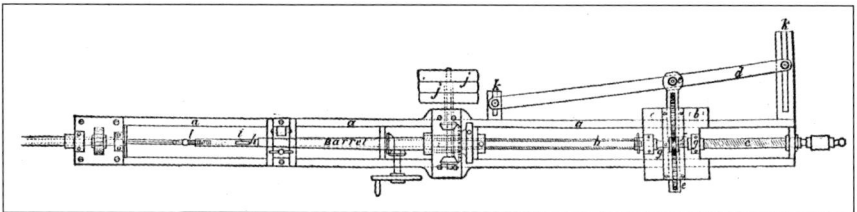

Fig. 7.51. Plan view of the Enfield rifling machine. (Miles, 1860, p. 33)

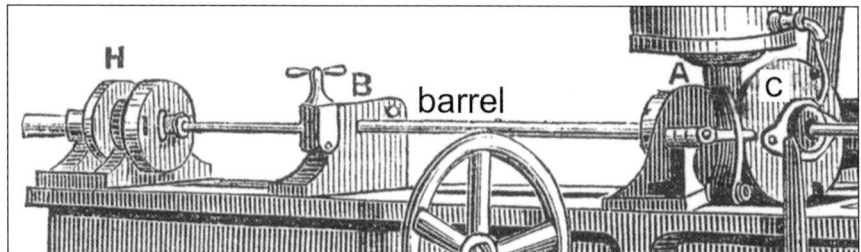

Fig. 7.52. Detail of Fig. 7.50 showing the barrel fitted into the housing, A, and clamped at B.

This plate had three equally spaced holes into which was fitted a peg carried on a detent arm attached to the bed. Un-clamping the barrel, rotating the spindle carrying it one third of a revolution so the detent registered with a hole in the division plate and re-clamping the barrel, positioned the barrel for the three grooves to be cut.

The rifling cutter head was carried on a rod, E, (Fig. 7.53) fitted at its right-hand end into a spindle mounted on the carriage, D, which was moved along the bed of the machine by a leadscrew positioned centrally beneath it and reversibly driven by the belting via two bevel pinions which could selectively engage with a crown wheel to drive the leadscrew in either direction.

The spindle, to which the rifling rod was attached, carried a pinion engaging with a rack. At its outer end, this rack had a fitting which allowed it to engage with and slide on a horizontally-inclined bar referred to as a 'sine bar', fitted at the rear of the bed. This is more clearly shown in Fig. 7.54.

The 'sine bar' was set at the appropriate angle and, as the carriage was moved along by the leadscrew beneath it, the rack was caused to move transversely,

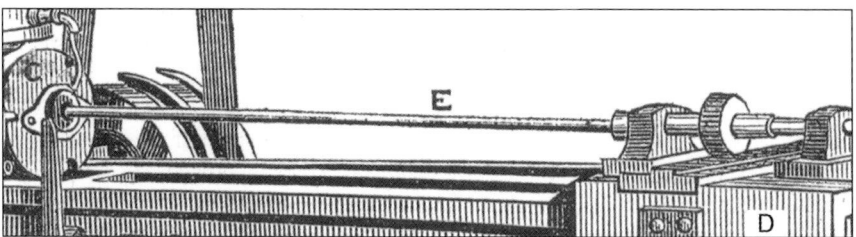

Fig. 7.53. Detail showing the rifling rod, E, fitted to the spindle mounted on the carriage, D.

MAKING THE ENFIELD PATTERN 1853 RIFLE-MUSKET

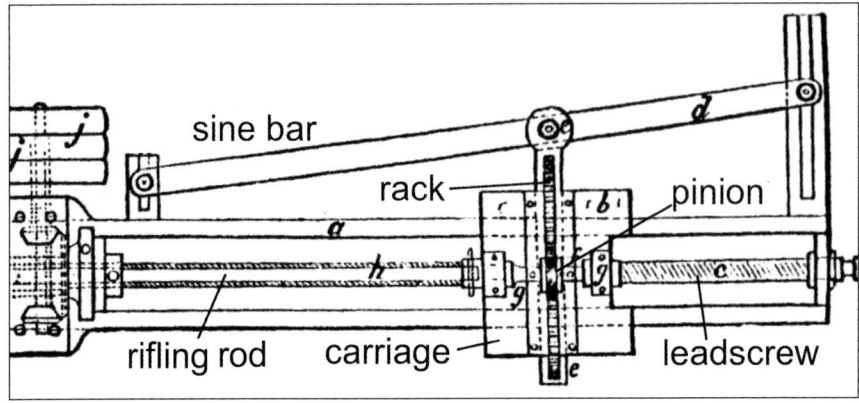

Fig. 7.54. Detail plan view of the machine showing the rifling rod, *h*, the leadscrew, *c* and the 'sine bar', *d*. (adapted from Miles, 1860, p. 33)

rotating the pinion and the rifling bar so that it made a half revolution in the length of the barrel. At the opposite end of the rifling bar, the 'cutter-head' was fitted with the hook-shaped cutter, *i*.

The depth of cut was regulated by the screw, *l*, at the end of a long rod which is free to slide through the housing on the left but is prevented from rotating. Aspects of this are more clearly shown in the schematic diagram, 7.56.

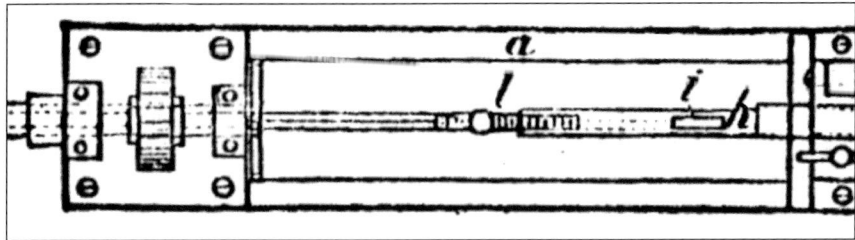

Fig. 7.55. Detail plan view showing the rifling rod, *h*, with its cutter, *i*. (From Miles, 1860, p. 33)

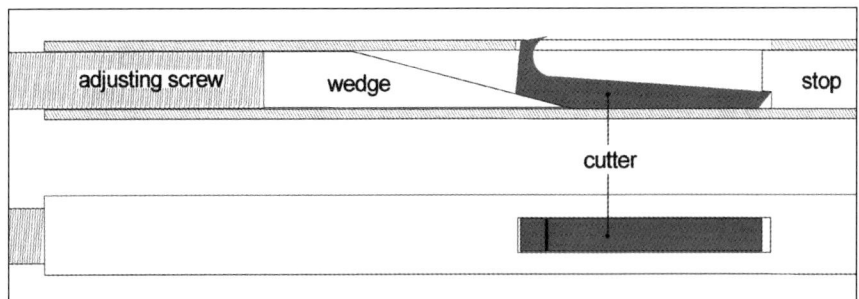

Fig. 7.56. Schematic representation of the construction of the 'hook-rifling' cutter head. © P. Smithurst)

MANUFACTURE OF THE BARREL

The adjusting screw bore upon a wedge whose tip was under the cutter. Rotation of the screw raised or lowered the cutter, the width of which was the same as that of the groove to be cut and whose cutting edge was at the tip of a hook, hence the term 'hook-rifling' being used to describe this method of cutting rifling grooves.

Rifling commenced at the muzzle. As the rifling bar began its longitudinal travel, it was rotated by the sine bar acting on the rack and rotating the pinion connected to the rifling bar. The rifling bar also rotated relative to the adjusting screw, causing the screw to be continuously advanced by a very small amount. This, acting on the wedge, caused its cutting edge to rise as it travelled along the bore, cutting a groove of progressively increasing depth towards the breech. The pitch of the screw combined with the angle of the wedge had to be such as to raise the cutter only 0.010 inch over a length of 39 inches.

The rifling process often left a burr along the lip of the groove and according to a contract document:

> *any burr raised in rifling to be removed by lapping according to Barrels approved from Enfield. Draw leading will be allowed to remove any little roughness or rivelling on the cuts which may be considered necessary* (Appendix 2, item 7)

'Rivelling' might be defined as pronounced striations left by a worn rifling cutter with an uneven cutting edge. 'Lapping' was usually performed using a short lead cylinder cast on the end of a brass or steel rod, machined to correct diameter, and moved up and down the bore whilst being rotated and fed with a small amount of fine emery/oil paste. This could only act on the lands, removing any burrs raised at their edges and polishing their faces. Draw-leading was a similar process but acted only longitudinally. The lead cylinder had ribs on its surface to fit the rifling grooves and could only be used with uniform, as opposed to progressive rifling which varied in depth.

With regard to the rifling machine, contradictory information surrounds it. On 24th July 1854, Robbins & Lawrence submitted a tender to the War Department, which was accepted on the same day, and included:

> *2 machines for rifling barrels 3ft. 3in. long.*[51]

Unfortunately, no illustration of this machine has been located, but there is a description of it:

> *In 1854 a rifling machine was designed by H. D. Stone at Windsor, Vermont, for the English government, which cut three grooves at a time, the twist being given by a vertical rack, actuated by a roller moving in guides.*[52]

The use of a *vertical* rack to impart rotation to the rifling rod makes it similar in this respect to Manceaux's machine, but in its ability to cut three grooves at once it differs from any machine described in the various accounts of operations at Enfield.

Even Major Mordecai of the U.S. Army, following a visit to Enfield in 1855/56, makes the comment:

> *The machine used at Enfield for rifling barrels with grooves of progressive depths, was obtained from Liège, being of the same kind as those ordered from there for our armories.*[53]

this again, confirms that some machines ordered from Robbins & Lawrence were never supplied.

Rear Sight

The manufacture of the rear sight required twelve or thirteen operations, by hand and machine, before it was ready for fitting, but absence of details precludes these being examined. According to contract details, it was fitted between the fine boring and rifling stages since it was to be *viewed for boring-up with the back sight soldered on* (see Appendix 1, item 5). On the other hand, '*The Mechanics' Magazine*' states that at Enfield it was fitted after rifling. It had to be positioned so that its centre line lay on a line between the centres of the front sight and the breech tang and coincident in the vertical plane with the axis of the bore. It also had to be at the correct distance from the front sight, otherwise the ranges shown upon it would be incorrect. It was soft-soldered in place and its position checked by the use of gauges.

'Viewing' and gauging

Viewing

Throughout its manufacture, the barrel, like other components, was subjected to various examination procedures. 'Viewing' was the term applied to a combination of a visual inspection and, for the barrel, an examination of the bore with gauges. The visual inspection was to ensure there were no defects which would not be detected by gauging. At its simplest it would check surface finish and for flaws in the metal, such as cracks, which might indicate forging/welding had been carried out at too low a temperature, or an imperfect weld in a barrel seam, all of which might have been hidden until revealed by machining.

The most rigorous visual examination was to check for straightness of the bore. This was carried out intermittently during manufacture but, on the finished barrel this viewing acquired a high degree of rigour and required great expertise. The barrel was held up and pointed towards a source of light and rotated slowly while the inspector looked down the bore. The shadow cast by the muzzle was examined to make sure its edge was a continuous line. Any diversion from that would indicate a slight bend which was corrected by laying the barrel across a channel on an anvil and striking at the correct point with a mallet.

External gauging

To ensure the barrel satisfied the external requirements for dimensions and, in some instances, shape, it was checked by the use of gauges.

MANUFACTURE OF THE BARREL

The barrel gauges supplied by the Ames Company noted earlier, comprised:

gauge for testing the counterboring of the breech
gauge for testing the tapping of the breech
gauge plug for testing drilling of cone [nipple] *seat*
gauge for screw-tapping
gauge, receiving, for testing cone-seat drilling
gauge, receiving, for breech
gauge, profile, for testing underside of tang
gauge, profile, for top of tang
gauge plate for testing diameter of breech tenon [taken to be the cylindrical body of the breech pin]
gauge plate for testing the barrel at six points
gauge plate for length and height of stud
gauge for testing the position of stud from breech
gauge nut for testing the thread of breech screw and tap

In the set of gauges once used at Enfield and now part of Royal Armouries' collections, only a few barrel gauges have survived. Their names, however, make their purpose, on the whole, self-evident.

One gauge in this collection tests the finished barrel at six points: butt [the wide breech end], 4, 6, 16, 26 and 35 inches from the butt, and one open and one ring gauge for muzzle.

What is noticeable on this gauge is the use of what are taken to be hardened steel inserts on the flanks of the openings to minimise wear and maintain dimensional accuracy. Such inserts have not been observed on any of the other gauges, raising the question of how they were produced. One clue exists in the comments of Richard Lawrence in relation to the contract for rifles where he states gauges were made and were 'carefully oilstoned' to bring them to accurate dimensions.[54] Precision grinding was only made possible by the development of the first successful artificial grinding wheel by Norton in 1877[55] and the first universal grinding machine produced by Brown & Sharpe in 1876.[56]

Two of the barrel gauges ordered from the Ames Company are described as:

Gauge, plate, for length and height of stud
Gauge, plate, for testing the position of stud from breech

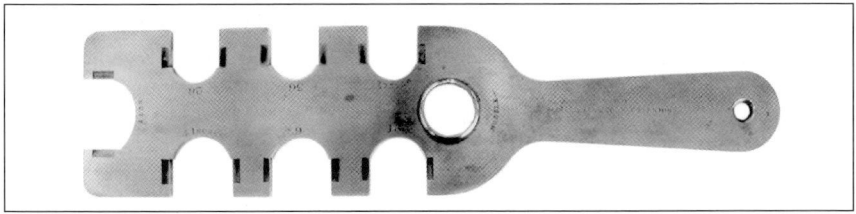

Fig. 7.57. Barrel external dimensions gauge. (Part of PR.10142 © Royal Armouries)

Left: Fig. 7.58. Barrel gauge applied to 'butt' and muzzle. (© Royal Armouries)

Below: Fig. 7.59. The sight, A, and bayonet stud, B. (Benton, 1878, p. 6)

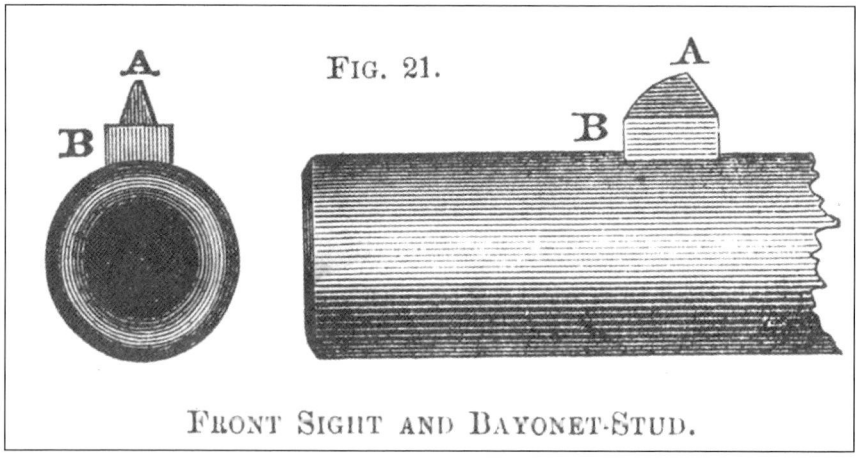

The use of the term 'stud' is unusual in this context since it is usually taken to refer to a small rectangular block beneath the muzzle of a musket which engages with the slot in the socket of a bayonet. However, Benton[57] refers to the front sight being composed of two elements, the sight proper, A, and the bayonet-stud, B.

Since both Ames and Benton were American, they very likely used the same terminology, and it is taken to have the same meaning with the Enfield rifle since this also used the front sight base as a 'stud' for attachment of the bayonet and there is nothing else which needed to have a specific size and distance from the breech.

None of the surviving set of Enfield gauges matches those described in the Ames contract, but there is one that mimics the bayonet socket and performs the functions of testing the size of 'stud' of the front sight and the diameter of the muzzle.

The only other gauge on the original list which may survive is the *Gauge, receiving, for breech*. That shown in Fig. 7.61, is for checking the result of the work on the finished bolster and breech pin and is dated 1857.

A barrel length gauge, Fig. 7.62, exists but forms part of a second set of gauges in Royal Armouries' collections which had possibly been used by The London Armoury Company. It is inscribed 'Length of Barrel 3 Feet 3 Inches' so is clearly for the Pattern 1853 Enfield rifle. The inscription is also flanked by two 'stars' in identical fashion

MANUFACTURE OF THE BARREL

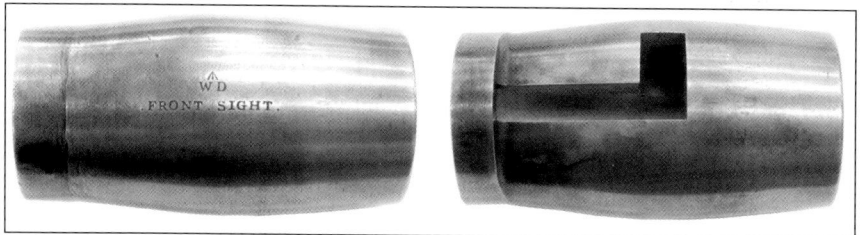

Fig. 7.60. Front sight / muzzle gauge. (Part of PR.10142 © Royal Armouries)

Fig. 7.61. 'Gauge, Receiving, Breech' in use checking form and alignment of breech pin tang and heel and the underside of the nipple bolster. (Part of PR.10142, © Royal Armouries)

Fig. 7.62. Barrel length gauge. (part of XIII.949 © Royal Armouries)

to gauges, such as the bridle gauge, Fig. 7.64, contained in the Enfield set which suggests it may once have been part of this set or from the same unknown source.

Another gauge in this collection is simply inscribed "L.A.C./ For Barrel Gauges / Standard", Fig.7. 65.

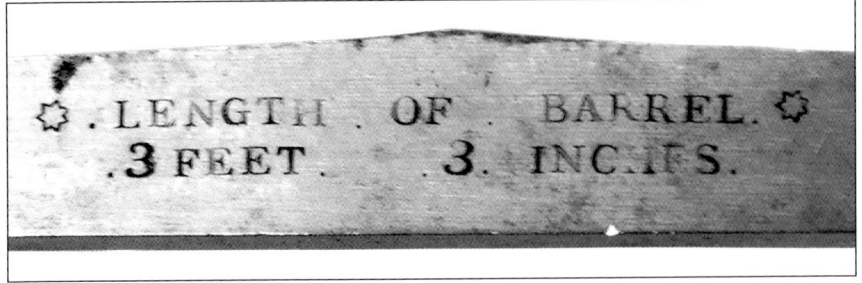

Fig. 7.63. Marking on barrel length gauge. (part of XIII.949 © Royal Armouries)

Fig. 7.64. Marking on bridle gauge. (part of PR.10142 © Royal Armouries)

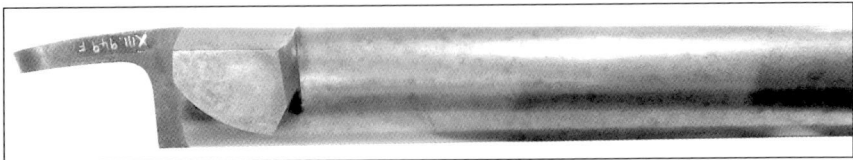

Fig. 7.65. Model of breech-end of a barrel marked 'For Barrel Gauges'. (XIII.949F © Royal Armouries)

Fig. 7.66. Detail of inscription – 'For Barrel Gauges / Standard'. (XIII.949F © Royal Armouries)

MANUFACTURE OF THE BARREL

It can be seen that it is fitted with a finished breech pin with correctly formed tang, but an unfinished nipple bolster/seating and, whilst not obvious from the photograph, it is only about 14 inches long. Considering that the barrel gauge previously examined gauged the diameter at several points along its length from 'butt' to muzzle, and had a ring gauge for the muzzle, it is difficult to comprehend the purpose of the item in Fig. 7.65 since it would have had limited use in checking the barrel gauge! Had it read 'For Breeching Gauge' its purpose would have been more understandable, particularly as it has the mating face between nipple bolster and lockplate finished, it would have been ideal for checking the accuracy of that gauge shown in Fig. 7.61 and for checking the work on the stock, as has been discussed in Chapter 5.

Internal Gauging - the bore and rifling

Once rifled, the rifling and bore were subjected to close inspection and gauging. All of the gauges mentioned so far provided either an upper or a lower limit on size, but not both. In the case of the bore and rifling, the concept of 'tolerance' is encountered for the first time. There is little to offer guidance in the application of these inspection processes beyond the specifications outlined at the beginning of this chapter and the little that does exist serves to highlight the fact that those early specifications were changed over time.

A number of recently discovered documents (Appendix 2) have a bearing on this. Some are private contracts relating to the 1853 Pattern Artillery Carbine, the 1856 Pattern Short Rifle and the 1856 Cavalry Carbine, East India Co. Pattern, all of which are rifled arms of ·577-inch calibre so it can be presumed that identical gauging procedures would have been applied to them. In addition are two memoranda on bore and rifling gauging.

The Artillery Carbine and Short Rifle both had limits of the acceptance of a plug gauge of 0.572-inch and rejection if the 0.577-inch plug entered the bore (Appendix 2, items 5 & 6). This is at odds with the specification quoted by Roads[58] where he states that the lower limit was 0.575 inches, but without a source being given. In the Cavalry Carbine contract (Appendix 2, item 7), there is no mention of limits on bore size, but a hand-written postscript, Fig. 7.67, clearly states that 'all

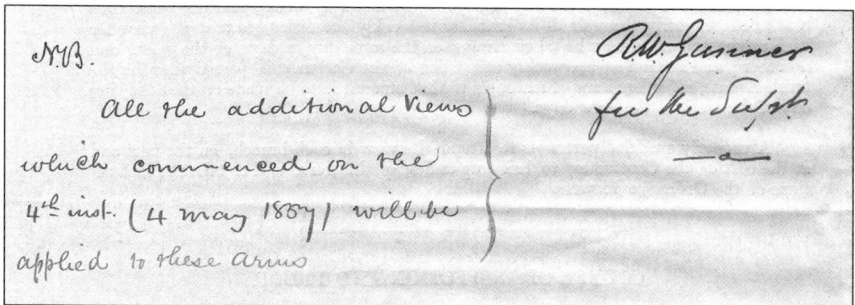

Fig. 7.67. A postscript on the cavalry carbine contract specification document by R. W. Gunner on behalf of the Superintendent. (author's collection)

the additional 'views' in force from 1857 shall be applied' and this would apply to the gauging process.

This simple note probably caused confusion among contractors and later in the same year a memorandum on Gauging the Bores of Barrels was issued by The Military Store Office in Birmingham on 11th November 1857, (Appendix 2, item 8) and added to the confusion. A transcript is provided here in full:

> *In order to ensure the barrels being perfect to guage [sic] at the muzzle, when the arms are finished and ready for transmission to Store. Mr. Shirley will guage [sic] each barrel separately, after the arm has undergone its Finishing View, with a ·577 inch plug, taking care that it passes freely from end to end of the barrel and he will also test the muzzle with the ·580 inch plug, allowing no toleration, but condemning every Barrel that takes the rejecting plug at all. He will place his mark for this view crosswise on the stock, between the back end of the guard strap and the contractor's name.*
>
> *The present plan of guaging [sic], examining, and marking the Barrel, in the view for 'Browning', will be continued. In guaging [sic], no toleration will be allowed at the muzzle beyond the difference between the receiving and rejecting plugs; but at the breech the rejecting plug may enter for one third of its length, but anything beyond this will cause the barrel to be condemned.*
>
> *Since a certain amount of toleration has inadvertently been permitted, in order not to throw on the Contractor's hands Barrels that have been Viewed for Boring whilst this toleration has been reckoned on by them, a new view mark for Boring will be put on all Barrels coming in after this date, and in the after views no toleration will be allowed to barrels thus marked.*
>
> *All Barrels that have been viewed for boring previous to this date, will be allowed a toleration of the rejecting plug entering the muzzle to the extent of one inch.*

Whereas a .577-inch plug was previously a *rejecting* gauge, it now had to pass freely along the full length of the bore and a .580-inch plug was not allowed to enter the barrel except under certain circumstances. Of particularly interest in this memorandum, despite its confusion, is the use of the term *toleration* which, since it is applied to dimensions, can be read as 'tolerance' and is the earliest reference to this concept to have been noted although it is implicit in the bore gauges.

Whilst the contractors at that time would have been totally familiar with regulations prior to this date, what had been the 'rejecting' plug had become the 'accepting' plug and something just over five-thousandths of an inch had been added to the acceptable bore size and new rules for the 'acceptance' and 'rejection' of barrels introduced. It can be imagined that the barrel contractors began searching through their previously over-sized rejected barrels to find ones that might comply

MANUFACTURE OF THE BARREL

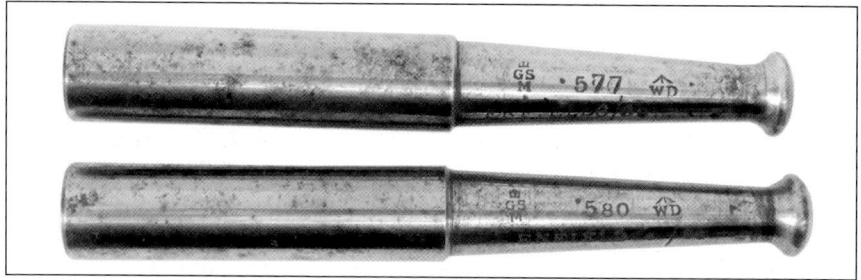

Fig. 7.68. A .577 'accepting' and a .580 'rejecting' plug gauge, dated 1868, corresponding with those described by Gunner in 1857. (author's collection)

with the new specification. It may have been this confusion which prompted James Burton, Chief Engineer at Enfield, to issue the following memorandum in 1858 (Appendix 2, item 9):

> *In order not to exceed the depth of .015, which is the figure representing the depth of groove at breech, calculated upon one turn over 6ft 6in: and commencing at the muzzle with .005, it is absolutely necessary to have a plug to test the rifling, slightly under .015 at the breech.*
>
> *A plug for this purpose measuring .013 of an inch is to be employed for the reception plug, and one measuring .016 for the rejecting.*
>
> *At the muzzle, where the depth is intended never to exceed .005 of an inch, two plugs will be used in the Rifling View in order not to allow of any groove being accepted which may be deeper than .005.*
>
> *These two plugs will be respectively .005 for acceptance and .008 for rejection. The width of the groove in the rifle barrel is to be .235 of an inch.*

Burton, in stating the need for and specifying the sizes of gauges for the rifling, taken in conjunction with the absence of any previous references to gauges for this purpose, suggests that none had been used. His memorandum concludes with a detailed listing of relevant dimensions and limits. The first four items relate to gauges for the bore and are distinctly larger than any sizes specified hitherto – a full ten-thousandths of an inch greater in upper and lower limits. It has to wondered why such a change was made and no other reference to it, or explanation for it, has been found. The last four items in the list relate to the gauges for rifling depth limits at breech and muzzle. The set of bore gauges in Royal Armouries' collections, once forming part of the Pattern Room collections and dating to 1862, contains rifling gauges in addition to the standard bore gauges, but is incomplete.

MAKING THE ENFIELD PATTERN 1853 RIFLE-MUSKET

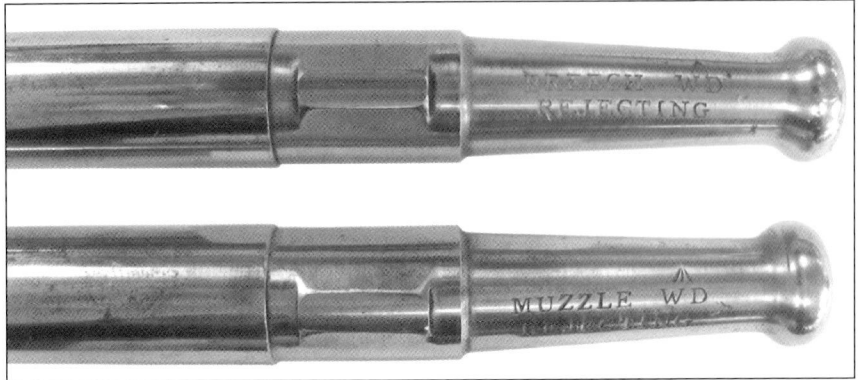

Fig. 7.69. Rifling rejecting plug gauges for Breech (top) and muzzle (Bottom). (Part of PR.10144 © Royal Armouries)

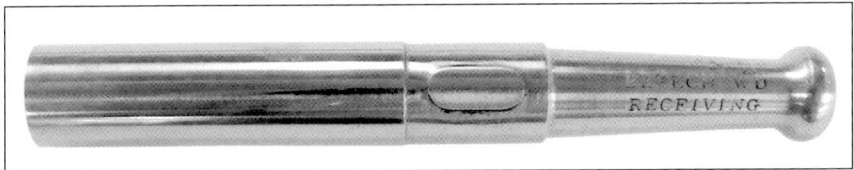

Fig. 7.70. Rifling receiving plug gauge for Breech. (Part of PR.10144 © Royal Armouries)

Proving

The most vital and demanding test of a barrel was 'proving' in which it was fired to confirm its ability to withstand the explosion within it. According to a specification by Pellat on behalf of the Inspector of Small Arms:

> *the barrel **before percussioning** to be proved with seven drachms of small arm powder, and an elongated service ball, with a cork wad over the ball.*[59] [author's highlights]

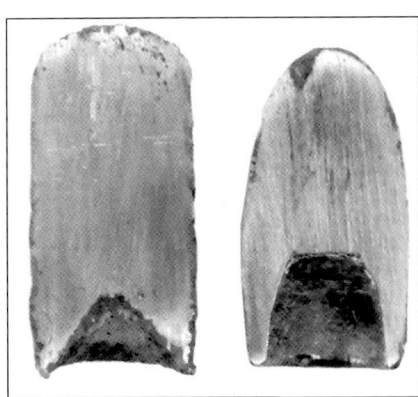

The standard 'service' charge was 2½ drachms (1.5g.) of powder with the service bullet of 530 grains (34.5g.). This was an extreme test, using nearly three times the normal service charge of powder.

What is believed to be an 'elongated service ball' is shown in Fig. 7.71,

Fig. 7.71. A sectioned 'proof ball', left, compared with a sectioned standard 'service ball', right. (author's collection)

216

compared with a standard service 'ball'. This 'elongated ball', from the weight of a 'half' section, weighs approximately 710 grains, almost 30% more than a service ball.

It is also longer and cylindrical and its increased inertia would serve to increase the pressure generated by the explosion momentarily until it began moving. There was an obvious logic to 'proving' at this early stage since if the barrel failed after percussioning, all the extra time and effort would have been wasted, but the only way of accomplishing it before a vent had been drilled was by the use of a special plug with a vent.

This early proving is also supported in contract specifications issued to Joseph Smith for the manufacture of the East India Company cavalry carbine adopted for British service use in 1856.[60]

The fact that it states that the barrel underwent a second proof after percussioning, implies that it underwent its first proof before percussioning. However, an earlier contract document issued in 1853 for the Pattern 1853 Artillery Carbine clearly states percussioning took place before the barrel was proved.

This same contract also shows that for proof – no second proof is mentioned – the 'elongated service ball' was replaced by a standard service ball.

Prior to 1860/61, carbines and short rifles were produced only by contractors and such variations in specifications reflect the confusion pervading the contract system.

COURSE OF INSPECTION AND PROOF.

1st. The Barrel will be viewed for Percussioning and Machine Jointing.

6th. The Barrel to be sent in for the following views :—

1st. Second Proof.

Fig. 7.72. Abstract regarding proof from contract to Joseph Smith for 1856 cavalry carbine, E.I.Co. Pattern. (Appendix 2, item 7, author's collection)

SPECIFICATIONS and Rules to be attended to in the Supply of Rifled Artillery Carbines ·577 bore, and in the Delivery and Inspection of the same, Weekly, under the general conditions of the Contract to be annexed hereto, and to form part hereof.

COURSE OF INSPECTION AND PROOF.

Barrels.

The Barrel to be sent down for Proof and inspection, percussioned, jointed, and properly breeched ; the bore is to be tested with two plugs, viz.,

---

Fig. 7.73. Abstract re proof from contract to Joseph Smith for the 1853 artillery carbine. (see Appendix 2, item 5, author's collection)

> They are to be proved with the regulated proof charge of 7 drachms of powder (FG) and a conical service bullet of 530 grains weight, and to stand at least twenty-four hours after proof before being inspected.

Fig. 7.74. Another abstract regarding proof from contract to Joseph Smith for the 1853 artillery carbine. (Author's collection)

Browning

After all the inspection processes, the barrel was 'browned' which actually gave it a deep blue-black colour. It inhibited rusting and prevented reflection. The polished and de-greased barrel was rubbed over two or three times a day for four or five days in a room at 70°F with a solution of

6 oz. spirits of wine (ethyl alcohol)
6 oz. tincture of steel (Iron III Chloride in alcohol)
2 oz. corrosive sublimate (Mercury II Chloride)
6 oz. sweet spirits of nitre (Ethyl Nitrite in alcohol)
3 oz. Nitric Acid
Made up to 1 gallon with water

until the desired colour was produced[61] and an experiment has shown this to be effective. The result of the application of all this technology was a masterpiece of engineering with an occasional input of high quality craftsmanship:

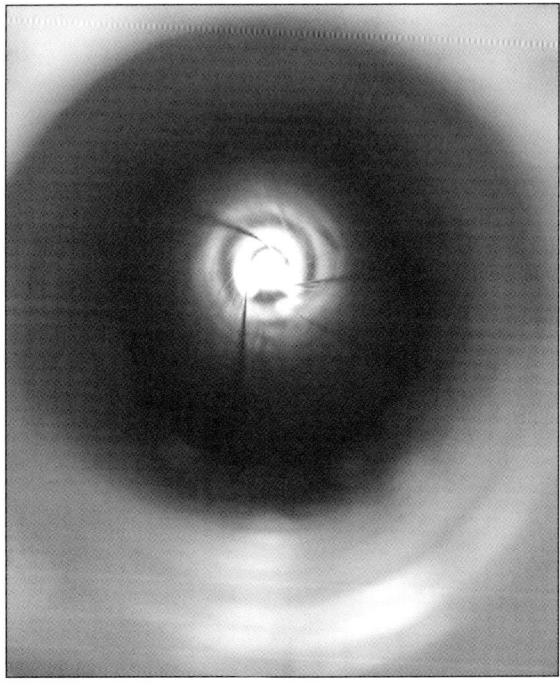

Fig. 7.75. The perfectly finished bore (1855.5.731 courtesy Birmingham Museums Trust)

MANUFACTURE OF THE BARREL

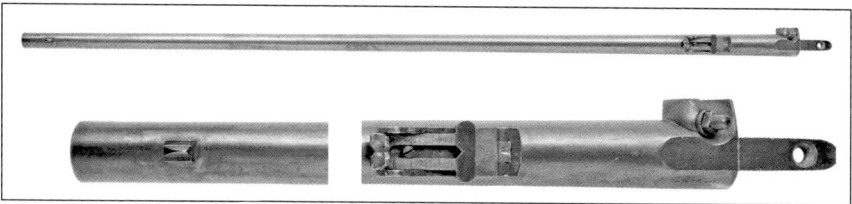

Fig. 7.76. The perfect finished barrel. (1855.5.731 courtesy Birmingham Museums Trust)

> *After this process the bore is a most perfect cylinder, beautifully polished, and the light is reflected to such an extent as to cause an appearance, on looking through it, as if the farther extremity were a thin film of light concentric with the circle which is nearest the eye.*
> (*Engineer*, 1859, p. 348)

7.4 Conclusion

The technology employed has been examined more extensively and in greater depth than in any previously published account. The barrel was finished to high standards of accuracy requiring a series of complex operations, details of which were omitted from many of these earlier accounts and which have been examined and supplemented with the aid of previously unstudied documents and unique items in the collections of various museums. Particular attention has been given to the manufacture and fitting of the breech pin, superficially a simple operation, and the machining of the complex geometry of the nipple seat bolster. The first encountered use of the concept of 'tolerance' as applied to an examination procedure has also been noted.

Chapter 8

Conclusion

An attempt has been made to show that in Britain the adoption of a new rifle for military service in 1853 was quickly followed by a dramatic revolution in the method of its procurement and manufacture so that by 1857 it was being manufactured almost entirely by machines with a precision previously unparalleled. There is no doubt that the Enfield Pattern 1853 Rifled Musket produced after 1857 by using that technology was an outstanding masterpiece of both design and manufacturing finesse, yet it has been so widely ignored in the published history of 19th century firearms! You might say I am biased in my opinion – so let someone who was intimately involved at the time speak of it. In the Archives at Yale University are the papers of James Burton, the man largely responsible for the setting up and operation of the Enfield factory for the rifle's production. Amongst them is a copy of a letter of Nov. 13th, 1860, by which time Burton had returned to the United States, addressed to a Captain Randolph concerning details of the new model rifle to be adopted by the State of Virginia;

Richmond VA
Nov 13th 1860

Sir,
With reference to the subject of the Model Rifle Musket to be adopted by the State of Virginia as a pattern by which to govern the manufacture of arms in the State Armory, I beg to submit, in compliance with your request, the following suggestions for your consideration in the hope that they may be of some service to you and the other members of the Commission having charge of the question.

I have carefully examined the US Rifle-Musket placed in my hand by you for that purpose, and am of the opinion that it can be much improved upon in matters of detail as I shall endeavour to describe.

1st the barrel.

I would recommend that the U.S. length and diameter of bore be retained but the external form and dimension of the breech of the Enfield rifle be adopted. In the latter named arm the cone or nipple

CONCLUSION

is inserted so as to establish an almost direct communication with the bore of the barrel, and thus sure fire is secured, which is a most necessary quality in a military firearm. This is not the case in the U.S. model and should not be imitated in establishing the new model. With reference to the waist and fore end of the Enfield barrel I would recommend that about 4 ounces of weight of metal be added to the barrel, and properly distributed in such a way as to strengthen this part, which has been found rather too weak to effectively resist the rough usage of service, and at the same time preserve the shooting qualities of the arm. This modification was under consideration by the proper authorities of the War Department in England when I left them a few weeks since, and has been most likely adopted and put into practice. The nose end of the barrel should be rounded [crowned] as in the Enfield arm. This renders loading easier and the corners of the grooves and the lands of the rifling are protected from injury. I would also strongly recommend that the barrel be browned like the Enfield barrel for two reasons – first the glitter of a bright barrel distracts the aim of the soldier using it: second, the bright barrel is more likely to be seen by the enemy under any circumstances, and thus discover to him the locality of the rifleman, whether acting singly or in masses. Besides these reasons, the barrel is more likely to rust than the browned barrel. The form and twist of the grooves in the U.S. barrel are not open to much improvement and may be copied with safety.

2nd The Lock

I would strongly recommend that the Maynard primer lock be abandoned for the reasons that the lock is not in itself a good one, encumbered as it is with the extra machining necessary to the use of the primer, and the action of the mainspring in connection with the tumbler and hammer is not uniform. The mainspring is weak when the hammer is down on the nipple, and much stronger when the hammer is at full-cock. This should not be the case and may be remedied by a proper adjustment relatively to each other parts of the lock. The lock of the Enfield Rifle is a good one and I would recommend that it be copied in every particular, as much experience has proved its reliability and efficiency.

3rd Furniture

I would recommend the adoption of the convex adjustable screw bands after the style of those on the Enfield arm and that they be blued in the same manner to correspond with the browned barrel – in this case the three band springs will be dispensed with. The advantage of the screw bands are too apparent to render it necessary to describe them and I regard these bands as embracing one of the best features in the Enfield arm. I would recommend the separate arrangement of the trigger plate and Guard-plate, as in the Enfield Rifle be copied as it offers better facility of adjusting the trigger to the sear of the lock, and that the Butt-plate, Trigger-plate, Guard-plate, Side cups and the nose cap be made of brass, as in that arm. The experience of the English service has proved that iron furniture, being in contact with the wood of the stock is very liable to rust badly when exposed in service, which is not the case with brass furniture and hence the adoption of the last named metal for this purpose in the Enfield Rifle. The cost of the brass furniture is also less than the iron. I would also recommend the adoption of a third screw at the angle of the butt-plate, which screw will protect the latter from injury in "grounding arms".

I would recommend that the "swell" ramrod of the U.S. arm be abandoned, and a plain straight ramrod substituted instead, of a somewhat larger diameter, the head to be of iron and of cylindrical form in part, so as to permit of a hole being bored through it not less than 1/10th inch in diameter. This hole will be useful as offering a facility of obtaining a good grip or purchase in cleaning and wiping out the barrels, by passing through it a nail, a bit of iron etc. as may come handiest to the soldier. The "jagged" head of the Enfield ramrod is of no special use and should not be copied. In abandoning the "swell ramrod", it will be necessary to introduce a 'rod spring' to hold the ramrod, as in the Enfield arm, in which respect it may be copied. The plain ramrod can be made at less cost, too, than the 'swell' ramrod. I would also recommend that the nosecap be secured to the stock by means of a screw, instead of a rivet as in the U.S. arm.

4th Stock

The stock of the Enfield Rifle is of better form and proportions than that of the U.S. arm, but can be improved upon at the butt by reducing the depth about ¼ inch and thinning the 'toe' further forward so as to render the heel or angle of the buttplate more prominent; at the same time the butt end of the stock may be dressed flat with advantage, instead of concave as in the Enfield Rifle.
Experience in the English Service has proved the desirability of making this alteration in the butt of the stock as a means of preventing the splitting off of the wood at the 'toe', which point, from its ??? and too great prominence, has been found liable to this casualty: and the alteration had been decided upon before I left England by the authorities. This modification of the stock of the Enfield arm involves also a corresponding alteration of the form of the Butt-plate

CONCLUSION

to one of more simplicity and consequent facility of manufacture, and I therefore recommend the adoption of the Enfield stock with the above described alteration.

5th Rear Sight.

I would recommend the adoption of a "block and three leaved sight" for 100, 200, 300 and 400 yards ranges instead of the elevating sight as on the Enfield Rifle, the utility of which is much questioned by military men, both in the English Service and on the continent of Europe. This "leaf sight" should be firmly attached to the barrel by soldering. This will prevent it from its being laterally displaced by an accidental blow or fall, which is liable to frequently occur in service.

*

The above suggestions cover the chief features of the arm as I should establish them, and I shall be most happy if they may prove [to] be of any service to you, as I feel very desirous that the State of Virginia should establish the manufacture of an arm of the very best model and quality, and now is the time to consider and decide upon all the details of its construction and capabilities.

If I can be of further service to you do not hesitate to call on me at any moment.

I am, Sir
Your most obedient servant,
Jas. H Burton.

* 6th Bayonet

The bayonet of the Enfield Rifle is a good one in form and proportions, and in my opinion cannot well be improved upon, and I therefore recommend it as a good model to be copied.

It is also worth noting the regard in which the Pattern 1853 rifle was held in that it probably saw greater service in the United States than in British service. Roughly half a million were supplied to the Union and Confederate armies during the Civil War. Indeed, the Confederacy effectively adopted it as their 'official' infantry rifle and James Burton, late of Enfield, was appointed superintendent of Confederate armouries to manufacture them.[1]

The nature of the machines used at Enfield, what they achieved and how they achieved what they did has been presented here in greater detail than before. The shortfalls in detail and errors in the few contemporary accounts that exist have been amplified and corrected to hopefully provide a greater understanding of their mechanical attributes. Anyone with an understanding of the traditional gunmaking processes based extensively on hand-craft techniques will appreciate the virtuosity applied in creating machines that replaced the greater majority of those hand skills. This virtuosity is apparent in the barrel rolling process, the profile milling machines for the lock components and stock making machines. All of these machines and others had to be designed and constructed with meticulous care because, while producing very different components, all those components had to be produced to the same standard of accuracy so that they could be combined to create a finished rifle. An even greater achievement, not seen before in Britain, was that they were able to interchange with the same components on other rifles.

However, it should be noted that not all the Enfield Pattern 1853 series of firearms were manufactured at Enfield after 1857 – Carbines and Short Rifles continued to be made by contractors for a while and were not interchangeable.

There is work still to be done in this field. The metrology study was only able to concentrate on pre-1857 rifles. The same needs to be applied to post-1857 rifles to gain some idea of the acceptable tolerances in a gauging system which only set an upper size limit. However, herein lies a problem. It has to be assumed that post-1857 long rifles were mostly, if not entirely, converted to the Snider system, thus transferring interchangeability to the new weapon. This view is supported by an unlikely source. On May 18th, 1876, a writ of summons was served on Richard Brown Roden by Thomas Wilson for infringement of his, Wilson's, patent.[2] Evidence was given by Major General William Manley Hall Dixon, CB, who at that time was Secretary and Manager of The Birmingham Small Arms Company Limited. In his evidence he stated:

> *"That from the year 1855 to the year 1872 he was Superintendent of the Royal Small Arms Factory at Enfield. That he was fully and minutely acquainted with the circumstances connected with the adoption of the Snider gun into the British Service and had the active management of perfecting the so-called Snider Improvements and the application of such improvements to Enfield rifles. The first application of such improvements was to the transforming the new Enfield guns in-store into breech loaders. The whole of the guns so transformed were new and had not been issued to the army or used. Subsequently, some portion of the Enfields returned to store were also transformed. When the whole of the Enfield guns in store had been exhausted in or about the year 1868 . . ."*

Thus, simply searching for Pattern 1853 long rifles with post-1857 dated locks inscribed 'Enfield' may therefore prove futile. Out of the 152 'standard issue' long rifles in Royal Armouries collections, only 1 has a post-1857 dated 'Enfield' marked lock. However, should anyone wish to pursue this line of enquiry it is reasonably safe to assume that any Snider rifles designated as Pattern (or Mark) I, I*, II*, II** will have the original Enfield rifle locks. The same should apply to Pattern (Mark) III rifles which, although manufactured from new, should have identical locks to the earlier converted rifles.

Appendix 1

The Percussion Cap

Copy of a document once in the collection of Springfield Armory and now in the State Archives of Massachusetts. (Only the relevant portion included, with transcript and highlighted). A patent, *"Improvement in the percussion gun"*, was issued June 19th, 1822, to Joshua Shaw of Philadelphia.

Remarks Relative to Mr. Shaw's Detonating Gun.
July 11th, 1822, directed to the Editor of the *Hampton Patriot*

Mr. Joshua Shaw of Philadelphia has made an improvement in firearms which promises to be of considerable public utility. It is termed "the Detonating Gun". It differs from the common gun by having attached to the barrel a kind of pan called the Antechamber which is screwed into and communicates with the chamber in the barrel, to this is attached what is called the priming tube which is a piece of case-hardened iron or steel about half an inch in length and about ¼ of an inch in diameter next to the pan and a little smaller at the other end where it receives the cock with a small hole through it lengthwise into the antechamber, **on this tube is placed the priming cap which is a small piece of copper made in a form so as to exactly fit the tube, and has on the upper part of the inside a small quantity of oximuriate of potash and some other ingredients, which on the percussion of the cock explodes and communicates to the charge in the barrel with the rapidity of lightning**. [Author's emphasis]

(Oximuriate [sic] of potash is potassium chlorate and mixed with sulphur is easily detonated.)

Appendix 2

Joseph Smith

Joseph Smith of 28, Loveday Street, Birmingham, was a government contractor, c. 1851-1860. His work seems to have been varied, acting mainly as a 'setter-up' but later referring to himself as a 'gunmaker'. This small archive consists of a variety of documents issued to him in his various capacities.

Some are official Board of Ordnance, or after 1854, War Department printed contracts, whilst others are manuscript documents notifying changes to contracts, specifications or inspection procedures, etc.

With the exception of a letter from the secretary of the 'combination' of 'setters-up', and two contracts relating to the 1851 'Minié' rifle, all encompass arms within the Pattern 1853 series and whilst the '3-band' rifle musket is only mentioned twice, the details regarding various components and inspection criteria applied to the other arms in the Pattern 1853 series are equally applicable to the rifle, except proof charges where specified.

MAKING THE ENFIELD PATTERN 1853 RIFLE-MUSKET

1. 1851 – contract for 'jegging breech pins' and 'filing the lumps' of Minié barrels.

Summer Lawn OFFICE OF ORDNANCE,
27 October 1851.

Terms and Conditions of Contract proposed to be entered into by The Board of Ordnance for *jegging the Breech pins and filing up the lumps of 23,000 Barrels for Rifle Musquets.*

THE Contractor to undertake to supply the Articles, or to perform the Work described in the annexed Specification, which are required for Her Majesty's service, or such portion thereof as The Board may agree for, and to deliver the same in regular weekly proportions on stated days during the continuance of the Contract.

The articles and workmanship to be equal and according in all respects to the Patterns and Specifications, which may be inspected on application at the Small Arms Office at the Tower or at Birmingham; and the workmanship and quality of the several parts will be viewed, inspected, and tested by the rules inscribed in the Specification annexed to this Contract; and the articles that may be found faulty in workmanship or defective in quality will be rejected; and the Contractor is to remove the same at his own expence on notice being given to him, without any allowance being made to him for such rejected articles.

If the Contractor should fail in the regular delivery of the articles contracted for at the periods of time agreed on, The Board are to be at liberty after the lapse of one of those periods, to purchase or procure the articles of other persons, and to charge the Contractor with the difference between the cost of such articles and the Contract prices; which difference may be deducted from any bills that may be then due to the Contractor, or the amount may be demanded of him to be paid within fourteen days in such manner as The Board may direct. And if the Contractor should not properly fulfil his agreement the Contract may be wholly terminated by The Board.

The party whose tenders shall be accepted, to name, if so required, two sureties, to be approved by The Board, to be bound with him in such sum as The Board may name, for the due performance of the Contract.

Payment is to be made monthly on the certificates of the Inspector of Small Arms, and the Contractor will, on receiving notice from the proper office, and not before, be authorized to draw a Bill for the amount of his account at three days sight, or to attend personally at the Paymaster General's Office to receive payment.

One-half the expense of the Stamp on the Contract and Bond to be paid by the Contractor, should The Board direct a stamped Contract and Bond to be prepared.

The Contract to be declared void should the Contractor, or any person employed by him, pay or offer to pay any gratuity or reward to any clerk, viewer, or other person under the employ of the Ordnance, for anything done or to be done by such clerk, viewer, or other person concerning the execution of this Contract, or inspection of the work delivered, or for the passing such work, or discharging any imprest or debenture for the payment of money in pursuance thereof.

JOSEPH SMITH

., No member of the House of Commons in himself or through other persons to be admitted to any part or share of this Contract, or to any benefit to arise therefrom, and the Principal Officers of Her Majesty's Ordnance are not to be answerable in their private capacities or in their own persons or estates for anything connected with or arising out of this Contract.

The Board of Ordnance reserve to themselves the power of rejecting the whole or any of the tenders.

The persons whose tenders may be entertained, will be called on to send in a sample of their workmanship or materials before their offers are finally accepted, and after the Contract is entered into, the prices will not be reconsidered or augmented under any circumstances whatever.

This Form of tender to be properly filled up, signed, and returned, so as to be received on or before Tuesday, the 4th November, addressed to the Secretary to the Board of Ordnance, Pall Mall London, and marked on the outside tender for "Jigging Breechpins"

Specification & Rules to be attended to in jigging the Breechpins & filing up the dumps thereof of 23000 Barrels for Rifle Muskets, and in the inspection of the same under the general Conditions of Contract annexed hereto, forming part hereof —

The Barrels after being cut and Recessioned by Machine will be delivered to the Contractor in fixed numbers, not exceeding 800 P[er]ch, on a day to be named in each week, & the same number to be returned into the View Rooms within the following week with the Breech-pins jigged & the dumps filed up — and to insure regularity in the weekly deliveries, the Contractor to be liable to a fine of three pence for every Barrel passed over that shall be short of the weekly proportion agreed for; which fine is to be deducted from his monthly account. —

> The work is to be filed up in a workmanlike manner, spoon, and to fit the Jigs & Gauges adapted to it; and any Barrel that may be injured by the carelessness or unskilfulness of the Contractors workmen will be rejected; the marks struck out, and the full value of the Barrel deducted from his account.
>
> ———
>
> The Party to state the price Per Barrel, and the number of Barrels he will undertake to deliver weekly, viz:
>
> Price per Barrel ———— at
>
> Barrels to be delivered weekly — N°.

the undersigned have carefully considered the Terms and Conditions set forth in the foregoing form of Contract, with the Specifications and Rules annexed, and   hereby agree and bind   to fulfil and abide by the same; and to perform the Work or supply the Articles above mentioned, or such portion thereof as The Board of Ordnance may agree for, at the price or prices specified by   against each Article.

Signature _____

Address _____

Witness.

Signature _____

Address _____

JOSEPH SMITH

2. 1852. Documents relating to cock [hammer] stamping.

This Form of Tender to be properly filled up, signed, and returned, so as to be received on or before *10 June* addressed to the Secretary to The Board of Ordnance, Pall Mall, London, and marked on the outside Tender for *Stamped Musket Cocks*

OFFICE OF ORDNANCE,
4 June 185*2*.

Terms and Conditions of the Contract

PROPOSED TO BE ENTERED INTO BY THE

Honorable Board of Ordnance.

THE CONTRACTOR will supply the Articles specified in the accompanying Schedule, which are required for Her Majesty's Ordnance Service, to which he has affixed prices, or such part thereof as The Board may be pleased to order of him, within the time specified for the completion of the Order.

The Articles to consist of the qualities and sorts described, and to be equal in all respects to the Patterns or Specifications, which may be inspected

On application at the Principal Storekeeper's Office, at the Tower.

Duplicate Patterns will be delivered by the Ordnance, on payment of their value, when application is made for them.*

The Articles are to be delivered by the Contractor at the

The Tower,

but previous to their being received into Store, they are to be surveyed, and if found inferior, or defective in quality, they will be rejected, and the Contractor is to remove the same at his own expence, within ten days after he is required so to do, without any allowance being made to him for such rejected Articles.

If the Contractor neglects or refuses to supply the Articles contracted for, The Board are to be at liberty to purchase or procure the same of other persons, and to charge the difference between the cost of such Articles and the Contract prices to the Contractor, which difference may be deducted from any Bills which may be due to the Contractor, or may be demanded of him, to be paid within Fourteen days, for service of the Ordnance Department, in such manner as The Board may direct; or if the Contractor does not properly fulfil his Agreement, The Board of Ordnance to have the power wholly to terminate it.

In case the Orders given by The Board for the provision of any Articles are not fulfilled within the period specified in the Order or Warrant for the supply, the same will be considered to be cancelled, and no Article remaining unsupplied upon the Order will be allowed to be delivered without the specific order of The Board.

* Parties tendering are, for their own security, particularly recommended to inspect the Patterns and Specifications, as the Contract is entirely governed thereby. And they are also requested to observe that the reference number on the Pattern, and the number inserted in the Specification agree in every instance.

3. Though undated, from the designation of 'new Pattern' rifle muskets, this document is believed to relate to the Pattern 1853 rifle, rather than the Pattern 1851.

SPECIFICATIONS AND RULES to be attended to in setting up new Pattern Percussion Rifle Musquets, and in the delivery and inspection of the same weekly, in addition to the general conditions of Contract annexed hereto.

The Musquets are to be set up with the Barrels, Locks, and Materials to be received by the Contractor from the Ordnance Stores, and none others are to be substituted, under any pretext whatever. If any of the materials should turn out faulty in working, they are to be exchanged from Store, but no allowance made to the Contractor for the labour of replacement. No cracked or glued stock will be passed in any branch. If any part of the materials be made waste through the carelessness or want of skill of the workmen, or from any accident, that article to be replaced from Store, and the value deducted from the Contractor's account.

Barrels, Locks, and Materials, in complete sets, to the amount of six times the number agreed to be set up weekly, are to be issued to the Contractor at the commencement of the agreement, on a day or days of the week, to be named; and the weekly number of Musquets agreed for are to be delivered for view, in the rough-stocked state, on the eve of that same day or days, in the following week; and the work thus continued regularly, through all its branches, in weekly succession; so that the screw-together view, the barrel and the finish view, may be taken always on the same day or days of the week, and no other; and as soon as one weekly proportion of Musquets is finished and passed, a further issue of materials to that amount to be made on the following day of the week to the view days; so that there shall always be six times the number of sets of materials to the musquets agreed for weekly in the hands of the Contractor, up to the closing of his Contract; but no more than the numbers agreed for are to be passed, in either branch, in any one week.

The Contractor to find two good and sufficient sureties to enter into a bond with him for such sums as the Board of Ordnance shall name, to cover the value of the materials issued to him, and the penalty for non-performance of the Contract.

Also, to ensure regularity in the deliveries for view, the Contractor is to be liable to a fine of 2s. 6d. for every finished Musquet passed view that shall be found short of the weekly number agreed for, which fine is to be deducted from his account monthly.

A properly qualified person is to attend the view in each branch, on the part, and at the cost of the Contractor, to bring up and strip the work, and to take the instructions of the Ordnance Viewers.

COURSE OF INSPECTION AND PROOF.

1st.—The Musquet will be viewed in the rough-stocked state. The Stock to be sent in with the Ordnance Marks on the Butt; to be clean cut, to fit the standard and guages, and the Barrel and Lock neatly fitted and well bedded, to ascertain which they are to be taken out of the wood.

2nd.—The Musquet will be viewed in the screwed-together state, percussioned, cock filed, springed, and with the bayonet, rammer, and all its parts neatly and firmly fitted, and well bedded; after examination of the exterior, to be stripped for view by the person attending on the part of the Contractor.

3rd.—The Barrel to be examined internally for rifling by plugging, and viewed for sighting, to see that the sight be fitted square, upright, and of due elevation; to be then proved for the second time, and afterwards examined internally and externally.

4th.—The Barrel to be viewed for smoothing and browning; in this stage, the barrel to be sent in with the breech-pin loose, and to be retained in the view-room long enough to ascertain that the browning has been perfectly killed.

5th.—The Stock to be viewed for making off, oiling and cleansing with the lock-plates in, and to see that it fits the guages.

6th.—The Musquet to be viewed in the finished state, to be sent in completely made off, oiled, and finished in a workmanlike manner, according to the pattern and quality of the Arm agreed for. In this state the fitting and play of the Springs to be tried; the Barrel to be taken out and plugged; the Lock to be taken off, and its interior, and the side nails examined; and the whole Musquet carefully viewed at all points, when stripped, by the person attending on the part of the Contractor according to the custom of the Service.

JOSEPH SMITH

4. Contract for setting-up Pattern 1853 rifles

28 Loveday St
Birmingham
31 July, 1853

The Hon'ble Board of Ordnance
To Joseph Smith Dr.

			£	s	d
	B O 12 June 1853 $\frac{0}{2033}$				
49	New Rifle Musquets setting up patt. 1853 @ 28/-		68	12	0

Joseph Smith

MAKING THE ENFIELD PATTERN 1853 RIFLE-MUSKET

5. Contract specifications for the Artillery Carbine Pattern 1853.

SPECIFICATIONS and Rules to be attended to in the Supply of Rifled Artillery Carbines ·577 bore, and in the Delivery and Inspection of the same, Weekly, under the general conditions of the Contract to be annexed hereto, and to form part hereof.

The whole of the parts of the Carbine which are to be submitted for view previously to their being worked (with the exception of the stocks, which are to be supplied from store), are to be provided by the Contractor, and are to be of the best description of material and quality of workmanship, in accordance with the patterns and specifications, viz. :—

Barrels.

The Barrels to be rolled from wrought iron moulds of the best quality, and to contain no injurious greys or flaws either inside or outside. The bore of the Barrel is to be free from rings or tears, and must shade straight from end to end; the exterior to be ground, struck up and filed in a proper workmanlike manner, to be gauged in three places, viz., at the breech end, at the middle, and at the muzzle: the breeching is to be executed in a sound workmanlike manner, to have a full and clear screw inside the chamber, and on the breech pin, and to joint at top and bottom; the percussion lump (of the best wrought iron) is to be soundly and neatly welded to the barrel; the front sight and sword bar are to be firmly brazed with pin clippings, and to be in accordance with the standard gauges and pattern barrel.

Locks and Cocks.

The Locks and Cocks to be filed, fitted, and finished in a sound workmanlike manner, agreeably to the Lock of the pattern arm, and the standard tests and gauges; the main-springs, sear-springs, and all the screws (except the tumbler screws), are to be made of the best cast steel, and the rest of the parts (with the exception of the swivel, which is to be made of specially prepared steel of a very tough and mild description), are to be made of the very best wrought iron, free from flaws and greys.

Sword Bayonet and Steel Scabbard.

The blade to be made of the best clear cast steel, properly hardened and tempered, and the tang to be made of the best wrought iron, and to be shut on at the shoulder of the blade. The weight of each blade, when ground and polished, to be from 1 lb. 2 oz. to 1 lb. 2½ oz. The cross piece and pommel to be made of the best wrought iron, and to be neatly and firmly fitted and brazed. The scales for the grip to be made of the best neat's butt leather, to be made in two pieces as far as possible, and in no case in more than three pieces. The parts to be securely and firmly joined together, and the chequering to be well brought up in properly made dies.

The Scabbards to be made of the best cast sheet steel, to be neatly jointed and brazed with pin clippings, and to be ground and polished in a workmanlike manner, in accordance with the pattern and gauges, and to weigh from $13\frac{1}{2}$ oz. to $14\frac{1}{2}$ oz.

Rammers.

The Rammers to be made of the best cast or shear steel, with the exception of the head and two inches below, which is to be made of the best wrought iron, tempered, and finished in every way equal to the rammer belonging to the pattern arm, agreeably to the standard gauges and tests.

Bands.

The Bands to be made of the best and toughest wrought iron; the front Band to be fitted with a swivel, and to be filed and finished in every respect equal to the Bands belonging to the pattern carbine, agreeably to the standard mandrils and gauges.

Nipples.

The Nipples are to be made of the best cast steel, to be properly tempered and finished agreeably to the standard tests and gauges.

Small Materials.

The whole of the small work, viz., rammer-springs, wood, screws, &c., to be made of the best description of materials of their various kinds, and to be finished agreeably to the same article belonging to the pattern carbine, and in accordance with the jegs and gauges.

Furniture.

The Mountings to be made of pure tough brass, equal to those of the pattern carbine, agreeably to the jegs and gauges.

Setting-up.

The Setting-up will comprise rough stocking, screwing together, fitting sword bayonet, sighting, freeing, and adjusting the pull off of the lock, engraving and hardening lock, smoothing and browning barrel, making off stock, and finishing complete, which is to be executed in a sound workmanlike manner, in every respect equal to the pattern arm, and in strict accordance with the standard jegs and gauges.

A properly qualified person is to attend the view at the cost of the Contractor, to bring up and strip the work, and to take such instructions as may be found necessary.

COURSE OF INSPECTION AND PROOF.

Barrels.

The Barrel to be sent down for Proof and inspection, percussioned, jointed, and properly breeched; the bore is to be tested with two plugs, viz., one of the diameter of ·572 for the receiving, and one of ·577 for the rejecting plug. They are to be proved with the regulated proof charge of 7 drachms of powder (FG) and a conical service bullet of 530 grains weight, and to stand at least twenty-four hours after proof before being inspected. The bores are to be tested and examined for boring and straightness. The outsides to be closely examined and gauged, and the sword bar and front sight struck, and gauged, and the breeching to be closely tested. They will be examined for greys or flaws, and are to be filed and finished equal to the pattern.

Locks.

The Locks with Cocks fitted to be sent in for view in the filed state, and to be examined closely to jeg and gauge. The whole of the Locks to be stripped, and the fitting and filing of each part separately examined. The main-springs to draw from 13 lbs. to 15 lbs. at half-cock. The sear spring to weigh from 6 lbs. to 8 lbs. The quality and workmanship to be equal in all respects to the Lock of the pattern carbine agreeably to the standard jegs and gauges.

Sword Bayonet and Steel Scabbard.

1. The blade to be sent down for view in the ground state; to be gauged in two places, and to fit accurately to trough, in accordance with the pattern; to be struck on the back and edge on a block of wood in the usual manner, and to be sprung twice on each side at the centre of the blade over a bridge that will raise the blade at the shoulder 4 inches from the level.

2. The blade to be brought down for view, with the cross piece and pommel brazed on and filed up, and the spring fitted. The tang and mountings to be properly gauged and tested, and the blade to be also tested at the shoulder, in order to ascertain that it has not been softened in brazing on the cross piece.

3. The scales for the grip to be brought down and examined previously to mounting

4. The Sword Bayonet to be brought bown mounted, polished, and finished complete with the Scabbard, to be finally examined as to its being finished and mounted in every way equal to the pattern. The Sword Bayonet, when complete, without Scabbard, to weigh from 1 lb. 10 oz. to 1 lb. 12 oz.

Rammers.

The Rammers are to be viewed in the polished and finished state. They will be gauged in three places, at the middle and at each end, and to be sprung over a bridge in the centre of the stem $4\frac{1}{2}$ inches high; to be struck fair lengthways on an oak or hard wood bench, and across a block at the end of the shut near the head. They must be ground, polished, and finished in every respect equal to the rammers of the pattern arm.

Bands.

The Bands to be viewed in the filed state with swivel on the front band; to be gauged and tested to the standard tests and gauges, and filed and finished in all respects equal to the bands of the pattern arm.

The small materials and mountings are to be examined in the Setting-up inspection. A departmental mark will be placed upon the various materials as a guarantee that they have been examined by the Viewer of the Small Arms' Department, and have been accepted as fit to be worked; but these marks are to have no other signification. They must, however, remain on the various parts in all the stages of viewing.

Nipples.

The Nipple will be tested for soundness and being properly tempered, by being placed in a barrel when proved, or struck with a hammer on the top; to be closely gauged in all parts, and the screw strictly tested as to its correct size and pitch.

SETTING-UP.

Carbines.

The Carbines to be viewed in the rough stocked state; the stocks to be sent down with the Government marks on the butt; to be clean cut to fit the standards and gauges correctly; and the barrels, locks, and bands to be neatly fitted and well bedded, to ascertain which they are to be taken out of the wood. Should the arm pass view in this stage, the Viewer will transfer the Government mark to a part of the stock, where it will remain permanently.

2nd. The Carbine to be sent down in the screwed-together state complete, viz., Sword Bayonet, and Rammer and Rammer Spring fitted; Cock and Lump filed; mountings all properly fitted and filed; all these points will be carefully viewed and marked.

3rd. The barrels to be viewed for boring up with the back sight soldered on.

4th. The barrels in the rifled and sighted state to be sent down for verifying proof; and inspection for rifling and sighting, to be proved with a charge of 4 drachms of (F.G.) powder, and conical service bullet, and to be closely tested for rifling and sighting, with standard plugs, and gauges, agreeably to the pattern arm.

5th. The barrels in the browned state to be sent down to the view room two clear days previous to viewing. They will then be examined on the outside for browning and smoothing, and in the inside for straightness, &c. Such barrels as pass this view, are to be marked and retained in the view room, until presented for final view as a finished arm.

6th. The arm will lastly be viewed complete, and carefully examined.

MAKING THE ENFIELD PATTERN 1853 RIFLE-MUSKET

6. Specifications for the Short Rifle, Pattern 1856.

SPECIFICATIONS and RULES to be attended to in the supply of SHORT RIFLES, ·577 *bore, complete, for Rifle Corps, and in the delivery and inspection of the same, weekly, under the general conditions of the Contract to be annexed hereto and to form part hereof.*

THE whole of the parts of the Rifles, which are to be submitted for view previously to their being worked (with the exception of the Stocks, which are to be supplied from Store), are to be provided by the Contractor, and are to be of the best description of material and quality of workmanship, in accordance with the Patterns and Specifications, viz.:

Barrels.

The Barrels to be rolled from wrought-iron moulds of the best quality, and to contain no injurious greys or flaws, either inside or outside. The bore is to be free from rings and tears, and must shade straight from end to end; the exterior to be ground, struck up and filed in a proper workmanlike manner, and to be gauged in three places, viz., at the breech end, at the middle, and at the muzzle; the breeching is to be executed in a sound workmanlike manner, to have a full and clear screw inside the chamber, and on the breech-pin, and to joint at top and bottom: the percussion-lump, breech-pin, and front-sight are to be made of the best wrought-iron; the percussion-lump is to be soundly and neatly welded to the Barrel; the front-sight is to be firmly brazed with pin-clippings, and to be in accordance with the standard gauges and pattern Barrel.

Locks and Cocks.

The Locks and Cocks to be filed, fitted, and finished in a superior and sound workmanlike manner; the tumblers and sears to be freed top and bottom, agreeably to the Lock of the pattern Arm, and the standard tests and gauges; the main-springs, sear-springs, and all screws (except the tumbler screws) are to be made of the best cast-steel, and the rest of the parts (with the exception of the swivel, which is to be made of specially prepared steel of a very mild description) are to be made of the very best wrought-iron, free from flaws and greys.

Sword Bayonet. *& Scabbard*

1st. The Blade to be made of the best clear cast-steel, properly hardened and tempered, and the tang to be made of the best wrought-iron, and to be shut on at the shoulder of the blade; the weight of each blade when ground and polished to be from 1 lb. ½ oz. to 1 lb. 1½ oz.; the cross piece and pommel to be made of the best wrought iron, and to be neatly and firmly fitted and brazed; the scales for the grip to be made of the best Neat's butt leather, to be made in two pieces as far as possible, and in no case in more than three pieces; the parts to be securely and firmly joined together, and the chequering to be well brought up in properly made dies.

The Scabbards to be made of the best description of Leather, and to be fitted with steel mountings in accordance with the Pattern both as regards size and form.

Rammers.

The Rammers to be made of the best cast or sheer steel, with the exception of the head and two inches below, which is to be made of the best wrought-iron, tempered, and finished in every way equal to the Rammer belonging to the pattern Arm, agreeably to the standard gauges and tests.

Bands.

The Bands to be made of the best and toughest wrought-iron; the front band to be fitted with a swivel and sword-bar, and to be filed and finished in a superior workmanlike manner, in accordance with the bands of the pattern Arm, agreeably to the standard mandrils and gauges. The sword-bar is to be forged solid with the band.

Elevating Back-Sights.

The Beds to be made of the best wrought-iron, free from greys and flaws, and the rest of the parts to be made of the best sheer or cast-steel; the sight to be filed, pierced, and finished in a superior workmanlike manner, in accordance with the pattern, agreeably to the standard jegs, gauges and tests.

Furniture.

The Mountings to be made of the best wrought-iron, in accordance with those of the pattern Arm, agreeably to the standard jegs and gauges, and to be case-hardened, when finished.

Nipples.

The Nipples are to be made of the best cast-steel, to be properly tempered and finished, agreeably to the standard tests and gauges.

Small Materials.

The whole of the small work, viz., rammer-springs, triggers, nails, wood, screws, &c., to be made of the best description of material of their various kinds, and to be finished agreeably to the same article belonging to the pattern Rifle, and in accordance with the jegs and gauges.

Setting-up.

The setting-up will comprise rough stocking and screwing together, fitting sword bayonet, rifling and sighting, freeing and adjusting the pull off of the lock, engraving and hardening lock, and case-hardening furniture, smoothing and browning barrel, making off stock and finishing complete, which is to be executed in a sound workmanlike manner, in every respect equal to the pattern Arm, and in strict accordance with the standard jegs and gauges.

A properly qualified person is to attend the view, at the cost of the Contractors, to bring up and strip the work, and to take such instructions as may be found necessary.

Course of Inspection and Proof.

Barrels.

The Barrels to be sent down for proof and inspection, percussioned, jointed, and properly breeched. The bore is to be tested with two plugs, viz., one of the diameter of ·572 for the receiving and one of ·577 for the rejecting plug; they are to be proved with the regulated proof charge of 7 drachms of powder(F. G.), and a conical bullet, 530 grains weight, and to stand at least 24 hours after proof before being inspected; the bores are to be tested, and examined for boring and straightness; the outsides to be closely examined and gauged, the front-sight struck and gauged, and the breeching to be closely tested. They will be examined for greys and flaws, and are to be filed and finished equal to the pattern.

Locks.

The Locks and Cocks fitted to be sent in for view in the filed state, and to be closely examined to jeg and gauge. The whole of the locks to be stripped, and the fitting and filing of each part separately examined. The main springs to draw from 13 lbs. to 15 lbs. at half-cock. The sear springs to weigh from 6 lbs. to 8 lbs.; the quality and workmanship to be equal in all respects to the lock of the pattern Rifle, agreeably to the standard tests and gauges.

Sword Bayonet.

The Blade to be sent down for view in the ground state; to be gauged in two places, and to fit accurately to trough, in accordance with the pattern; to be struck on the back and edge over a block of wood, in the usual manner, and to be sprung twice on each side, at the centre of the blade, over a bridge that will raise the blade at the shoulder 4 inches from the level.

2nd. The blade to be brought down for view with the cross piece and pommel brazed on, and filed up, and the spring fitted; the tang and mountings to be properly gauged and tested, and the blade to be also tested at the shoulder, in order to ascertain that it has not been softened in brazing on the cross piece.

3rd. The scales for the grip to be brought down and examined previously to mounting.

4th. The Sword Bayonet to be brought down mounted, polished and finished complete, and to be finally examined as to its being finished and mounted in every way equal to the pattern. The Sword Bayonet when complete (without Scabbard) to weigh from 1 lb. 10 ozs. to 1 lb. 12 ozs.

Rammers.

The Rammers are to be viewed in the polished and finished state. They will be gauged in three places, at the middle and at each end, and sprung over a bridge in the centre of the stem, $4\frac{3}{4}$ inches high; to be struck fair lengthways on an oak or hard wood bench, and across a block at the end of the shut near the head. They must be ground, polished, and finished in every respect equal to the rammer of the pattern Arm.

7. Contract for the Pattern 1856 Cavalry Carbine (East India Co. carbine)

SPECIFICATIONS AND RULES to be attended to in setting up Carbines Cavalry, Rifle E. I. C. pattern (last approved), bore .577 ; and in the delivery and inspection of the same weekly, under the general conditions of the Contract, to be annexed hereto, and to form part hereof.

The Carbines are to be set up with Stocks, Barrels, Locks, Elevating Sights, Ribs with Rings, Rammers with Swivel and Rammer Springs, Bands, and Brasswork, to be delivered to the Contractor from the Military Store Office, and none others (except the small materials which are to be provided, according to the Tender, by the Contractor himself, and which are to be of the best quality) are to be substituted under any pretext whatever. If any of the materials found by the Department should turn out faulty, they are to be exchanged from Store, but no allowance is to be made to the Contractor for the labour of replacement ; no cracked or glued Stock will be passed in any branch. If any part of the material be made waste through the carelessness or want of skill of the workman, or from any accident, that article to be replaced from Store, and the value deducted from the Contractor's account.

The materials to be provided by the Department in complete sets, to the amount of thirteen weekly issues, according to the number agreed to be set up weekly, are to be issued weekly to the Contractor, on a day or days of the week to be named, and the weekly number of Carbines agreed for are to be delivered for view in each progressive state, on the eve of that same day or days in the following week ; and the work thus continued regularly through all its branches in weekly succession, so that the view for each progressive state may be taken always on the same day or days of the week, and no other ; and as soon as one weekly proportion of the Carbines is passed view in the finished state, a further issue of materials to that amount is to be made on the following day of the week to the view days.

A properly qualified person is to attend the view in each branch, on the part and at the cost of the Contractor, to bring up and strip the work, and to take the instructions of the Ordnance Viewers.

COURSE OF INSPECTION AND PROOF.

1st. The Barrel will be viewed for Percussioning and Machine Jointing.

2nd. The Barrel will be viewed for jegging and filing up the percussion lump, which is to be executed in a workmanlike manner, without the slightest alteration of the top square at the breech end, and to fit to jeg and guage correctly.

3rd. The Carbine will be viewed in the rough stocked state, the Stock to be sent in with the Government marks on the butt, to be clean cut, to fit the standard and guages, and the barrel, lock, and bands neatly fitted, and well bedded, to ascertain which, they are to be taken out of the wood.

4th. The Carbine will be viewed in the screwed together state, cock filed up, and rammer fitted, and all its parts neatly fitted and firmly bedded ; after examination of the exterior, to be stripped for view by the person attending on the part of the Contractor.

5th. The Barrel to be viewed for boring up for Rifling to an easy .577 plug, to be rejected if the .580 plug enters either end of the Barrel ; the Backsight to be soldered on, and to be tested for soundness in fixing, and also to its being correctly on the Barrel.

6th. The Barrel to be sent in for the following views :—

    1st. Second Proof.

    2nd. Rifling, graduating and filing up front Sight, and to be tested in the Government machines, and barrel smoothing ; the burr raised in Rifling

MAKING THE ENFIELD PATTERN 1853 RIFLE-MUSKET

2

to be removed by lapping according to Barrels approved from Enfield. Draw leading will be allowed to remove any little roughness or rivelling on the cuts which may be considered necessary.

7th. The Barrel to be viewed for browning; in this stage the Barrel to be sent in with the breech pin loose, and to be retained in the view room long enough to ascertain that the browning acid has been perfectly neutralized, and the inside clean and uninjured.

8th. The Stock to be viewed for making off, oiling, and cleansing, with lock plates in, and to see that it fits the guages.

Lastly. The Carbine to be viewed in the finished state, to be sent in completely made off, oiled, and finished in a workmanlike manner, according to pattern and quality of arm agreed for; in this state the Barrel to be taken out and plugged, the lock taken off and its interior examined and freed, and the whole Carbine carefully viewed at all points, when stripped by the person attending for the Contractor, according to the custom of the Service.

NB. All the additional views which commenced on the 4th inst. (4 May 1857) will be applied to these arms

R.W. Gunner
for the Supt.

8. 1857 Memorandum issued from Military Stores Office, Birmingham, relating to bore gauging, inspection, and tolerances. (Enlarged and spread over three pages for clarity.)

> Guaging and Viewing Bores of Barrels.
>
> In order to ensure the Barrels being perfect to guage at the muzzle, when the Arms are finished and ready for transmission to Store, Mr Shirley will guage each Barrel separately, after the Arm has undergone its Finishing View, with a ·577 inch plug, taking care that it passes freely from end to end of the Barrel, and he will also test the muzzle with the ·580 inch plug, allowing of no toleration, but condemning every Barrel that takes the rejecting plug at all —

In order to ensure the barrels being perfect to guage [sic] at the muzzle, when the arms are finished and ready for transmission to Store. Mr. Shirley will guage each barrel separately, after the arm has undergone its Finishing View, with a ·577 inch plug, taking care that it passes freely from end to end of the barrel and he will also test the muzzle with the ·580 inch plug, allowing no toleration, but condemning every Barrel that takes the rejecting plug at all.

He will place his mark for this view crosswise on the stock, between the back end of the guard strap and the contractor's name.

The present plan of guaging, examining, and marking the Barrel, in the view for 'Browning', will be continued. In guaging, no toleration will be allowed at the muzzle beyond the difference between the receiving and rejecting plugs; but at the breech the rejecting plug may enter for one third of its length, but anything beyond this will cause the barrel to be condemned.

> Since a certain amount of toleration has inadvertently been permitted, in order not to throw on the Contractor's hands Barrels that have been Viewed for Boring whilst this toleration has been reckoned on by them, a new view mark for Boring will be put on all Barrels coming in after this date, and in the after views no toleration will be allowed to barrels thus marked.
>
> All Barrels that have been viewed for boring previous to this date, will be allowed a toleration of the rejecting plug entering the muzzle to the extent of one inch.
>
> Military Store Office
Birmingham 11 Nov. 1857

9. 1858 Further memorandum issued by James Burton changing the specifications of bore sizes.

> *In order not to exceed the depth of .015, which is the figure representing the depth of groove at breech, calculated upon one turn over 6ft 6in: and commencing at the muzzle with .005, it is absolutely necessary to have a plug to test the rifling, slightly under .015 at the breech.*
>
> *A plug for this purpose measuring .013 of an inch is to be employed for the reception plug, and one measuring .016 for the rejecting.*
>
> *At the muzzle, where the depth is intended never to exceed .005 of an inch, two plugs will be used in the Rifling View in order not to allow of any groove being accepted which may be deeper than .005.*
>
> *These two plugs will be respectively .005 for acceptance and .008 for rejection. The width of the groove in the rifle barrel is to be .235 of an inch.*

> Rifling Plug Guages
>
> Receiving, Breech, '603
> Rejecting, Do. '609
> Receiving, Muzzle, '587
> Rejecting, Do. '593
> Receiving, Breech '013
> Rejecting Do. '016
> Receiving Muzzle '005
> Rejecting Do. '008
>
> (Signed) Jas. H. Burton
> Ch. Engineer
>
> R.S.a.F.
> ap. 6 1858

10. Minute from a meeting of the contractors for 'setting-up', chaired by J. D. Goodman Esq. and signed by the secretary, Joseph Bourne, shows that 'combinations' did exist.

> That this meeting is of opinion that the Setting Up price of 28/ is indivisible, and it will not recognize the right of any member

At a meeting of Contractors for Setting Up held at the Stork Hotel, Thursday, March 27th, 1856, J. D. Goodman Esq. in the Chair.

(see transcript below)

That this meeting is of opinion that the Setting Up price of 28/- is indivisible, and it will not recognise the right of any member of this body of Contractors for Setting Up, either by himself, or his partners, or indirectly through any connexion of his with whom he may have pecuniary interest or advantage to tender for, or to accept any offers, or make any terms for any the least significant position or branch of the work for which in the aggregate 28/- is paid by the War Department, and this meeting further pledges itself not to countenance, but to oppose most strongly, under any form or pretext, a departure from such principle, and hereby bind each member of this body individually and collectively to seek personal interests by consulting the general interests of the entire body of Contractors for Setting Up.

[signed]
Joseph Bourne
Secretary

To Mr Joseph Smith

Appendix 3

James Burton

Transcripts and abstracts of relevant documents contained in the archives of Yale University (copies of actual documents too poor to be included) and in the Public Record Office in London

---

Springfield, Mass.

May 16th [?] 1854

To
    Col. Burns, H[er] B[ritannic] M[ajesty's] Ordnance Board.

Sir,

Understanding that the English Government contemplates establishing a manufactory of small arms and also the introduction therein of machinery and tools similar in character to those in use at the U.S. armories for facilitating the manufacture of military small arms, I have thought that it might be deemed expedient to secure the services of some mechanical engineer familiar with the construction and use of such machinery etc. Should this be the case, I beg leave to tender my services in that capacity, having been in the service of the U.S. Ordnance Department for the last ten years at the Harper's Ferry Armory in various positions immediately connected with the manufacture, and for the last four and a half years in the responsible position of Master Armorer having charge of the entire mechanical department of the Armory. I have testimonials from Colonel Benjamin Huger and Major Thos[?] Symington, both members of the U.S. Ordnance Board and have been in command of the Armory, and should it be necessary, can obtain more from other officers of the Department.

    For reasons which it is not necessary here to explain, I am desirous of obtaining an appointment elsewhere and would therefore be pleased to enter the service of your Government in the above mentioned capacity, provided a sufficient inducement were offered, say, not less than 400£ per annum, and expenses out for myself and my family.

    The consideration of this application is respectfully solicited at as early a day as may suit your convenience.

    I am
    Very Respectfully
    Your Obedt. Servt.

MAKING THE ENFIELD PATTERN 1853 RIFLE-MUSKET

London,

May 31st 1855

J Burton Esqu.
U.S. Armory

Sir,

I have the authority of the Hon Board of Ordnance for informing you that should you now be willing to engage yourself to the British Government as Engineer of the new Armory for 5 years or permanently, you will receive as pay four hundred pounds a year clear of taxes & a house will be found you.

I have to request that you will answer this immediately, as should you not be willing to accept this offer, it is necessary someone else should be appointed without delay.

    I have the honour to be Sir
    Yr obedient servant
    J Picton Warlow
    Capt. R. A.

Office of Ordnance

20th September 1855   S
10086

Sir,

Mr Monsel with the assent of the Lords Commissioners of her Majesty's Treasury having approved of your engagement as Chief Engineer of the Royal Small Arms Factory at Enfield, on the salary of £400 per annum, exclusive of Income Tax, commencing from the 15th June last and with the understanding that you shall be provided with a suitable house and that the expenses incurred by yourself and family in travelling to England shall be defrayed by the Government.

I am directed by Mr Monsell to notify the same to you, adding that authority has been given for the payment of your salary etc. accordingly.

    I am
    Sir
    Your most obedient servant,

    J H Burton Esq
    Care of Captain Dixon

JAMES BURTON

Royal Manufactory, Enfield

Sept 10th 1855

Capt. Dixon, R.A.
Inspector of Small Arms

Sir,

I beg leave respectfully to submit for your information the accompanying brief statement, exhibiting the numbers, description and probable cost of the machines yet required for this establishment, in order to complete the system of machinery for the manufacture of small arms. No facilities having as yet been provided for the finishing of the barrels.

List of Machines for finishing the Barrel of the Enfield Musket – 250 per day

8 rough boring machines	@ $475 = $3,800
4 Second " "	@ $285 = $1,140
6 Finish " "	@ $475 = $2,830
12 Turning lathes	@ $450 = $5,400
2 Double Hand lathes	@ $400 = $800
12 Milling machines	@ $300 = $3,600
fixtures for do.	@ $525 = $525
4 Drill presses	@ $250 = $1,000
2 Milling machines for breech screw	@ $200 = $400
2 Counterboring machines	@ $200 = $400
2 Screwcutting machines	@ $175 = $350
2 Tapping cone seat	@ $115 = $230
2 Milling machines for muzzle	@ $400 = $800
1 Machine for grinding	@ $100 = $100
1 Machine for inside polishing	@ $400 = $400
4 do outside do	@ $500 = $2000
	$23,795.00
add 7½ per cent for packing etc	$1,784.62
	$25,579.62
for gauges, small tools and fixtures	$4,420.38
	$30,000.00

[This differs in a few small but important details from the document in Public Record Office – see below]

MAKING THE ENFIELD PATTERN 1853 RIFLE-MUSKET

[Transcribed from memorandum in Public Record Office, Cat No. T 1/5975.]
List and estimated cost of machines &c. necessary for finishing 250 barrels per day.

		£	£
8 machines for rough boring at		101	= 808
4 do for second boring at		61	= 244
6 do for finish boring at		101	= 606
12 lathes for turning exterior at		90	= 1080
2 do double hand for milling at		86	= 172
12 machines for milling butt at		65	= 780
Fixtures for do do at		113	= 113
4 drill presses, vertical, at		54	= 216
2 machines for milling breech screw		43	= 86
2 do for counterboring for do		43	= 86
2 do for cutting screw of breech		38	= 76
2 do for tapping cone seat		25	= 50
2 do for clamp milling muzzle		86	= 172
1 do for grinding exterior		20	= 20
1 do for polishing interior		86	= 86
4 do do exterior		107	= 428
For gauges, small tools, fixtures &c.			900
For contingencies			577
64 machines	total cost		£6,500

Signed Jas. H. Burton
Civil Engineer

Royal Manufactory
Enfield
September 10th 1855

Royal Small Arms Factory,   Enfield Lock,
Nov. 14th 1855

Messrs Robbins & Lawrence,
Windsor, Vt, USA

Gentlemen,

I am authorized by Capt. Dixon, the superintendent of this Armory, to inform you that your proposals for furnishing the machinery, tools, &c &c, necessary for finishing 250 Rifle barrels per day, of the pattern at present adopted by this Govt., are accepted with some modifications and additions.

Annexed you will find a schedule as modified of the machines &c, the prices being in accordance with your tender.

Upon examining it you will observe that several machines not embraced in your proposals have been added to it, which you will, it is expected, furnish - at fair remunerative prices, and which you will be paid enough to make known to me as early as convenient. There are 2 machines for clamp milling the muzzle of the barrel for the bayonet socket, and 1 machine for grinding the exterior of the barrel; that is the lever, press[?] and block used for forcing this barrel up to the grindstone. The clamp milling machines, as also the lever press for grinding barrels – are desired to be made similar to those in use at the Springfield Armory for the same purposes, and which you will have reference previous to their construction.

The 7th item in your proposal, viz; 15 machines and fixtures for milling the barrel & sight is superseded by your last tender, July 17th 1855, for 12 milling machines, fixtures, cutters &c for the same purpose.

The time specified in your proposal for the delivery of this machinery is within fifteen (15) months of the date of order. Inasmuch as you now have all these machines developed and the patterns made, it is expected that you will be able to deliver them in much less time, which you will please specify in writing by return mail, fixing it at as early a date as possible as the machinery is much wanted.

The 16 machines for rifling the barrel are entirely omitted, inasmuch as we have machines here for the same purpose that are preferred to the American machines. You have, however, two to furnish upon the order you are now engaged in filling. We have also lathes for turning the barrel, which are considered superior machines, but as we wish to test your improvements in barrel lathes, one is included in the accompanying lot.

A model barrel will be sent to you as soon as it can be prepared, but you need not wait for this to put the machines in hand. Any further information you may require upon this subject, I will cheerfully give you upon addressing me as above.

I am Sirs
Your most Obedt. Servt.
J H B
Chief Engineer.

Schedule of Machines, fixtures &c &c to be furnished at the prices annexed by the Robbins & Lawrence Co. of Windsor, Vermont, U.S.A. for the British Government, the same to be delivered on shipboard at New York, well packed and boxed, within [blank] months of the date of order; an allowance of 7½ per cent upon the prices annexed to be added for boxing and delivery at New York. Said machinery to be adapted to the fabrication of the British Rifle Musket barrels.

MAKING THE ENFIELD PATTERN 1853 RIFLE-MUSKET

No 1	8 Rough boring machines for the first cut after the barrel is welded, of 3 spindles each, and each to be furnished with one set of tools called nut augurs.	Each $475
No 2	4 Second boring or reaming machines of 1 spindle each, and each to be furnished with one set of reamers.	Each $285
No 3	6 Finish boring machines of 2 spindles each, each to be furnished with one set of reamers.	Each $475
4th	1 Lathe for turning the barrel whole length, with one set of tools.	$245
5th	2 Double hand lathes, each to contain two head and tailstocks, with all the tools &c for squaring & cutting the barrel to exact length.	Each $400
6th	12 Milling machines for milling the barrel (breech &c)	Each $300
7th	17 sets of fixtures for the above 12 milling machines, and 21 sets of mills or saws for the various cuts upon the barrel &c	$525
8th	4 Drill presses for drilling & counterboring the tang & cone seat; each to be furnished with drills and fixtures for holding the barrels.	Each $250
9th	2 Machines for milling the end of the breech screw to diameter & length, with tools, cutters & fixtures for holding the tang.	Each $200
10th	2 Machines for counterboring the breech of the barrel for breech screw; each with tools and fixtures for holding the barrel.	Each $200
11th	2 Machines for cutting the thread upon the breech screw, each to be furnished with one set of Dies and taps.	Each $175
12th	2 machines for tapping and countersinking the cone seat, with b??? and taps complete.	Each $115
13th	1 Machine for polishing the interior of the barrel; to be furnished with one lot of rods, connecting rod, balance wheel and all the overhead works complete.	Each $400
14th	4 Machines for polishing the exterior of the barrel of 5 spindles each, with all the overhead works complete.	Each $500
15th	2 Machines for clamp milling the muzzle of the barrel for bayonet socket, with dies and reamers complete.	Each $ #
16th	1 Machine for grinding the exterior of the barrel, complete	Each $ #

[70 machines]

JAMES BURTON

Woolwich   26th January 1857

Dear Mr Burton,

I am truly delighted to hear that the musket is now complete.

It has been uphill work but let us thank God that it is now an accomplished fact and that success has crowned the efforts that have been made. I will not say how proud I shall be to have a complete Enfield Musket. I look upon it as the most mechanical triumph of the age.

On Wednesday I shall of course say a few words on the application of machinery to small arms, and should feel obliged for the loan of as many parts as can be got ready for me, mainly to lie on the table for examination.

I shall feel glad if I could have the box into which you enclose the lockplate while being drilled. Could you bring that with you if you come, if not can you send it.

Do not disturb yourself regarding the ?? arrangements, you will accomplish all in due time.

With very best regards
I am my dear Mr Burton,
Yours sincerely,

John Anderson

6th October 1858

My dear Burton,

Mr. ??? has informed me that Genl. Peel has highly approved of my proposal for increasing the production of the factory and he has sent out the papers to the Treasury. There will be no demur and I think you may also consider that you will be ordered to America to see to them

Dixon

3rd December 1858

My dear Burton,

You will be glad to hear that we are going on very well here. For the last two weeks we have done over 1200 arms.

Dixon

Appendix 4

R.S.A.F. Enfield Orders To Greenwood & Batley, Leeds

(Transcribed from order books contained in the Greenwood & Batley Archives, West Yorkshire Archives Service.)

Order No	Details	
	Col Dixon, Enfield Lock 26th May 1856	
7	12 small drilling machines [delivery commenced with 3 on 26 July and final 3 sent 23 Aug]	
	Col Dixon, Enfield Lock 8th Aug 1856	
56	7 vertical drilling m/c, in all respects like the twelve in Or No 7	£126
	Col Dixon, Enfield 22nd Oct 1856	
80	2 lathes for making sight spring screws with cross slide carrying the top two longitudinal slides with levers for milling the wire and threading the screw and 1 cross slide for cutting off the screw.	£64
	5 milling m/c for finish milling the head of the screw. Placed upon baseplate upon [???] same as the American m/c.	£190
	Col Dixon, Enfield 14th July 1857	
226	2 clamp milling m/cs	£90
	Royal Small Arms Factory, Enfield 3rd Sept 1857	
259	1 milling m/c for screws as before except bedplate 2" longer	£35
	2 clamp milling m/cs same as Or No 226	£90
	1 m/c to slit screws same as Or No 82	£30
	1 tapping m/c as before with 2 pullies [sic] and double clutch	£15
	Royal Small Arms Factory, Enfield 5 Sept 1857	
265	2 large grindstone troughs and spindles	£80

266	4 m/cs for rifling	£400
267	1 m/c for leading out after rifling	£40
268	1 m/c for tapping cone seat	£30
269	1 m/c for threading breech screw	£30
270	1 m/c for edge milling	£60
271	2 m/cs for cupping head of ramrod	£20
272	1 polishing bench, double shaft and pullies	£120
	Col Dixon, Enfield 16th Oct 1857	
284	2 wood screw threading m/cs	£30
	Col Dixon, Enfield 31st July 1858	
424	2 edge milling m/cs, single spindle as before	£120
	1 pair edge rollers 20" die with gearing for driving them, complete	£55
	2 grindstone troughs for stones 3 feet die 5 in thick, with screw plates, fast and loose pulleys 14" dia.	£30
	1 m/c for screwing the ends of ramrods	£20
	Col Dixon, Enfield 10th Nov 1858	
502	1 m/c for leading interior of barrels same as Or No 267	£40
	1 cast iron frame, spindle etc. for large grindstone for barrels, same as Or No 265 with reduced length of pillow blocks to allow of more wear of stone	£40
	2 m/cs for clamp milling and cupping head of ramrod, same as before, Or No 271, for standing on ordnance bed.	£60
	2 m/cs for mortice drilling head of ramrod, same as before [order number not given], on ordnance bed	£60
	1 cast iron frame, spindle etc. for large grindstone for ramrods, same as for barrels	£40
	1 cast iron bench 20ft long for polishing heads, with shafting, pullies etc. same as before Or No 272	£80
	7 sets of headstocks and spindles for polishing wheels, fitted to the above bed and same as now in use	£70
	1 m/c, double head, for milling small screws for bands, same as Or No 259, fitted on bed to rest on ordnance bed	£35
	1 m/c for slitting heads of guard screws, same as Or No 261	£26
510	Or No 7	£18
503	1 m/c, vertical, 1 spindle, similar to mortice drilling m/c minus the feed gears, for milling sight notches.	£25

	Col Dixon, Enfield **4th Feb 1859**	
570 – 576	lineshafting, pulleys etc.	
	Col Dixon, Enfield **21 Mar 1859**	
590	4 m/cs for copy milling the backs of butt plates of arms to pattern butt plate and specification	£240
	Col Dixon, Enfield **28th September 1859**	
826	2 m/cs for first boring barrels with self-acting feed motion and same as special m/c at Enfield	£160
	1 m/c for second boring barrels with self-acting feed motion and same as special m/c at Enfield	£40
	total	**£2609**

Appendix 5

Enfield Expenditure Memorandum 1855

These are loose documents in Royal Armouries' archive and while their exact provenance is not known, they are 'official'.

Expenditure on account of the Royal Small Arm Factory Enfield in the 1855-6 to 30th November.

Name	Service	Amount £ s d
Captain Warlow	Machinery	891 19 8
Do	Travelling	431 14 4
Wm Wood	Dining hall	296 . .
Smith Beacoll & Co	Machinery	441 - -
Platt Bro. & Co	Do	1,425 . .
C. Gunn	Freight of Machinery	70 - 7
Col. S. Colt	Blower	50 . .
C. Gunn	Freight of Machinery	45 7 11
H Carter	Fixing hearths	90 . .
Fairburn & Son	Iron Work	8,000 . .
Holgate & Co	Millbands	33 5 3
Fairbairn & Son	Iron Work	5,000 - -
Do	Works	5,000 - -
Do	Do	4,000 - -
J. Whitworth	Machinery	74 - -
W. Wood	Building Cottage	252 . .
J. Nasmyth	Steam drop Stamps	350 - .
Lieut F. Beaumont	Travelling	10 17 0
E. A Sme	Erecting Police Station	150 - -
Lieut. Col. Foster	Travelling	9 5 -
Wm Wood	Works	266 1 11
Sanderson Bro.	Cast Steel	12 17 -
Wm Jackson	Lead	17 13 4
Shaledge Howell & Co	Cast Steel	46 9 6
J & F. Hepburn	Mill bands	8 8 4
J. Frith Sons	Circular Cutter	46 13 6
	Carried over	£ 27,231 17 9

Name	Service	Amount
		£ s d
	Brought forward	27,231 17 9
Fairburn Greenwood & Co	Iron Work	3,125 — —
Hobbs & Co	Screw Machines	233 6 8
E. A. Smee	Builders work	180 — —
G. Lloyd	Patent noiseless Fan	33 — —
E. Jones	Stationery	— 7 4
W. Bardwell	Plans of Cottages	12 — —
S. Austin	Advertising	2 — 6
W. M. Cooper	Works	1,000 — —
Captain Collinson	Travelling	9 9 9
W. Wood	Artificers Work	132 6 5
Do	Erecting Gas Works	1,000 — —
Captain Jervis	Machinery	51 9 5
E. A. Smee	Artificers Work Police Station	150 — —
W. Higgins	Milling Machine	600 — —
Bootz Stinger	Tallow	18 7 2
Beckeby & Co	Engraving Plates	8 15 6
Wm Jackson	Lead	22 10 —
J. McMahon	Shovels	9 18 0
C. Goods	White Rope	12 17 11
A. Morewood & Co	Galvanized Iron	17 14 3
J. Porter	Wood Screws	10 14 1
G. Winter	Iron chain	26 12 1
Sanderson Bro:	Cast Steel	89 12 3
J. Buckingham	Patent Composition	44 17 1
Wm Jackson	Tin	22 6 6
W. Fairburn & Son	Steam Engine	2,460 — —
W. Wood	Carpenters Work	389 15 6
J. & J. Hepburn	Mill bands	11 14 4
A. W. Cooper	Building Factory	500 — —
J. Fenn	Tools	421 4 7
		34,827 17 1

ENFIELD EXPENDITURE MEMORANDUM 1855

Name	Service	Amount
		£
	Brought forward	34,827 17 1
J. Russell & Son	Turning Chisels	7 5 1
G. Knight & Sons	Hammers	108 6 -
W. C. Haddon	Casks of Ungent	62 4
Sanderson Bro:	Cast Steel for bayonets	54 12 10
Wm Jackson	Lead	23 10 -
J. Russell & Son	Tubes	107 11 3
J. Frith & Son	Swedish Iron	14 10 3
Wilson Hawkesworth & Co	Cast Steel Cutter	14 2 -
J. Nasmyth & Co.	Machinery	65 - -
A. Handyside	Do	672 10 -
J. Windsor	Pay of Clerks of Works. June 2r.	124 15 -
Captn. Jarvis	Machinery U S	2118 3 3
		£38,200 7 3

Wm Thomas
11th Dec. 1855

Appendix 6

Official Returns

1858

ENFIELD FACTORY.

RETURN to an Address of the Honourable The House of Commons, dated 19 April 1858;—*for*,

"STATEMENTS of the TOTAL SUM Expended at ENFIELD FACTORY, from the 1st day of January 1854 to the 31st day of March 1858; the Amounts under the following Heads to be separately specified, viz., Building, Machinery, Stores of all kinds, Salaries, Wages, Miscellaneous Expenses :"

"And, of the Number of GUNS delivered into Store, up to the 31st day of March 1858, from *Enfield*."

War Office, 19 April 1858.

J. PEEL.

STATEMENT of the TOTAL SUM Expended at ENFIELD FACTORY, from the 1st day of January 1854 to the 31st day of March 1858; the Amounts under the following Heads being separately specified, viz., Building, Machinery, Stores of all kinds, Salaries, Wages, Miscellaneous Expenses.

PERIOD.	Buildings.	Machinery.	Stores.	Salaries.	Wages.	Miscellaneous Expenses.	TOTAL.
	£. s. d.	£. s. d.	£. s. d.	£. s. d.	£. s. d.	£. s. d.	£. s. d.
1 January 1854 to 31 March 1854.	85 – –	– – –	1,117 3 9	94 10 7	3,708 3 6	40 2 5	5,945 – 3
1 April 1854 to 31 March 1855.	15,690 13 9	25,478 19 4	4,468 15 –	672 9 2	15,762 16 6	237 5 6	62,310 19 3
1 April 1855 to 31 March 1856.	35,433 18 –	24,092 10 7	7,603 11 5	1,838 6 5	16,969 3 11	301 1 8	86,228 12 –
1 April 1856 to 31 March 1857.	24,776 16 9	13,105 15 11	14,081 6 4	2,111 – 5	31,512 7 4	414 11 5	98,061 18 2
1 April 1857 to 31 March 1858.	15,632 7 10	3,985 16 6	21,421 4 1	2,332 2 5	67,180 5 –	445 2 1	110,996 17 11
£.	91,618 16 4	68,653 2 4	48,692 – 7	7,048 9 –	135,132 16 3	1,438 3 1	352,583 7 7

Note.—The above amounts show the gross expenditure at Enfield during the specified period; but the whole of the expenditure from the 1st January to the 31st March 1854 (amounting to 5,045*l*. 0*s*. 3*d*.), the expenditure on stores, salaries, and wages in 1854–55 (amounting to 21,141*l*. 6*s*. 7*d*.), and the sums expended under the same heads in the first quarter of 1855–56 (amounting to 4,838*l*. 15*s*. 11*d*.), are not chargeable to the present Royal Small Arms Factory, and these sums amount to 31,025*l*. 2*s*. 4*d*.

STATEMENT of the Number of GUNS delivered into Store, up to the 31st day of March 1858, from *Enfield*.

Musket Rifles (pattern 1853) made by machinery, complete - - - 26,739.

In addition to the above number of arms which have been delivered into store, there are, in various stages of progress, parts of arms and materials, equal to about 10,000 finished rifles.

OFFICIAL RETURNS

1859

ENFIELD FACTORY.

RETURN to an Address of the Honourable The House of Commons, dated 7 April 1859;—*for*,

" STATEMENTS of the TOTAL SUM Expended at ENFIELD FACTORY, from the 1st day of April 1858 to the 31st day of March 1859 inclusive; the Amounts under the following Heads to be separately specified, viz.:—Building, Machinery, Stores of all kinds, Salaries, Wages, and Miscellaneous Expenses:"

" And, of the Number of GUNS delivered into Store, from the 1st day of April 1858 to the 31st of March 1859 inclusive, from *Enfield*."

STATEMENT of the TOTAL SUM Expended at ENFIELD FACTORY, from the 1st day of April 1858 to the 31st day of March 1859 inclusive; the Amounts under the following Heads being separately specified, viz., Building, Machinery, Stores of all kinds, Salaries, Wages, and Miscellaneous Expenses.

PERIOD.	Buildings.	Machinery.	Stores.	Salaries.	Wages.	Miscellaneous Expenditure.	TOTAL.
	£. s. d.	£. s. d.	£. s. d.	£. s. d.	£. s. d.	£. s. d.	£. s. d.
1 April 1858 to 31 March 1859	9,356 6 9	2,908 18 2	41,241 15 1	2,514 15 8	94,526 6 10	2,669 19 3	153,218 1 9

STATEMENT of the Number of GUNS delivered into Store, from the 1st day of April 1858 to the 31st day of March 1859 inclusive, from *Enfield*.

Musket Rifles (pattern 1853) made by machinery, complete -   -   - 57,256

War Office, 18 April 1859.   J. PEEL.

1860

RETURN to an Address of the Honourable The House of Commons, dated 30 March 1860;—*for,*

" STATEMENTS of the TOTAL SUM Expended at ENFIELD FACTORY from the 1st day of April 1859 to the 31st day of March 1860 inclusive; the Amounts under the following Heads to be separately specified, viz.:—Building, Machinery, Stores of all description, Salaries, Wages, and Miscellaneous Expenses:"

" And, of the Number of GUNS delivered into Store, from the 1st day of April 1859 to the 31st day of March 1860 inclusive, from *Enfield* (in continuation of Parliamentary Paper, No. 249, of Session 1, 1859)."

War Office, 26 April 1860.

S. HERBERT.

STATEMENT of the TOTAL SUM Expended at ENFIELD FACTORY, from the 1st day of April 1859 to the 31st day of March 1860 inclusive; the Amounts under the following Heads being separately specified, viz., Building, Machinery, Stores of all kinds, Salaries, Wages, and Miscellaneous Expenses.

PERIOD.	Buildings.	Machinery.	Stores.	Salaries.	Wages.	Miscellaneous Expenditure.	TOTAL.
	£. s. d.	£. s. d.	£. s. d.	£. s. d.	£. s. d.	£. s. d.	£. s. d.
1 April 1859 to 31 March 1860	20,107 9 1	12,252 11 –	55,931 10 5	2,281 9 6	120,700 6 11	4,303 16 9	215,577 3 8

STATEMENT of the Number of GUNS delivered into Store, from the 1st day of April 1859 to the 31st day of March 1860 inclusive, from *Enfield*.

Musket Rifles (pattern 1853) made by machinery, complete - - - 87,405

War Office, 25 April 1860.

S. *Herbert.*

Appendix 7

Enfield Pattern 1853 Lockplate Hole Metrology Data

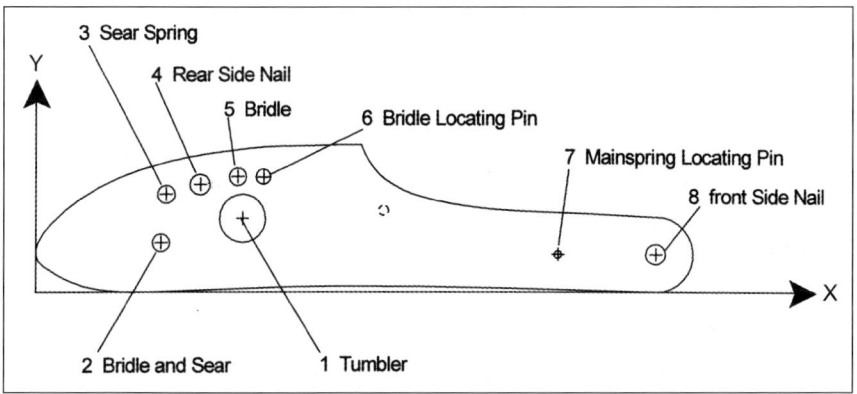

Tower 1855

Hole coordinates (mm) taking the centre of Hole 1 (tumbler arbour) as Origin.
(Highlighted entries show main data studied relating to the bridle.)

Acc. No	Hole 2 X	Hole 2 Y	Hole 3 X	Hole 3 Y	Hole 4 X	Hole 4 Y	Hole 5 X	Hole 5 Y
XII.7235	-18.49	-4.40	-16.69	4.85	-8.88	6.94	-0.17	8.80
XII.8999	-18.68	-4.46	-16.99	4.86	-8.77	7.13	-0.15	8.80
XII.9000	-18.71	-4.50	-16.92	4.78	-8.58	7.04	-0.11	8.65
XII.9058	-18.58	-5.07	-17.04	4.42	-8.85	7.00	-0.09	8.44
XII.9061	-18.58	-4.39	-17.01	4.71	-8.88	7.05	-0.09	8.80
XII.9070	-18.84	-4.72	-17.33	4.50	-9.02	7.36	-0.04	9.25
XII.9074	-18.22	-4.61	-16.60	4.30	-8.34	7.20	-0.70	8.86
XII.9076	-18.46	-4.90	-17.30	4.56	-9.02	7.32	-0.58	9.33
XII.9078	-18.23	-4.68	-16.68	4.68	-8.50	7.35	-0.76	8.67

MAKING THE ENFIELD PATTERN 1853 RIFLE-MUSKET

Acc. No	Hole 2 X	Hole 2 Y	Hole 3 X	Hole 3 Y	Hole 4 X	Hole 4 Y	Hole 5 X	Hole 5 Y
XII.9079	-18.95	-4.58	-17.48	4.99	-9.13	7.66	-0.79	9.96
XII.9088	-17.97	-6.92	-17.46	2.43	-9.79	6.30	-1.09	8.98
XII.9090	-17.79	-6.81	-17.24	2.53	-9.83	6.16	-1.58	9.42
XII.9091	-17.87	-6.88	-17.65	2.59	-9.91	6.01	-1.40	9.19
XII.9095	-17.66	-6.57	-17.36	2.86	-9.54	6.38	-1.57	9.69

Acc. No	Hole 6 X	Hole 6 Y	Hole 7 X	Hole 7 Y	Hole 8 X	Hole 8 Y
XII.7235	5.83	9.60	67.07	-8.52	89.03	-8.86
XII.8999	5.68	9.46	66.86	-8.27	89.49	-8.32
XII.9000	5.83	9.16	67.11	-8.79	89.66	-8.20
XII.9058	5.78	9.22	66.72	-8.42	88.67	-8.50
XII.9061	5.62	9.00	67.08	-8.63	88.81	-8.58
XII.9070	5.57	9.55	66.52	-8.68	89.22	-8.65
XII.9074	block	block	66.44	-8.72	88.85	-8.77
XII.9076	block	block	66.10	-8.34	88.77	-8.43
XII.9078	block	block	66.61	-9.05	88.87	-8.74
XII.9079	block	block	66.47	-8.02	89.04	-8.07
XII.9086	3.75	10.24	65.95	0.00	89.63	1.25
XII.9088	block	block	69.16	-8.98	89.92	3.04
XII.9090	block	block	66.79	-8.00	89.53	2.5
XII.9091	4.16	10.30	66.97	-8.45	89.50	2.62
XII.9095	3.66	11.34	67.20	-8.23	90.25	2.52

Hole diameters (mm)

Acc. No	Hole 1	Hole 2	Hole 3	Hole 4	Hole 5	Hole 6	Hole 7	Hole 8
XII.7235	10.10	4.02	4.00	4.38	3.99	3.60	2.72	4.36
XII.8999	10.11	3.87	3.80	4.31	3.81	3.65	2.71	4.45
XII.9000	10.13	3.86	3.81	4.50	3.82	3.69	2.75	4.44
XII.9058	10.06	4.04	3.97	4.45	4.00	3.47	2.62	4.53
XII.9061	10.07	4.15	4.10	4.55	4.08	3.47	2.64	4.35
XII.9070	10.18	3.98	4.02	4.41	3.93	3.63	2.74	4.42
XII.9074	10.07	4.01	3.96	4.42	4.00	block	2.69	4.43
XII.9076	10.26	4.09	4.04	3.94	4.12	block	2.93	4.43
XII.9078	10.05	4.05	4.04	4.53	4.01	block	2.58	4.47
XII.9079	10.09	3.96	3.90	4.25	3.95	block	2.80	4.44

ENFIELD PATTERN 1853 LOCKPLATE HOLE METROLOGY DATA

Lock makers/markings

Acc. No	
XII.7235	Brazier
XII.8999	Brazier
XII.9000	Brazier
XII.9058	WC (William Corbett)
XII.9061	J & EP (Partridge)
XII.9070	J Steatham / SSP / T&T
XII.9074	WS
XII.9076	GP / HS
XII.9078	RA / WW
XII.9079	------

Belgian Contract 1856

Hole coordinates (mm) taking the centre of Hole 1 (tumbler arbour) as Origin.

Acc. No	Hole 2 X	Hole 2 Y	Hole 3 X	Hole 3 Y	Hole 4 X	Hole 4 Y
XII.8428	-18.31	-4.54	-16.84	4.63	-9.00	6.73
XII.9102	-18.15	-3.98	-16.86	5.34	-8.71	7.51
XII.9105	-18.14	-4.11	-16.69	5.04	block	block
XII.9121	-18.23	-4.23	-16.90	4.89	-9.41	7.01
XII.9122	-18.38	-4.29	-17.02	4.73	-9.04	6.93

Acc. No	Hole 5 X	Hole 5 Y	Hole 6 X	Hole 6 Y	Hole 7 X	Hole 7 Y	Hole 8 X	Hole 8 Y
XII.8428	-0.51	8.98	5.54	9.23	66.25	8.83	88.15	8.98
XII.9102	-0.32	9.35	5.52	9.41	66.73	-8.60	88.59	-8.56
XII.9105	-0.22	9.35	5.87	9.43	66.67	-8.73	88.54	-8.65
XII.9121	-0.34	9.23	5.56	9.11	66.26	-8.75	88.51	-9.29
XII.9122	-0.67	9.34	5.52	9.26	66.50	-8.76	88.42	-8.49

Hole diameters (mm)

Acc No	Hole 1	Hole 2	Hole 3	Hole 4	Hole 5	Hole 6	Hole 7	Hole 8
XII.8428	10.00	3.68	3.83	4.52	3.66	3.74	2.72	4.50
XII.9102	10.03	3.70	3.68	4.45	3.69	3.39	2.64	4.50
XII.9105	10.09	3.78	3.77	blocked	3.79	3.80	2.75	4.42
XII.9121	10.03	3.76	3.78	4.412	3.73	3.72	2.67	4.44
XII.9122	10.04	4.04	3.96	4.39	4.00	3.67	2.77	4.43

Appendix 8

Messrs. Fox, Henderson and Co.'s Contract for rifles

Transcript of
Précis of Correspondence on Messrs. Fox, Henderson & Co.'s contract for rifles, 1861.

(London, National Archives, Catalogue reference: WO 33/10)

MESSRS. FOX, HENDERSON, AND CO.'S
CONTRACT FOR RIFLES.

On the 12th February 1855 the late Board of Ordnance accepted an offer from Messrs. Fox, Henderson, and Co. to supply, through a manufacturing firm in the United States of America, 25,000 rifles, at 70 s[hillings]. each, to be delivered at New York. The Board at the same time intimated its intention to enter into further agreements with Messrs. Fox, Henderson, and Co. for 75,000 additional rifles upon certain conditions depending upon the fulfilment of the first agreement.

On the 14th February 1855 the Treasury authorized the Board of Ordnance to issue to Messrs. Fox, Henderson, and Co. an advance of £21,875, being one fourth of the contract price of the 25,000 rifles, the object being to enable them to put in funds the manufacturing firm in America, Messrs. Robbins & Lawrence.

It should be remarked that no *contract* was made with Messrs. Fox and Henderson; their offer by letter being merely accepted; that they were under no penalty for nonfullfilment of agreement, and that no time was fixed for the completion of the agreement, although it was subsequently agreed that the limit for the delivery of the rifles should be February 1859.

In June 1855 Captain Jervis, RA, was appointed by the War Department to proceed to America with a staff of viewers, to superintend the delivery of the rifles, and with power to draw money. He was instructed to use his discretion in making any expenditure of money he might think proper, to meet sudden emergencies, without waiting for authority from home.

On the second of December 1855 Captain Jervis reported that Messrs. Robbins & Lawrence [the sub-contractors under Fox and Henderson] were in pecuniary difficulties, that the workmen had not been paid for three months, that but little work had consequently been done; and that it would probably be necessary for him to advance to Messrs. Robbins & Lawrence £2,000. Captain Jervis was informed in reply, by letter from the War Office of 17th December 1855, that no money should be paid on account of rifles until value should be received for the amount already advanced.

But in December 1856, Captain Jervis reported to Captain Dixon, the Superintendent of the Royal Small Arms Factory at Enfield, that the firm of Robbins & Lawrence had become insolvent, and that he had advanced £6,000 to them on the security of the plant, &c., to enable them to go on with the rifles.

As it was considered that Captain Jervis had not acted with sufficient judgment in this matter, he was instructed to return to this country; and Captain Warlow was ordered to replace him and was told to consult with the British Minister and the most eminent American lawyers recommended by him as to the steps which should be taken in reference to the security accepted by Captain Jervis.

On the 27th December 1855 Mr. Bannister, the solicitor for The War Department, reported that "Captain Jervis having on the part of the Government, advanced on security £6,000 to Messrs. Robbins & Lawrence, the Government must abide the security taken, and cannot make any claim upon Messrs. Fox and Henderson on account of that advance; at the same time the act of Captain Jervis was a beneficial one to Messrs. Fox and Henderson and cannot release them from the execution of their agreement.

Dismissing, therefore, the question of the *£6,000* advance, the solicitor considered the position of the War Office with Messrs. Fox and Henderson to be as follows:

> *"That the Department may, in the event of Fox and Henderson not supplying the whole of the rifles in accordance with their contract, bring an action against them to recover the difference between the £21,875 advanced and the value of such rifles as may be supplied."*

On the 17th December 1855, Captain Jervis forwarded, from the American lawyer employed, explanations of the reasons for the advance of £6,000 and the consequent arrangements made. The result of those arrangements were stated to be:

> *"That the contract for rifles, before in peril, can now proceed. That £20,000 already advanced under that contract, without security, and otherwise certain to be lost to a great extent, if not entirely, is now secured; and that that sum, and $30,000 (£6,000) further advance, is now fixed by way of first charge on property worth at least from $150,000 to $200,000, and is therefore substantially secured."*

On 31st January 1856, Captain Jervis reported that he had made a further advance to Messrs. Robbins & Lawrence of $6000 (£1,266 4s. 2d.)

In April 1856, the following arrangements were entered into with Messrs. Fox and Henderson for repayment of the advances made. These advances were:

£	s	d	
21,875	0	0	One fourth part of the whole contract
6,000	0	0	Advanced by Captain Jervis to Robbins & Co.
1,266	4	2	Do Do
6,185	11	3	Advanced by Fox and Henderson to Robbins & Lawrence, and repaid by War Office to Fox. and Co.
109	4	6	Lawyers charges incurred by Captain Jervis
35,435	19	11	

These advances were to be thus repaid.

As regards the sum £21,875, Messrs. Fox and Henderson to deduct from their bills one fourth of the value of each delivery until the whole advance be repaid.

As regards the £6,000, a deduction of 4s. 9½d. to be made on each weapon delivered, until 10,000 are delivered.

As regards the £1,266 4s. 2d., a deduction of one dollar per weapon to be made until that sum be discharged.

As regards the £6,185 11s. 3d., after the discharge of the foregoing sums, a deduction of 4 dollars per weapon to be made.

The £109 4s. 6d. for lawyer's bill to be deducted from the first payment to Fox and Henderson.

On the 30th October 1856, the following was the state of the contract or agreement for rifles:

Number of rifles agreed for	-	25,000
Number delivered	-	10,000
Remaining to be delivered	-	15,000

	£	s	d
Amount advanced	35,550	10	4
Repaid by deduction from the several claims	11,369	11	7
Balance due	24,180	18	9

May and November 1856. Messrs. Fox and Henderson having in the meantime become insolvent, arrangements were made respecting the payment of bills drawn in America on the securities held by Fox and Henderson from their sub-contractors in America were transferred to Colonel Rowan (who had succeeded Captain Warlow) as trustee for the Secretary of State for War.

In November 1856 Colonel Rowan was instructed by the Solicitor for the War Office to exercise forbearance towards Messrs. Robbins & Lawrence, with a view to the ultimate completion of the contract, taking care not to release Messrs. Fox and Henderson, nor to prejudice the securities.

But on 12th November 1856 Colonel Rowan reported that Messrs. Robbins & Lawrence had stopped work, and that the agent of Messrs. Fox and Henderson had taken steps to rescind the sub-contract with Robbins & Lawrence.

On 23rd December 1856 the following was the state of the contract:

10,400 rifles had been delivered
£24,180 was still due on account of advances to Fox and Henderson.

Security of the premises and plant of Messrs. Robbins & Lawrence had been assigned to the [War] Department as security for that debt.

Both the firms of Fox and Henderson and Robbins & Lawrence had suspended payment.

The work of the establishment in America was stopped, and no further deliveries of rifles could be expected, unless some general arrangement could be made, with the concurrence of the creditors. If no arrangement could be made the agent of the War Office (Colonel Rowen [sic]) would realize to the fullest extent the securities of the property and plant in America, and Messrs. Fox and Henderson would be held responsible for the final payment of the balance (if any).

The firm in America had made proposals for continuing the contract under certain conditions, one of which was that they should have a new contract for 25,000 additional rifles. These proposals were considered at a meeting of War Office Authorities, and it was decided that it was not advisable to accept them; also, that Colonel Rowan and Mr. Whittaker (the American lawyer employed in the case) should be instructed to give any reasonable time for the realization of the securities without putting them in jeopardy.

In October 1857 the agreement was taken over by a new company called the "Union Arms Company" upon much the same terms as those of the previous firm.

But this company also soon got into difficulties and ceased to make any deliveries after 30th June 1858. A certain number of rifles were, however, delivered, and the debt was reduced, as shown by the following statement:

Total Rifles delivered to 30th June 1858 -	16,000
Due on account of advances when the "Union Arms Company" commenced work -	£24,180
Since obtained by deductions from bills -	7,925
Balance due on 30th June 1858 -	£16,255
Since obtained by sale of machinery -	662
	£15,593

MAKING THE ENFIELD PATTERN 1853 RIFLE-MUSKET

The new Company having refused to continue their deliveries of rifles, except upon more favourable terms than those previously agreed to, and having made fresh propositions with a view to the continuance of the deliveries, those propositions were, in August 1858, declined by the Secretary of State for War, upon the following grounds:

1st. Colonel Dixon, the superintendent of the Royal Small Arms Factory, Enfield, reported that the arms were of such inferior quality, compared with those obtained in this country, that he considered it desirable that none should be received.

2nd. That since the original agreement with Messrs. Fox and Henderson was made, the price at which the Department could obtain rifles had decreased from the price named in that agreement by about 1/- per rifle, so that a completion of the deliveries would have involved a loss of about £9,000 irrespective of the circumstance of the inferiority of the arms.

2472.   E. & S-5/61

Bibliography

Abbott, J. (1852). *The Armory at Springfield,* Harper's New Monthly Magazine, Vol. V. No. XXVI, July, 145-161. Reprinted Ontario; Fortress Publications Inc., Research Document No.6, 1-17. (n.d.)

Alder, K. (1997). *Engineering the revolution: arms and enlightenment in France, 1763 – 1815,* Princeton, N.J.; Princeton University Press.

Anderson, J. (1857) *On the application of machinery in the War Department,* Journal of the Society of Arts, Vol V, No. 213, pp. 155-166. London.

Anonymous. (1857). *Repeating Firearms - A day at the armory of Colt's Patent Fire-arms Manufacturing Company.* United States Magazine, Vol. IV, No 3, March, 221-249.

Anonymous. (1864). *The Norwich Armories,* Harper's New Monthly Magazine, March, 450-465. Reprinted Ontario; Fortress Publications Inc., Research Document No.7, 1-16 (n.d.)

Anonymous. (1977). *The American system is born,* American Machinist, 100th Anniversary Issue, November. New York, McGraw-Hill Inc.

Babbage, C. (1835). *On the economy of machinery and manufactures*, London; Charles Knight.

Bagust, H. (2006) *The Greater Genius: a biography of Marc Isambard Brunel*, Hersham, Surrey; Ian Allen.

Battison, E. (1984) *Eli Whitney and the milling machine.* Smithsonian Journal of History, (1), (1966, revised 1984), pps 9-34.

Beasley, W. (1852). *Manufacture of metal tubes, &c.* U.K. Patent, No. 14,163.

Bedford, C. (1969). *The Forsyth percussion system,* Bulletin No. 19, American Society of Arms Collectors, 4-13.

Bentham, S. (1791). *Machinery for cutting and planing wood*, U.K. Patent No. 1838.

Bentham, S. (1793). *Methods of, and machinery and apparatus for, working wood, metal and other materials,* UK Patent No. 1951.

Benton, Lt. Col. J. G. (1878). *Fabrication of Small Arms,* Ordnance Memorandum No. 22, Washington, Government Printing Office. Reprinted by Benchmark Publishing Company Inc., Glendale, New York.

Blackmore, H. (1961). *British Military Firearms 1650 – 1850*, London, Herbert Jenkins.

Blanc, H. (1790). *Mémoire important sur la fabrication des armes de guerre*. [Important memoire on the manufacture of weapons of war]. A L'assemblee Nationale, Paris, printed L M Cellot, rue des Gr.-August.

Blanchard, T. (1819), *Machine for turning gunstocks, tackle and shipping blocks*, U.S. Patent Sept. 6, altered by Act of Congress to Jan. 20, 1820. (US patents not numbered prior to 1836)

Bodmer, J. (1869) Obituary, *Proceedings of the Institute of Civil Engineering, Vol. 28*, 573-608.

Bond, Lt. Col. H., RA. (1884). *Treatise on Military Small Arms and Ammunition*, London. Her Majesty's Stationery Office.

de Borda, de la Place, Coulomb and le Roy (1791). *Rapport Fait à L'Academie Royale des Sciences D'Un Mémoire Important, De M. Blanc, sur la fabrication des Armes de guerre*. [Report to the Royal Academy of Sciences of an Important Memoir, by M. Blanc, on the manufacture of Weapons of War]. Le Samedi 19 Mars 1791. Extrait des Registres de l'Academie Royale des Sciences, Paris, Du 19 Mars 1791.

Bradley, J. (1811). *Machinery for manufacturing gun skelps*, U.K. Patent No. 3437.

Bradley, J. (1990). *Guns for the Tsar*, Northern Illinois University Press, Dekalb, Illinois.

Bramah, J. (1802). *Machinery for forming gunstocks &c.*, UK Patent, No. 2652.

Brown, W. (1885). UK Patent, No. 5265.

Brunel, M. (1801). *Ship's Blocks*, UK Patent, No. 2478.

Buckingham, E. (1921), *Principles of interchangeable manufacturing,* New York, The Industrial Press; London, The Machinery Publishing Co. Ltd.

Buckle, J. (1822). *Machinery for shaping or cutting wood &c.*, UK Patent, No. 4652.

Burton, 1855. A collection of the original letters and other correspondences of James Burton are held in the archive of Yale University and copies/transcripts of those referred to are included in Appendix 3 of this thesis.

Burton, J. (1860) *Improvement in the manufacture of gun barrels*, US Patent No. 27,539, March 20; reissue No. 4,686, December 26, 1871.

Charles I, (1631). *Arms and armour: Orders in regard to Patterns*, Patent Rolls, Chancery, T/Charles I, Part 20 (C66/25791)

Colt, Col. S. (1851). *On the application of machinery to the manufacture of rotating chambered-breech fire-arms and the peculiarities of those arms*. Proceedings, Institution of Civil Engineers, November 25th, London, 30-68.

Cooper, C. (1991). *Shaping Invention: Thomas Blanchard's machinery and patent management in nineteenth century America,* New York, Columbia University Press.

Cotty, H. (1806). *Mémoire sur la Fabrication des armes portatives de Guerre* [Memorandum on the Manufacture of Portable Weapons of War]. Paris, Magimel. Reprinted by Facsimile Publisher, Delhi, n.d. but printed on demand 2017.

Dale, Capt. M, (1810). *Mémoire sur le montage du fusil,* [Memorandum on the assembling of the musket]. Liège. (Reprint published on the initiative of

BIBLIOGRAPHY

Fabrique Nationale Herstal on the occasion of the inauguration of its collection of industrial archeology as part of the fourth Colloquium of Industrial Archeology. Liege 1977.)

Daumas, M. (1969). *A History of technology and invention: progress through the ages*, Crown Publishers, Vol. II.

Deyrup, F. (1970) *Arms making in the Connecticut Valley – a regional study of the economic development of the small arms industry 1798-1870*. First published in 1948 as *Arms makers of the Connecticut Valley* in Volume XXXIII of the Smith College Studies in History; reprinted in limited edition under this new title by George Shumway, York, Pennsylvania.

Diderot, D. (1779) Diderot et D'Alembert, *L'Encyclopédie : Fabrique des armes escrime, Fusil de Munition*, Paris.

Eadon, T. (1856) *An improvement in the manufacture of band saws and other endless bands or hoops of metal*. U.K. patent no. 1628, 1856.

Edwards, W. (1953). *The Story of Colt's Revolver*, Harrisburg, PA, Stackpole Publishing.

Engineer [Hobbs]. (1859). *Hobbs' Lock Manufactory*. The Engineer, London, 188-190; 203-204.

Engineer, (1859). *The Royal Small-Arm Manufactory, Enfield*, The Engineer, London, 204, 258, 294, 295, 348, 384, 422, 423.

Exhibition, (1851). *Illustrated exhibitor; sketches by pen and pencil of the principal objects of the Great Exhibition of the Industry of all Nations*, London, John Cassell.

Fitch, C. H. (1882). *Report on the Manufacture of fire-arms and ammunition*, Extra Census Bulletin, Washington. Reprinted by American Archives Publishing, Topsfield, Massachusetts, 1970.

Forsyth, Revd. A. (1807). *Apparatus for Discharging Artillery, &c. by Means of Detonating Compounds*, U.K. Patent, No. 3032.

Freeland, W. (1918). *Mass production at the Winchester shops,* Iron Age, 101.

Gamel, J. (1826). *Описание Тульского оружейного завода с учетом исторических и технических аспектов, [Description of the Tula Weapon Factory in regard to historical and technical aspects]*. Moscow: August Semen Press. Translated reprint, E. A. Battison (Ed.), published for the Smithsonian Institution Libraries and The National Science Foundation, Washington D.C. by Amerind Publishing Co. Pvt. Ltd., New Delhi, 1988.

Gilchrist, M. (2003). *Patrick Ferguson – a man of some genius*, Edinburgh, NMS Enterprises Ltd.

Gooding, J. and J. Morton (2009) *Honoré Blanc and Interchangeable Gunlocks*' Journal of the Arms and Armour Society, Vol. XIX No.5, 208 – 218.

Goodman, J. D. (1866). In *The Birmingham gun trade*, in *Birmingham and the Midland hardware district*, S Timmins, (Ed.), Originally published by Robert Hardwicke, London, 1866; reprinted by Frank Cass & Co Ltd, 1967

Gordon, R. (2010). *The Armoury Gauging System and Interchangeable Manufacture*, Arms & Armour, *Vol. 7, No. 1*, 40-52. Royal Armouries, Leeds.

Greener, W. W. (1884). *The Gun and its development.* 2nd ed., London, Paris & New York, Cassell & Company Ltd.

Greenwood, T. (1862) *Machinery for the manufacture of gunstocks*, Proceedings, Institute of Mechanical Engineers, 328-340.

Hall, J. (1811). *Breech loading rifle*, U.S. Patent, (later numbered as No. 1515X.)

Hamilton, A. (1791) *Report on Manufactures*, 6th ed., Philadelphia, 1827.

Heywood, G. (1814). *Machinery for the manufacture of gun and pistol barrels.* U.K. Patent, No. 3813.

Hobbs, A. (1859) *Hobbs' lock manufactory,* The Engineer, London, p. 190.

Hounshell, D. (1984). *From the American System to mass production, 1800 - 1932*, Baltimore and London, Johns Hopkins University Press.

Hubbard, G. (1922). *Pioneer machine builders; the mechanics of the Windsor region and their industries.* Windsor, Vermont; Town School Board.

I.LN. (1851). *The manufacture of gun barrels at Birmingham,* Illustrated London News, Supplement, *Feb. 1.* 85.

I.L.N. (1855). *The manufacture of the Enfield rifle*, Illustrated London News, Supplement, April 28. 410-411.

I.LN. (1861). *The manufacture of the Enfield rifle*, Illustrated London News, Sept. 21. 304-305.

Jervis, Capt. J-W (1854). *The rifle musket*, London, Chapman Hall.

Lewis, J. (1996) *The development of the Royal Small Arms Factory (Enfield Lock) and its influence upon mass production technology and product design C1820 - C1880* Ph.D. thesis, Middlesex University

LoC. *List of Changes in War Matériel and of Patterns of Military Stores,* London, War Department (various dates).

Manceaux, F. J. (1852). *Machinery for rifling firearms, etc.*, U.K. Patent No. 13,934.

Marks, E. (1903). *The manufacture of iron and steel tubes*, Manchester, Technical Publishing Company.

Maugham, S. *The moon and sixpence* (various dates and publishers)

Mayr, O. and Post. R. (Eds.) (1984). *Yankee enterprise: the rise of the American system of manufacture.* Papers from a symposium held in 1981, Washington, D.C., Smithsonian Institution Press.

McNeil, I. (1968). *Joseph Bramah: a century of invention, 1749-1851,* New York, Augustus M. Kelley.

Mechanic. (1861). *Leviathan workshops: the Enfield Small Arms Factory,* The Mechanics' Magazine, August 23 to September 6, No. 11. 110-111; 127-128; 144-145.

Memoria, (1850) *Memoria De la fabricacion del fusil de nuevo modelo en la Real Manufactura de Armas de Lieja en los anos de 1847 y 1848.* Madrid, Imprenta y fundicion de Don Eusebio Aguado. *[Memoire on the manufacture of the new model musket in the Royal Manufacture of Weapons in Liege in the years of 1847 and 1848].* Madrid, Printing and foundry of Don Eusebio Aguado.

Miles, H. (Ed) (1860). *Construction of the Enfield Rifled musket,* Book of Field Sports, Div. 1, Vol. 1. London, pps.29 – 35.

BIBLIOGRAPHY

Mordecai, Major A. (1860). *Military Commission to Europe in 1855 and 1856*, War Department, Washington, DC.

Morse, S. A. (1863) *Bit,* U.S. Patent 38,119, April

Nash, S. (1818). *Turning Musket Barrels,* U.S. Patent, (later numbered No. 2939 or 2939X)

Osborn, H. (1813). *Machinery for tapering gun barrels &c.* U.K. Patent, No. 3740.

Osborn, H. (1817). *Machinery for welding barrels of fire-arms and other cylinders,* UK Patent, No. 4105.

Ordnance (1857). *Proceedings of the Ordnance Select Committee,* 1st January – 31st March, 1857, Minute No. 1169, London, Ordnance Select Committee.

Ordnance, (1860) *Proceedings of the Ordnance Select Committee,* 19th October, Minute No 2058, Report No. 1186. London, Ordnance Select Committee.

Pam, D. (1998). *The Royal Small Arms Factory, Enfield, and its workers.* Published by the author, Enfield, Middlesex.

Petrie, Capt. M. (1866). *Equipment of Infantry*, London, Her Majesty's Stationery Office.

Pettibone, D. (1814). *Boring gun barrels,* U.S. Patent, (later numbered No. 2064X).

Précis (1861). *Précis of Correspondence on Messrs. Fox, Henderson & Co.'s contract for rifles, 1861*: London, National Archives, Catalogue reference: WO 33/10

Pritchard, R. (2014) *The English connection*, Thomas Publications, Gettysburg, Pennsylvania.

Report (1823). *Report of the Secretary of War: Expenses of the Ordnance Department, 1816-1822,* Washington, D.C.; Ordnance Department.

Report, (1852) *Report of Experiments with Small Arms carried on at the Royal Manufactory, Enfield.* London; Board of Ordnance.

Report (1854). *Report from the Select Committee on Small Arms*, London; House of Commons.

Return (1858). *Return to an address of the Honourable House of Commons, 13th April, 1858.*

Return (1859). *Return to an address of the Honourable House of Commons, 7th April, 1859.*

Return (1860). *Return to an address of the Honourable House of Commons, 20th March, 1860.*

Roads, C. (1961) *The history of the introduction of the percussion breech loading rifle into British military service, 1850 – 1879,* a thesis submitted for the degree of Doctor of Philosophy in the University of Cambridge, Trinity Hall, February 1961.

Roads, C. (1964). *The British Soldier's Firearm, 1850 – 1864*, London; Herbert Jenkins.

Robinson, C. (1944). *Explosions, their anatomy and destructiveness*, New York & London, McGraw-Hill Book Company Inc.

Roe, J. W. (1912). History of the first milling machine, *American Machinist, 1912, No 36.* 1037-8.

13. Osborn patent, 1813
14. Osborn patent, 1817
15. Heywood's patent, 1814
16. Fitch, 1882, p. 10
17. Hall patent, 1811
18. Smith, 1977, p. 235
19. Fitch, 1882, p. 26
20. Smith, 1977, p. 212
21. Smith, 1977
22. Cooper, 1991
23. Dale, 1810
24. Memoria, 1850
25. Hounshell, 1984, p. 38
26. Bagust, 2006, p. 20
27. Hamilton, 1791, p. 22
28. see also Cooper, 1991, p. 139
29. Gamel 1826, p. 155, footnote 1
30. Smith, 1977, p. 125
31. Benton, 1878
32. Mayr, 1981, pps. 58-60
33. Smith, 1843 edn, p. 4
34. Babbage, 1835, p. 120
35. Babbage, 1835, p. 268
36. Babbage, 1835, p. 66
37. Gordon, 2010
38. Mayr & Post, 1984, p. 104
39. Daumas, 1969
40. Rolt, 1965, p. 89
41. Roe, 1916, p. 212
42. Rolt, 1965, p. 177
43. Lewis, 1996, pps. 16, 17
44. Petrie, 1865, p. 32

Chapter 2: Basic Technology of the Enfield Rifle

1. Forsyth, 1807
2. Robinson, 1944, p. 2
3. Bedford, 1969, p. 9
4. Cotty, 1806, pps. 75-87
5. Alder, 1997, pps. 197, 199
6. Bond, 1884, p. 197
7. Gilchrist, 2003, p. 20
8. Blackmore, 1961, p. 113

REFERENCES

9. Blackmore, 1961, p. 118
10. Report, 1852, p. 31
11. Roads, 1964, p. 23
12. Roads, 1964, p. 24
13. Roads, 1964, p. 25
14. Greener, 1884, p. 123
15. Roads, 1964, p. 25
16. Roads, 1964, p. 24
17. Ordnance, 1857, Minute 1169
18. Ordnance, 1860, Minute 2058
19. Boucher, J. 1858

Chapter 3: Origins and Procurement

1. Roads, 1964, p. 33
2. Report, 1852, p. 3
3. Report, 1852, p. 35
4. Charles I, 1631
5. Smith, 1843 edn, p. 4 et seq
6. Report, 1854, p. iii
7. Report, 1854, p. 286
8. Report, 1854, p. 292
9. Report, 1854, p. ix
10. Report, 1854, pps. 451, 452
11. Report 1854, p. 451
12. Roads, 1864, p. 86
13. Exhibition, 1851, p. 4
14. Hubbard, 1922, pps. 70-73
15. Satterlee, 1940
16. Smithurst, 2014, 2015
17. Anonymous, 1857
18. Rosenberg, 1968, pps. 91-97
19. Rosenberg, 1968, p. 94
20. Report, 1854, p. ii
21. Anderson, 1857
22. Report, 1854, pps. 26-38
23. Report, 1854, p. 27
24. Report, 1854, p. 29
25. Report, 1854, p. 91
26. Colt, 1851, pps. 44, 45
27. Report, 1854, pp. xxiv-xxv
28. Rosenberg, 1968, p. 90
29. Rosenberg, 1968, pps. 91-97

30. Rosenberg, 1968, pps. 181–184
31. Pam, 1998, p. 52
32. Rosenberg, 1968, p. 184
33. Rosenberg, 1968, pps. 187-190
34. Rosenberg, 1968, p. 181
35. Rosenberg, 1968, p. 191
36. Rosenberg, 1968, p. 180
37. Greenwood, 1862
38. Rosenberg, 1968, pps. 187-190
39. Lewis, 1996, pps. 16, 17
40. Engineer, 1859, p. 205
41. Report, 1854, p. 286
42. Fitch, 1882, p. 10
43. McNeil, 1968, pps. 199-200
44. Hobbs, 1859, p. 190
45. Spencer, 1873
46. Rolt, 1965, p. 165
47. Roe, 1916, p. 143
48. Engineer, 1859, pps. 384, 422
49. Mordecai, 1860, p. 107
50. Bond, 1884, p. 207
51. Sword, 1986, p. 49
52. Pritchard, 2014, p. 58
53. Petrie, 1866, p. 32
54. Roads, 1964, p. 83

Chapter 4: Manufacture of the Lock

1. Gamel, 1826, p. 131
2. Report, 1854, pp. xxiv-xxv
3. Mechanic, 1861, p. 144
4. Engineer, 1859, p. 384
5. Engineer, 1859, p. 384
6. Engineer, 1859, p. 384
7. Engineer, 1859, p. 384
8. Engineer, 1859, p. 384
9. Hubbard, 1922, p. 71
10. Engineer, 1859, p. 385
11. Engineer, 1859, p. 385
12. Rosenberg, 1968, p. 188
13. Rosenberg, 1968, p. 188
14. Engineer, 1859, p. 385
15. Rosenberg, 1968, p. 188

REFERENCES

16. Engineer, 1859, p. 385
17. Rosenberg, 1968, p. 182
18. Engineer, 1859, p. 422
19. Steeds, 1969, p. 141

Chapter 5: Manufacture of the Stock

1. Dale, 1810
2. Memoria, 1850
3. Blanchard, 1819
4. Cooper, 1991, p. 81
5. Buckle, 1822
6. Greenwood, 1862, p. 340
7. Cooper, 1991, p. 92
8. Report, 1823, p. 97
9. Cooper, 1991, p. 92
10. Cooper, 1991, p. 91
11. Fitch, 1882, p. 16
12. Rosenberg, 1968, p. 180
13. Greenwood, 1862, pps. 328-9
14. Roads, 1964, p. 75
15. Greenwood, 1862, pps. 328-9
16. Engineer, 1859, p. 258
17. Engineer, 1859, p. 258
18. Eadon Patent, 1856
19. Engineer, 1859, p. 259
20. Greenwood, 1862, p. 329
21. Engineer, 1859, p. 258
22. Engineer, 1859, p. 258
23. Benton, 1878, p. 111
24. Greenwood, 1862, p. 328
25. Engineer, 1859, p. 258
26. Engineer, 1859, p. 258
27. Greenwood, 1862, p. 329
28. Benton, 1878, p. 111
29. Engineer, 1859, p. 259
30. Engineer, 1859, p. 258
31. Engineer, 1859, p. 259
32. Greenwood, 1862, p. 328
33. Engineer, 1859, p. 294
34. Greenwood, 1862, p. 332
35. Greenwood, 1862, p. 332
36. Greenwood, 1862, pps. 328-9

37. Engineer, 1859, p. 294
38. Greenwood, 1862, pps. 332-334
39. Greenwood, 1862, p. 230
40. Benton, 1878, p. 112
41. Engineer, 1859, p. 348
42. Engineer, 1859, p. 348
43. Engineer, 1859, p. 348
44. Benton, 1878, p. 113

Chapter 6: Manufacture of the Barrel

1. Osborn, 1813
2. Osborn, 1817
3. Heywood, 1814
4. Report, 1854, p. 491
5. Jervis, 1854, p. 3
6. Miles, 1860, p. 31
7. Goodman, 1866, p. 388
8. I.L.N., 1855, p. 410
9. I.L.N., 1861, p. 304
10. Mechanic, 1861, p. 127
11. Engineer, 1859, p. 348
12. I.L.N., 1855, p. 410
13. I.L.N., 1855, p. 410
14. Fitch, 1882, p. 10
15. Fitch, 1882, p. 9
16. Russell, Patent No. 4892, 1824
17. Miles, 1860, p. 31
18. I.L.N., 1861, p. 304
19. Greener, 1884, p. 253
20. I.L.N., 1855, p. 410
21. Mechanic, 1861, p. 127
22. Engineer, 1859, p. 348

Chapter 7: Manufacture of the Barrel

1. Williams et al, 2018, pps. 22-57
2. Abbott, 1852, p. 149
3. I.L.N., 1855, p. 410
4. Mechanic, 1861, p. 127
5. Engineer, 1859, p. 348
6. Benton, 1878, p. 25

REFERENCES

7. Pettibone, U.S. Patent, 1814
8. Engineer, 1859, p. 348
9. Mechanic, 1861, p. 127
10. Benton, 1878, plate XVII
11. Fitch, 1880, p. 11
12. I.L.N., 1861, p. 304
13. Benton, 1878, p. 25
14. Mechanic, 1861, p. 127
15. Benton, 1878, p. 25
16. Engineer, 1859, p. 348
17. Engineer, 1859, p. 348
18. Mechanic, 1861, p. 127
19. Nash, 1818
20. I.L.N., 1855, p. 410
21. I.L.N., 1851, p. 85
22. Diderot, 1779, Pl. 3
23. Greener, 1884, p. 273
24. Jervis, 1854, p. 6
25. I.L.N., 1855, p. 410
26. Engineer, 1859, p. 348
27. Jervis, 1854, p. 5
28. Jervis, 1854, p. 5
29. Miles, 1860, p. 32
30. Mechanic, 1861, p. 127
31. Miles, 1860, p. 32
32. Mechanic, 1861, p. 127
33. Gamel, 1826, Figs. XI
34. Memoria, 1850, pps. 12, 13
35. Rosenberg, 1969, p. 189
36. Rosenberg, 1969, p. 189
37. Miles, 1860, p. 32
38. Benton, 1878, pps. 26-28
39. Benton, 1878, p. 26
40. Mechanic, 1861, p. 127
41. Engineer, 1859, p. 348
42. Jervis, 1854, p. 7
43. Fitch, 1882, p. 12
44. I.L.N., 1855, p. 410
45. Jervis, 1854, p. 47
46. Fitch, 1882, p. 12
47. Engineer, 1859, p. 349
48. Miles, 1860, p. 33
49. I.L.N., 1861, p. 304
50. Report, 1854, p. xxiv

51. Rosenberg, 1968, p. 184
52. Fitch, 1882, p. 13
53. Mordecai, 1860, p. 107
54. Hubbard, 1922, p. 71
55. Rolt, 1965, p. 179
56. Rolt, 1965, p. 183
57. Benton, 1878, p. 6
58. Roads, 1964, p. 102
59. Report, 1854, p. 491
60. Roads, 1964, p. 109
61. Miles, 1860, p. 33

Chapter 8: Conclusion

1. Vandiver, 1952, p. 79.
2. Andrew Appleby, Cape Town, personal communication